Video
Engineering

Other McGraw-Hill Books of Interest

Video Engineering

Andrew F. Inglis

Arch C. Luther

Second Edition

McGraw-Hill

New York San Francisco Washington, D.C. Auckland Bogotá
Caracas Lisbon London Madrid Mexico City Milan
Montreal New Delhi San Juan Singapore
Sydney Tokyo Toronto

Library of Congress Cataloging-in-Publication Data

Inglis, Andrew F.
    Video engineering / Andrew F. Inglis, Arch C. Luther.—2nd ed.
        p.   cm.—(Video/audio engineering series)
    Includes index.
    ISBN 0-07-031791-7 (hardcover)
    1. Television. 2. Digital video. I. Luther, Arch C. II. Title.
III. Series.
TK6630.I54   1996
621.388—dc20   96-10913
                                                            CIP

McGraw-Hill

A Division of The McGraw-Hill Companies

    3 4 5 6 7 8 9 0 DOC/DOC 9 0 1 0 9 8 7

ISBN 0-07-031791-7

*The sponsoring editor for this book was Stephen S. Chapman, the editing
supervisor was Stephen M. Smith, and the production supervisor was Donald
F. Schmidt.*

Printed and bound by R. R. Donnelley & Sons Company.

McGraw-Hill books are available at special quantity discounts to use as pre-
miums and sales promotions, or for use in corporate training programs. For
more information, please write to the Director of Special Sales, McGraw-Hill,
11 West 19th Street, New York, NY 10011. Or contact your local bookstore.

This book is printed on acid-free paper.

Contents

Preface   xiii

Chapter 1. Video System Fundamentals   1

   1.1 Introduction   1
   1.2 Video Systems   2
   1.3 Scanning   4
   1.4 Resolution   6
   1.5 Pixels, Sampling, and Quantizing   7
   1.6 Choosing Scanning Standards   11
   1.7 Worldwide Video Scanning Standards   14
   1.8 Television Signal Waveforms   17
   1.9 Digital Video Signals   25
   1.10 Analog-Digital and Digital-Analog Conversion   26
   1.11 Summary Comparison of Analog and Digital Systems   27
   1.12 The Audio Component   27
   1.13 Summary   29

Chapter 2. Color Video Fundamentals   31

   2.1 Introduction   31
   2.2 The Properties of Color—A Summary   31
   2.3 Color Specifications   33
   2.4 The Primary Colors   34
   2.5 Dominant Wavelength and Saturation   36
   2.6 Luminance and Color Difference Components   37
   2.7 The CIE Chromaticity Coordinates   37
   2.8 Camera Colorimetry   40
   2.9 Display Device Primary Standards   42
   2.10 Colorimetry of Computer Displays   46

Chapter 3. Introduction to Digital Technology   47

   3.1 Digital   47
   3.2 Number Systems   47
   3.3 Analog-to-Digital Conversion   50
   3.4 Serial and Parallel Formats   58
   3.5 Binary Arithmetic   60

3.6 Digital Signal Processing 63

3.7 Digital Video Transmission 65

3.8 Digital Error Protection 70

3.9 Digital-to-Analog Conversion 74

3.10 Video Bandwidth Reduction 75

3.11 Digital Data Compression Technologies 76

3.12 JPEG and MPEG Compression 81

3.13 Summary 85

Chapter 4. Elements of Image Quality 87

4.1 Overview 87

4.2 Basic Image Quality Criteria 89

4.3 Image Defects 90

4.4 Characteristics of Human Vision 92

4.5 Image Definition 94

4.6 Aperture Response 95

4.7 Sampling and Aliasing 101

4.8 Limiting Resolution 104

4.9 NTSC Television System Aperture Response 107

4.10 HDTV System Aperture Response 112

4.11 Image Definition—Broadcast Systems 115

4.12 Image Definition—HDTV Systems 116

4.13 Visual Perception of Broadcast and HDTV Images 118

4.14 Image Gray Scale 119

4.15 Signal-to-Noise Ratio 125

4.16 Image Defects (Continued from Section 4.3) 128

4.17 Comparison of Film and Television Picture Quality 132

Chapter 5. Analog Video Systems 139

5.1 Introduction 139

5.2 Standardization 139

5.3 Video Signal Formats for Color 140

5.4 Analog Color Television System Configurations 142

5.5 Broadcast Signal Components 143

5.6 Subcarriers 145

5.7 The NTSC System Subcarrier 145

5.8 The PAL System Subcarrier 149

5.9 Comparison of NTSC and PAL 152

5.10 Color Signal Distortions and Artifacts 153

5.11 Subcarrrier Cross Talk Reduction 155

5.12 Component Signal Formats 156

5.13 Multiple Analog Component (MAC) Systems   156

Chapter 6. Digital Video Systems—HDTV   159

6.1 Introduction   159
6.2 HDTV System Objectives   161
6.3 The Grand Alliance HDTV System   162
6.4 The Picture Formats Layer   164
6.5 The Compression Layer   166
6.6 The Transport Layer   168
6.7 The Transmission Layer   172
6.8 GA HDTV Audio   177
6.9 Summary   178

Chapter 7. Digital Video Systems—Computers   179

7.1 Introduction   179
7.2 The Personal Computer   179
7.3 Software   185
7.4 Computer Video Formats   188
7.5 Computer Video Standards   193
7.6 Digital HDTV on Computers   194

Chapter 8. Video Cameras   195

8.1 Introduction   195
8.2 Photoconductive Storage Tubes   199
8.3 Charge-Coupled-Device (CCD) Sensors   200
8.4 CCD Imagers   205
8.5 CCD Imager Performance   212
8.6 Camera Chain Configurations   218
8.7 Optical Components   220
8.8 Analog Camera Electronics   222
8.9 Digital Camera Electronics   226
8.10 Camera Operating Controls   229
8.11 Camera Specifications   229
8.12 Television Film Camera Systems   230

Chapter 9. Professional Video Recorders   233

9.1 Introduction   233
9.2 Principles of Magnetic Tape Recording   236
9.3 Video Head Design   238
9.4 The Helical-Scan Principle   240
9.5 Tape Transport Configurations   241

9.6 Magnetic Recording Tapes   243
9.7 Tape and Record/Playback Head Performance   245
9.8 Analog Recording   249
9.9 SMPTE Type C Recorders   252
9.10 ½-inch Cassette Recorders   259
9.11 Digital Recording   261
9.12 Digital Recording Standardization Policy   262
9.13 The D-1 Component Format   264
9.14 The D-2 Composite Format   264
9.15 The D-3 Composite and D-5 Component Formats   265
9.16 The D-6 HDTV Component Format   268
9.17 Computer Hard Disk Video Recording   268
9.18 Optical Video Recording   270
9.19 Trends in Professional Video Recorders   272

Chapter 10. Home and Semiprofessional Video Recorders   275

10.1 Introduction   275
10.2 Comparison with Professional Recorders   275
10.3 Recording Drum Configurations   280
10.4 Home VCR Record Formats   280
10.5 Automatic Scan Tracking   285
10.6 Record Information Density   286
10.7 Color-Under Signals   287
10.8 S-VHS and Hi-8   289
10.9 Digital Home Recorders   290
10.10 Camcorders   291
10.11 Semiprofessional and Professional Camcorders   294

Chapter 11. Video Postproduction Systems   297

11.1 Introduction   297
11.2 Electronic Editing of Videotape   298
11.3 Tasks of Postproduction   304
11.4 Time Code   307
11.5 Digital Postproduction   308
11.6 Summary   312

Chapter 12. Television Receivers and Video Monitors   313

12.1 Introduction   313
12.2 Receiver Configuration   313
12.3 RF Amplifier and Channel Selector   314
12.4 IF Amplifier   315

12.5 Detection   318
12.6 Sync Separation and Scanning Generation   318
12.7 Subcarrier Demodulation and Video Amplification   318
12.8 Color Display Devices   319
12.9 Special Receiver Features   322
12.10 Receiver Performance Criteria   324
12.11 Professional Picture Monitors   326

Chapter 13. Digital Video Display Systems   327

13.1 Introduction   327
13.2 Computer Video Subsystem   327
13.3 Analog Video Overlay   332
13.4 Digital Television Displays   334
13.5 Computer Display Monitors   334
13.6 Flat-Panel Displays   335
13.7 Digital HDTV Display Processing   339

Chapter 14. Interactive Video Systems   343

14.1 Introduction   343
14.2 Ingredients of Interactivity   344
14.3 Communicating with the User   345
14.4 Video-Based User Interfaces   345
14.5 Computing Capability for Interactivity   349
14.6 Multimedia   350
14.7 Interactive Television   356

Chapter 15. Terrestrial Broadcasting Systems   359

15.1 A Brief History and the Future Outlook   359
15.2 Allocations and Assignments   361
15.3 The Standard Analog Broadcast Television Channel   365
15.4 The Visual Carrier   365
15.5 The Aural Carrier   368
15.6 Elements of a TV Transmitting System   369
15.7 FCC ERP/Antenna Height Limitations   371
15.8 Television Transmitters   371
15.9 Representative Transmitter Power Ratings   376
15.10 Transmitter Performance Standards   376
15.11 Transmission Lines   379
15.12 Waveguides   383
15.13 Properties of VHF and UHF Radiation   385
15.14 Antennas   391

15.15 TV Station Service Areas 396
15.16 Digital Broadcasting 398
15.17 HDTV (ATV) Transmission Standards 400
15.18 Display Formats for HDTV (ATV) Systems 401
15.19 HDTV (ATV) Channel Assignments 402
15.20 HDTV (ATV) Station Coverage 402
15.21 HDTV (ATV) Channel Utilization Alternatives 403
15.22 HDTV (ATV) Transmitters 403

Chapter 16. CATV Systems 405

16.1 A Brief History of CATV 405
16.2 Cable System Elements 405
16.3 Cable Television Channels 406
16.4 Coaxial Cable 408
16.5 Line and Bridging Amplifiers and Equalizers 410
16.6 Trunk Circuits 412
16.7 Reverse Path Circuits 413
16.8 Video Performance Standards 413
16.9 Head Ends 414
16.10 Head-End Equipment 414
16.11 Set-Top Converters 415
16.12 Pay-TV Systems 417
16.13 Wireless Cable 418
16.14 Digital Transmission and Fiber-Optic Cable 422
16.15 Future Cable Systems 424

Chapter 17. Satellite Video Communications 427

17.1 Satellites and Television 427
17.2 Satellite Communication Systems Elements 428
17.3 The Geosynchronous Orbit 428
17.4 Satellite Classifications 432
17.5 Satellite Construction 434
17.6 Earth Stations 437
17.7 Transmission Formats 442
17.8 FM Performance Characteristics 444
17.9 Digital DBS Services 449

Chapter 18. Fiber-Optic Transmission Systems 457

18.1 Overview 457
18.2 Fiber-Optic Construction and Operation 458
18.3 Single and Multimode Fibers 459

18.4 Attenuation and Dispersion 461
18.5 Signal Formats and Modulation Modes 463
18.6 Path Performance, Digital Format 465
18.7 Transmitters for Fiber-Optic Systems 465
18.8 Receivers for Fiber-Optic Systems 467
18.9 Passive Components 469

Chapter 19. The Future of Video Systems 471

19.1 Introduction 471
19.2 Technologies in Video Engineering 471
19.3 Video Systems 475
19.4 Communications 477
19.5 New Video Systems 478
19.6 Conclusion 484

Glossary 485

Bibliography 507

Index 511

Preface

The spectacular progress of electronic technology has made electronics a participant in nearly all aspects of modern life. From television or personal computers to automobiles or kitchen appliances, electronics is everywhere. Nowhere are the advances more evident than in the electronic reproduction of pictures—video. That is the subject of this book and, in the three years since the First Edition of this book was written, so much has happened in video engineering that a Second Edition has become mandatory. This Second Edition, therefore, includes most of the subject matter of the First Edition and is greatly expanded with descriptions of the technical developments of the past three years.

The same electronic progress applied to personal computers has made digital technology inexpensive and eminently useful to video systems. The result is higher performance, higher reliability, and lower cost. The newly developed digital HDTV system is an excellent example—high-resolution moving pictures with surround sound are transmitted in the same 6-MHz bandwidth that handles analog television and by using channels that could not previously be utilized for analog broadcasting because of interference. The result is an unprecedented merging of television and computer technologies.

This expanded scope of video engineering calls for an organization of this book that brings digital technology to the forefront. At the same time, analog technology cannot be left out because most of the video world is analog. Thus, this book has grown in size.

The potential audience for this book has also grown because analog video engineers need to learn about digital technology and digital engineers entering the video field need to learn about the analog principles that underly all video systems—even digital ones. And, of course, new video engineers need to learn about both. This book treats analog and digital video equally in the perspective required by future video engineers.

In-depth coverage of all the subjects embraced by video engineering would require many volumes of great detail, far beyond what is practical in one book. This book focuses on the fundamentals of video engineering in all aspects and details are presented only as examples. Complete coverage of details is left to other writers. This concept is supported by the references in the chapters and in the Bibliography.

Another reason for adhering to fundamentals is that the details of a technology change much more rapidly than its fundamentals. This causes rapid obsolescence of books that go into great detail in a fast-moving field such as video engineering and that is unfair to readers unless they actually need the detail. It is also unfair to writers who will find that their books quickly go out of print!

In addition to the authors, a project such as this book involves the contributions of many people, including those who helped with the First Edition and now the Second Edition. We especially want to acknowledge the following for their suggestions, comments, or reviews: Sidney Bendell, Walter Braun, John Christopher, Peter Dare, Marvin Freeling, Luigi Gallo, Charles Ginsburg, Jukka Hamalainen, Norman Hurst, Anthony Lind, Robert Neuhauser, Koichi Sadashige, John Smiley, Larry Thorpe, and Tom Vreeland.

Video engineering is a field that has an exciting future. This book will give its readers a full view of that potential.

Andrew F. Inglis
Arch C. Luther

Video
Engineering

Video System Fundamentals

1.1 Introduction

Everyone knows what video is—it is another word for television. For more than 40 years, television has been in our homes; it is an integral part of modern life. News, information, and entertainment are all received through "video."

However, an engineer would take a broader view of the definition of video—he or she would equate it with the use of a cathode-ray-tube display or other equivalent means of electronic display. Since this book is about video engineering, the engineer's viewpoint will be used:

> *Video is the technology for electronic capture, storage, transmission, and reproduction of images and motion pictures.*

That definition embraces much more than television. In particular, it includes the display technology used in personal computers (PCs), which now are present in a large percentage of our offices, schools, and homes. These two spheres of usage of video are the hubs from which many other applications have grown.

For many years, the two spheres—television and computer—although they both used video, were artificially separated by the fact that TV was analog and computers were digital. That era is now past—analog and digital technologies now are used wherever they suit and are cost-effective. However, because of continued advances in capability and cost reduction of digital video technology, the trend is for digital techniques to replace many video functions that previously were analog. The day is nearing when the signal path in television equipment will be all digital. In fact, the new HDTV standards that

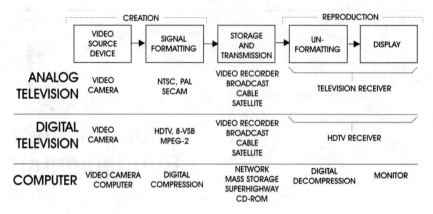

Figure 1.1 Elements of a video system

are near completion are all digital. This will accelerate the transition to digital television.

1.2 Video Systems

Video, as shown by Figure 1.1, requires a system that includes the elements of creation, storage, transmission, and reproduction. The figure shows the most common names for each of the parts of a system for both television systems and computer systems. Each of these system parts will be covered in later chapters.

Creation is the process of generating an electrical video signal, either from a live scene using a live camera, from motion picture film using a telecine camera, or artificially by a computer. Video cameras are covered in detail in Chapter 8. In signal formatting, the video signal is processed to enhance it and to conform it to a signal standard for use later in the system.

1.2.1 Video signal standards

Signal standards in a video system are extremely important because video signals pass through equipment of many different types and different manufacturers. Each equipment must know exactly the form of the signal it receives and must output a similarly precise signal. Otherwise, the many parts of a system could not work together. Signal formats are covered in Chapters 2 to 7.

Broadcast television systems have historically been standardized in terms of the signal radiated by a transmitter. This has provided a solid basis for the design of receivers, but as the industry has grown and become more diverse, there is also a need for a signal-generation standard. This is called a *produc-*

tion standard and it defines the video signal at its point of creation or assembly. Such a standard must be more demanding than a transmission standard because the signals will be stored, broadcasted, or presented in many formats, and the standard must allow for degradation that may occur in those processes.

1.2.2 Storage

Storage is another important part of a video system. The home video cassette recorder (VCR) is a typical video storage device for recording and replaying of video and audio. Video recorders of many types are covered in Chapters 9 and 10. Computers have their own forms of storage, including RAM memory and hard disks. These are covered in further detail in Chapters 3 and 7.

Because of the capability to store and replay video, one can assemble a video program from sequences taken at different times or different places. The result is stored as a complete program. This process is called *editing*. It allows program creation to be divided into two steps: the initial capture of video and audio, called *production*, and the editing and program assembly process, called *postproduction*. They are covered in Chapter 11.

1.2.3 Transmission

Transmission is the process of sending a video signal from one physical location to another. Usually it refers to transporting video in real time, although shipping tapes or disks is a valid alternative if the time of arrival of the video is not critical. The largest real-time video transmission system is the broadcast television system, although computer network systems are catching up. Broadcast transmission is covered in Chapters 15, 16, and 17. Computer transmission systems are covered in Chapter 3.

The reproduction part of a video system receives a signal and, by knowing its standard, unformats it to a form suitable for driving a display device such as a cathode-ray tube (CRT) or a solid-state display panel. The viewer sees the result. Displays are covered in Chapters 12 and 13.

1.2.4 Video system types

Figure 1.1 is an abstraction of a video system that emphasizes the common elements in all systems. Most of the time in this book, real systems are discussed—not abstractions. Three major system types that appear throughout the book are:

Television (TV)	The broadcast systems used around the world, including NTSC (U.S. and Japan), PAL (most of Europe and China), and SECAM (France and Russia) color video systems (see Chapter 2)
High-definition TV	The emerging higher-performance digital TV systems
Computers	The many forms of video used by personal computers

The electronic reproduction of video is not complete without sound. One has only to turn off the sound on a TV news program to appreciate how much information is carried by the sound channel. From an artistic point of view, the sound contributes as much as the picture to the overall psychological and emotional effect of a video presentation.

Often only part of a video system is evident, such as when viewing a television receiver, but the other parts of the system must exist (or have existed) for video to work.

1.3 Scanning

Natural scenes are three-dimensional (x, y, and time), but they can be converted to two dimensions using, for example, photography. A photograph presents millions of pieces of information at once, because all the points in the picture exist simultaneously. But an electrical signal has only one value at a time, so to represent a picture with a single electrical signal, one must use a technique that converts the picture to a one-dimensional signal vs. time. That process, known as *scanning*, is similar to the way one reads a page of English text. The reader starts at the upper-left corner of the page, reads across a line, moves down to the next line, and continues reading lines until the bottom of the page is reached.

Scanning is used by all video systems in the process of converting optical images into electrical signals. The invention of scanning was a major breakthrough and an indispensable first step in the development of video technology. Like many basic technical developments, it did not have a single inventor but appears to have occurred to a number of engineers simultaneously. Note that theoretically a picture could be reproduced electrically by creating a separate signal for each point (*pixel*) in the picture; however, the physical impracticability of having the many thousands of parallel channels that this approach requires precludes its use except inside of sensor devices such as CCDs (see Chapter 8.)

1.3.1 Scanning principles

Scanning operation in a television camera is illustrated in Figure 1.2. An image of the scene is optically focused on the sensitive surface of the camera's *imager*, where scanning takes place. In the scanning, an electronic sensing spot moves across the image in a pattern called a *raster*. The sensing spot converts each image point that it sees into an electrical voltage. Thus, the voltage output from the imager varies as the spot passes over different brightness areas in the image. The actual mechanism for moving the spot is different between tube imagers and solid-state imagers, but the raster pattern and the electrical output are virtually the same. Both television and computer video systems use a rectangular raster pattern. Scanning begins at the upper-left of the image and the spot moves rapidly in a horizontal direction across the image to form a scanning line. It then snaps back (called *retrace*) to begin another line. However, the spot also has a slower vertical motion downward, so that—after retrace—the spot is just below the previous line, ready to scan a new area of the image.

When the vertical motion reaches the bottom of the image, both horizontal and vertical retrace actions occur, bringing the spot back to the upper-left starting position. A complete scan of the image area is called a *frame*.

Scanning is also used to reproduce the image in a display device. In an analog video system, the display scanning process must be exactly the same as at the camera except that the moving spot is a spot of light whose intensity is controlled by the electrical signal from the camera. In a digital video system,

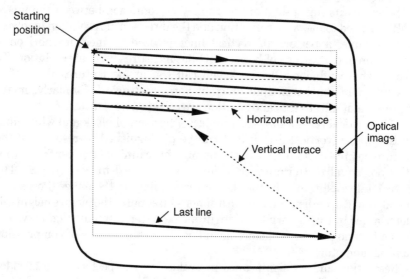

Figure 1.2 Scanning

the display scanning is not necessarily the same as at the camera (see Section 13.4).

Frames must be repeated often enough so that the viewer's eye will not see the moving spot or the lines in a reproduced image. This is possible because of persistence of vision in the eye—a suitable frame rate will produce an apparently continuous image without any visibility of the scanning motion.

1.4 Resolution

The *resolution* of a video system describes the system's ability to reproduce fine detail. The amount of resolution required in a video system depends directly on how the output of the system will be viewed. It is not necessary (or desirable) to have more resolution than the viewer will be able to see. The principal parameter of the viewing situation is the *viewing ratio,* which is the ratio between the distance between viewer and display and the height of the display screen. For analog television in the home, viewing ratios range from 4:1 up, whereas for computer screens, the ratio can be as low as 1:1.

In television, resolution is expressed as the maximum number of alternate black and white lines that can be distinguished in a dimension equal to the height of the picture (both black and white lines are counted). This number is expressed in *TV lines* (TVL), which can be measured in any direction on the image, although horizontal and vertical measurements are most common. The TVL measure of resolution is not often used with computers, but the way computer people express resolution is closely related.

When discussing resolution, there may be confusion between the image patterns used to measure resolution and the direction of resolution involved. For example, a pattern of vertical lines is used to measure horizontal resolution and a pattern of horizontal lines measures vertical resolution. To reduce the confusion, this book will call the image features that affect resolution in the horizontal direction *horizontal detail.* Similarly, image features that affect vertical resolution are called *vertical detail.*

The vertical resolution—the number of horizontal black and white lines in the image (vertical detail) that can be distinguished or resolved in the picture height—is a function of the number of scanning lines per frame (but it is not equal to the number of lines as explained in Section 6.8). The horizontal resolution—the number of vertical lines in the image (horizontal detail) that can be distinguished in a dimension equal to the picture height—is determined by the signal bandwidth in analog systems, but in a digital system it is a function of many factors including data rate, degree of compression applied, and others.

Resolution and its significance in evaluating the performance of video systems is described more fully in Chapter 4.

1.5 Pixels, Sampling, and Quantizing

Although an image in nature is a continuous analog function in space and time, the process of scanning creates a noncontinuous output in the form of lines and frames. This is known as *sampling*, which is a fundamental technique used to some extent in all video systems. This section discusses sampling and the related subjects of pixels and quantizing.

1.5.1 Sampling

The purpose of sampling is to reduce the infinite content of an analog moving image to a manageable number of values that can be handled by a video system. For example, it can be shown that 30 frames per second is a sufficient rate for the display of image motion, so there is no need to send frames any faster. Similarly, the human eye's resolving power and the viewing conditions allow the number of scanning lines to be set to a finite value. The values in the analog image that exist in space between the lines and in time between the frames can safely be eliminated.

A second purpose of sampling is to make possible a digital video system. Digital systems convey information as a series of numbers. By means of sampling in three dimensions (frames, lines, and horizontal), a series of discrete values is created to which a digital system can assign numbers. The process of assigning numbers is called *quantizing*. It is described below.

Although the parameters that are subject to sampling are not necessarily electrical (space and time), the analysis of sampling is often carried out by the use of electrical analogy. For example, frequency is the analog of TV line number. The powerful mathematical tools that have been developed for electric circuit analysis are then available. This analysis must be carried out in both the time and frequency domains, because both are needed for a full description of the sampling process.

1.5.1.1 Sampling analysis—time domain

The sampling process in the time domain is shown by Figure 1.3. In Figure 1.3(a), the sampling rate is more than twice the signal frequency and in Figure 1.3(b) it is less. A sampling rate of twice the highest signal frequency component is known as the *Nyquist frequency*. Similarly, the highest signal frequency that can be sampled without distortion for a given sampling frequency is known as the *Nyquist limit*. Sampling at lower rates than twice the highest signal frequency is known as *sub-Nyquist* sampling. Sub-Nyquist sampling generates a new frequency component equal in frequency to the difference between the sampling and signal frequencies [see Figure 1.3(b)]. This effect is called *aliasing*.

For vertical detail, the spatial sampling rate is determined by the number of scanning lines in the raster and ideally this would equal twice the highest

(a)

(b)

Figure 1.3 Sampling—time domain

line number of the vertical detail components in the optical image presented to the scanning imager. There would then be no scanning line aliasing. In modern cameras, this is usually not the case and vertical aliasing may be visible in scenes that have a strong vertical detail pattern.

A solid-state imager is inherently sampled both vertically and horizontally because its sensitive surface is divided into an array of cells that are scanned in a raster pattern to obtain the electrical output. The numbers of cells in the horizontal and vertical directions are primary determinants of the imager's resolution performance. However, the output signal is still analog in the sense that the output from each cell can have any value at all, depending on the optical brightness that falls on the cell.

1.5.1.2 Sampling analysis—frequency domain The pulses generated by sampling have characteristics similar to an RF carrier that is amplitude modulated by the video signal. The frequency spectrum of the pulses consists of a strong carrier component at the sampling frequency with sidebands that correspond to the frequency components of the video signal. (The sampling pulses themselves are rich in harmonics and the pattern of carrier and sidebands is repeated at harmonics of the sampling frequency.) The resulting spectrum for the baseband and sampled signals are shown in Figure 1.4.

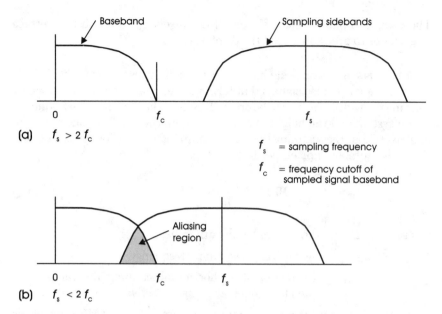

Figure 1.4 Sampling—frequency domain

Figure 1.4(a) shows the spectrum when the sampling rate is more than twice the highest video frequency and thus meets the Nyquist criterion.

Figure 1.4(b) shows the effect of a sub-Nyquist sampling rate. In this case, there is an overlap between the frequency components of the video and the sidebands of the sampling signal. This overlap cannot be removed by subsequent signal processing and it is the source of aliasing. The various cases of aliasing are discussed in Section 4.7.

1.5.2 Pixels

In a digital system, a picture is made up of a series of digital values that represent individual points along the path of scanning the image. Thus, the image is divided into discrete points, called *pixels*. If one examines a digital image closely, he or she will probably see little squares or rectangles of color—these are the pixels. In a well-designed system, the pixel pattern will be invisible for normal viewing. The process of dividing an image into pixels is horizontal sampling, which means that a series of equally spaced readings of the image color have been taken during the scanning process. The output from sampling is a series of individual values, rather than a continuously varying signal.

The resolution of a digital video system is determined by its pixel counts, horizontal and vertical. This is given by two numbers in the form: (horizontal) × (vertical). A typical system (computer VGA) will have 640 × 480 pixels.

The usual computer display shows all the pixels; none are lost to retrace intervals or to overscanning of the display.

1.5.2.1 Pixels in analog systems   It is convenient to use the concept of pixels even with analog video signals that have not been sampled. That is done by defining the width of a pixel to be one-half cycle of the highest video frequency in the system and the height of a pixel to equal the height of one scanning line. (Note that the scanning lines are a sampling process for vertical detail.) Thus, the number of pixels per line (PPL) is:

$$\text{PPL} = 2\,\frac{B_W}{f_H}\,(C_H) \tag{1.1}$$

where: B_W =   the system bandwidth in hertz

f_H =   the horizontal scanning frequency

C_H =   the fraction of the horizontal scanning interval devoted to signal transmission (active line factor)

For the NTSC television system with a bandwidth of 4.2 MHz, the pixels per line are:

$$2 \times \frac{4{,}200{,}000}{15{,}734} \times 0.84 = 448$$

The pixels per frame would then be:

$$448 \times 480 = 215{,}040$$

(The number 480 is the number of scanning lines that are actually visible. The rest of the 525 total lines are blanked out for the vertical retrace.) Pixels per frame is a measure of the maximum information content of a single frame for a system.

1.5.3 Quantizing

To create a fully digital signal, a second process, called *quantizing*, is necessary. In this process, each sampled value is forced to take a number corresponding to the closest position in a series of equally-spaced values. This series is determined by the number of bits per pixel in the digital standard being used. For example, a system that uses 8 bits per pixel is capable of a maximum of 256 (2^8) values. For more discussion of quantization, see Section 3.3.3.

1.6 Choosing Scanning Standards

Scanning is one of the properties of a video system that must be standardized. The choice of the number of scanning lines requires a tradeoff between the conflicting requirements of image resolution, signal bandwidth or data rate, and image flicker. Before examining this tradeoff, each of the requirements and a few other contributing factors will be discussed.

1.6.1 Image flicker and frame rate

In a video display, the action of scanning will become visible if the vertical-scan rate is too low. The common name for this effect is *flicker*. The visibility of flicker vs. vertical-scan rate is discussed in detail in Chapter 4, but it is sufficient to say here that it depends in complex ways on the characteristics of the human eye, the image brightness, and the viewing conditions. For television-style viewing, it has been determined that the vertical-scan rate ought to be 50 Hz or higher. For computer-style viewing, it may need to be even higher.

In the early days of television, it was important to consider the relationship between the vertical-scanning rate and the local power frequency because of serious problems in filtering all power-frequency interference out of a television receiver. As a result, early monochrome television systems synchronized vertical scanning to the power frequency—60 Hz in the United States, 50 Hz in Europe, etc. However, when color television was introduced, other considerations mandated that vertical scanning could not be locked exactly to the power line, but the frequency relationship was still kept close. These measures insured that power-line interference, if present, would cause a stationary or slowly-moving pattern in the picture, thus minimizing the visibility of such interference.

The *frame rate* is the rate at which all the lines in the picture are reproduced. In a simple scanning system, the frame rate is the same as the vertical-scan rate. However, in television systems, a technique called *interlaced scanning* is employed in order to make the frame rate lower than the vertical-scan rate.

1.6.2 Interlaced scanning

As will be explained shortly, higher frame rates lead directly to higher bandwidth requirements. In television systems, interlaced scanning permits the frame repetition rate, and hence the bandwidth, to be reduced by one-half with very little increase in flicker. This is done by scanning only half of the total lines in each vertical scan and designing the system so that scan lines on alternate vertical scans fall between the lines of the previous vertical scan.

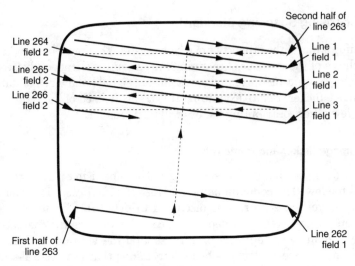

Figure 1.5 Interlaced scanning for 525 lines

Thus, each frame takes two vertical scans (fields), with the even and odd lines scanned on alternate fields. The perception of large area flicker is based on the field rate, but the bandwidth requirements are based on the frame rate, which is one-half the field rate. The use of interlaced scanning in television to reduce flicker is analogous to the use of a shutter in a film projector that displays the film with two light flashes per frame.

To generate interlaced scanning, each frame has an odd number of scanning lines as shown in Figure 1.5. The vertical-scan frequency is chosen so that half of the scanning lines are contained in each field. This causes the first line of alternate fields to begin in the center of the picture, and the lines are interleaved between fields. In the system employed in the United States there are 525 scanning lines in each frame and 262½ lines in each field.

In common with all techniques employed in television to reduce the bandwidth requirements, there is a price for the benefits of interlaced scanning. The technique is not fully effective because of the tendency of the scanning lines to pair in the receiver, i.e., to fail to interlace. This is a receiver design problem that is described in Chapter 12. When pairing occurs, the effective number of scanning lines and the vertical resolution are reduced by one-half, thus eliminating the advantage of interlacing.

A more fundamental problem is small area flicker. Fine vertical detail is repeated at the slower frame rate, and a sharp horizontal edge that is reproduced only by an odd or even field will appear to flicker. Another problem with interlaced scanning is the greater visibility of motion effects because the number of scanning lines is halved for moving objects.

Small area flicker and motion problems are not particularly noticeable in most standard broadcast transmissions because the sharpness of the edges and the fineness of detail produced by the camera signal are not sufficient to generate visible spurious signals at normal TV viewing distances. They become a significant problem in HDTV systems in which the images are necessarily sharper and the detail finer because of smaller viewing ratios. As a result, the issue of interlacing is being reconsidered for HDTV.

1.6.3 Sequential or progressive scanning

The problems of interlaced scanning described in the preceding section can be greatly reduced or eliminated by the use of sequential or progressive scanning in which every line is scanned on every field. This is an important consideration to HDTV because of the desire to reduce all kinds of picture impairments. The penalty, of course, is that twice the bandwidth is ordinarily required. However, the digital compression used in HDTV will allow sequential scanning to be available as an option for broadcasting HDTV. It may also be used in HDTV production systems or other wideband systems that employ alternative transmission modes, e.g., fiber optics.

Sequential scanning is commonly used in computers and it can be used in television receivers that employ a digital frame memory. The memory stores the broadcasted interlaced frames and outputs the lines in a progressive-scan format for display. HDTV receivers will always have a frame memory and that allows the receiver display scanning to be *decoupled* from the scanning assumed in the transmitted signal. That means, for example, that an HDTV receiver could employ sequential scanning for its display even though the incoming signal was interlaced.

1.6.4 Aspect ratio

Motion pictures have a wider screen than television or computers. This is determined by the system parameter called *aspect ratio*. For standard NTSC or PAL television and computers, the aspect ratio is 4/3 (1.33:1), which is the ratio of the width to the height of the picture. Wide-screen movies have a 16/9 (1.77:1) or higher ratio. At the time television standards were set, movies also were close to 4/3, but in the intervening years they have created new standards. Television developers plan to do the same thing with HDTV, which will have a 16:9 aspect ratio.

Computer displays generally use 4/3 aspect ratios. This is not to be the same as television, but because it is convenient at the small viewing ratios involved to have a display width that can be viewed without much movement of the head. Compatibility with television is not important—computer CRTs are different than television CRTs because of the need for higher resolution and flatter screens called for by the close viewing conditions.

1.6.5 Bandwidth and data rate

Bandwidth is a measure of an analog system's ability to transmit a range of frequencies. The scanning process generates both high frequencies and low frequencies, requiring a bandwith of at least 4.2 MHz (for NTSC television). Spectrum space for broadcast systems is limited, and it is desired to keep the bandwidth as low as possible. In a digital system, the equivalent parameter is data rate, which measures the speed at which the system can handle data.

1.6.6 The tradeoff

As an example of the tradeoffs in choosing scanning standards for a video system, the following equation describes how the factors discussed above combine (for analog television) to establish a system's bandwidth requirement:

$$B_W = \frac{1}{2}\frac{A_R\, F_R\, N_L\, R_H}{C_H}$$
(1.2)

where: B_W = the system bandwidth in hertz

A_R = aspect ratio

F_R = the frame rate in frames per second

N_L = the number of scanning lines per frame

R_H = the horizontal resolution desired in TVL

C_H = the fraction of time occupied by active horizontal scanning (after subtracting the horizontal blanking period)

The factor ½ is the ratio of the number of cycles to the number of lines resolved. The aspect ratio in the equation is required because the horizontal resolution is measured in lines per picture height.

Choosing the optimum combination of B_W, F_R, N_L, and R_H was a major concern of the original TV system designers, since it involves a complex series of tradeoffs. From the above it can be seen that a wide bandwidth is required to resolve fine detail while maintaining a high enough picture repetition rate to avoid objectionable flicker. This explains the huge spectrum requirements of television systems.

1.7 Worldwide Video Scanning Standards

1.7.1 Television scanning standards

Television scanning standards adopted by various countries around the world are listed in Table 1.1. The diversity of standards is the result of chronological,

Table 1.1 Worldwide television scanning standards

System	Aspect Ratio	Inter- laced	Frames/ second	Total/Active Lines	Lines/ second	Bandwidth (MHz)
United States						
Mono	4:3	2:1	30	525/480	15,750	4.2
Color NTSC	4:3	2:1	29.97	525/480	15,734	4.2
Color HDTV	16:9	no	60	750/720	45,000	6.0[1]
Color HDTV	16:9	2:1	30	1125/1080	33,750	6.0[1]
England						
Color PAL	4:3	2:1	25	625/580	15,625	5.5
Japan						
Color NTSC	4:3	2:1	29.97	525/480	15,734	4.2
France						
Color SECAM	4:3	2:1	25	625/580	15,625	6.0
Germany						
Color PAL	4:3	2:1	25	625/580	15,625	5.0
Russia						
Color SECAM	4:3	2:1	25	625/580	15,625	6.0
China						
Color PAL	4:3	2:1	25	625/580	15,625	6.0

1 Digital transmission. Scanning standards shown are typical—other variations are possible.

political, and technical factors that affected the decision differently in different countries. In the current move toward HDTV standards, there is a concerted worldwide effort to minimize the number of different systems.

Some of these TV standards are more than 50 years old. Since TV receivers are purchased by consumers, who are generally nontechnical and have limited financial resources, it has been important to set a single standard within a country and hold it unchanged for many years. This allows the lowest-cost receivers to be built and the consumers' investment in them pays back over a long useful life. The situation is quite different in the computer market.

1.7.2 Computer scanning standards

Until very recently, personal computer users were required to have considerable technical knowledge about their system. Although this has surely limited the market for PCs, the value of the technology has been great enough that large numbers of users have made the investment in learning their systems. This has caused the standards situation to be very different from TV.

The personal computer is inherently a programmable device. With software, one can theoretically program a PC to do anything. Although the software might be able to do anything, there are limits on what the hardware can do. The thrust in hardware, then, has been to make it as powerful and flexible as possible—consistent, of course, with a selling price that users can afford. Over the years of PC development, hardware capabilities have grown at a prodigious rate. For example, the processing power of the microprocessor chip that is the heart of a PC has increased by a factor of two every two years! That kind of growth has made product life in the PC market much shorter than it is in television.

Turning to the effect this growth environment has had on video display technology, computer displays are generally designed to operate with a variety of scanning standards, covering a range of line and field frequencies as well as a range of color formats. Often a single PC will operate with different scanning at different times, simply by changing its software. The display hardware must be able to accommodate such versatility. Thus, there is much less need for a single country-wide scanning standard for computers.

However, computers need to share data with one another and the data requirements are different when scanning standards are changed. Although data can theoretically be converted between standards with software, the process is not always perfect and it uses computer resources that may slow down the system operation. Therefore, the industry has made an effort to limit the number of different standards, but there is no reason to have only one. Table 1.2 lists some of the most popular scanning standards for IBM-compatible PCs.

The letters "VGA" in the table stand for *video graphics array*, which is a hardware video display standard originally developed by IBM and widely used. VGA defines many different resolution and color modes, although the one given in the table is the highest one. "SVGA" stands for super VGA, an enhanced version of VGA that supports higher-resolution and higher-color

Table 1.2 Computer scanning standards

Resolution Mode	Color Mode	Frames/ second	Lines/ frame	Lines/ second	Data Rate (MB/second)
640 × 480 VGA	4bpp	60	525	31,500	9.2
640 × 480 SVGA	8bpp	60	525	31,500	18.4
640 × 480 SVGA	RGB-16	60	525	31,500	36.8
640 × 480 SVGA	RGB-24	70	525	36,750	64.5
1024 × 768 SVGA	8bpp	70	800	56,000	55.0
1280 × 1024 SVGA	4bpp	70	1100	77,000	45.9

modes. SVGA has become the current standard because it supports video modes that do a better job of reproducing realistic images.

The Color column in the table tells how many colors the format can reproduce at once. "4bpp" refers to 4 bits per pixel and can only display 16 (2^4) colors on the screen at once. This is extremely limiting for displaying realistic images and is the reason that plain VGA displays are phasing out. An 8bpp color mode displays up to 256 colors at once, which is still too small a number for real pictures. However, an ingenious technique called *dithering* allows remarkably good reproduction of many images using only 8bpp (see Section 7.4.2.2).

The other color modes (RGB-16 and RGB-24) support up to 65,536 or 16,777,216 colors and give excellent reproduction of realistic images. These modes are sometimes called 16bpp and 24bpp modes. Color modes are discussed more completely in Chapter 5.

Notice that there is a range of horizontal and vertical scanning frequencies in the table and that the numbers are all higher than television numbers. This is because all the formats listed use progressive (noninterlaced) scanning.

The data rates listed in the table represent the raw data that must be supplied to update a display at the frame rates indicated. These rates normally occur only within the video display hardware of the computer. Data rates handled by the microprocessor and the storage of the computer are many times lower because of a variety of strategies for data rate reduction, called *compression*. This is covered in Chapter 3.

1.8 Television Signal Waveforms

The units of most television systems communicate by means of a composite signal, which is a special format that contains all the information needed to convey a picture (brightness, color, synchronization) on a single cable. The composite signal is standardized by specifying its waveform, which is a view of how the signal varies with time. Figure 1.6 shows the waveform diagrams for the brightness or luminance part of the NTSC color television system used in the United States.

It is important to remember that these waveform standards were set 50 years ago, when all television technology was analog and the hardware capabilities were not what they are today. It is remarkable that any standard in electronic technology should last so long! The standards contain features that are specific to the problems of the era when they were developed, some of these are not as important today. However, the following discussion explains the standards in terms of the environment in which they were developed. Later chapters will give the modern views of some of these issues.

Figure 1.6 NTSC luminance video waveform drawings

Views (a) and (b) of Figure 1.6 show the luminance waveform in the vicinity of the vertical retrace period of the scanning. The signal interval that accommodates vertical retrace is called the *vertical blanking interval* (VBI) and it takes about 7.5% of the frame time. The word "blanking" means that the video voltage during the VBI has a range of values that will always appear as black on a television receiver. This was necessary to insure that a video display will show no spurious information while vertical retrace is taking place. View (c) shows some details at the end of the VBI. View (d) shows details of the vertical synchronizing part of the VBI. Finally, view (e) shows details of the horizontal blanking interval, which is the time allowed for horizontal retrace of the scanning. Horizontal blanking takes 14 to 18% of a line time, so the total of blanking intervals is about 25% of a frame. That time is not available for picture transmission.

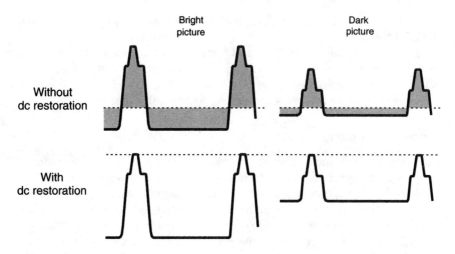

Figure 1.7 DC restoration

1.8.1 The dc component of a video signal

The drawings in Figure 1.6 show the signal in a black-positive polarity, with peak white in the picture represented as 7.5% signal, blanking level at 75% signal, and synchronizing pulses at 100% signal. This is the format at the output of a television transmitter, where the synchronizing pulses produce the highest power output. Notice that the average voltage of the signal will vary with the average brightness of the picture—dark pictures will have a higher average voltage than bright pictures. This means that there is a direct-current (zero-frequency) component in a video signal.

It is not common practice to transmit the dc component throughout an analog video system. At most places in the system, video voltages are ac-coupled, which means that there is no absolute voltage consistently associated with any part of the signal. As shown in Figure 1.7, the absolute voltages of the signal vary with the average brightness of the picture. However, there are places where the dc component is required for correct operation. The TV transmitter is one place; other places are where any gray-scale operations take place and at the video display device. Fortunately, it is easy to restore the dc component wherever required by using a circuit that continually sets some level in the signal to a specific value. This is called *dc restoration* and it operates by setting all sync pulse tips to the same level as shown in Figure 1.7.

In a digital video system, the dc component is retained throughout the system because the digital data carries absolute values for the pixels. When a system involves conversion from analog to digital formats, dc restoration must be used at the conversion point to make sure the digital data does contain the dc component.

The video waveforms in Figures 1.6 and 1.7 are shown as black-positive, which applies to the output amplitude from a television transmitter. However, at most other places in the system, the signals are white-positive. This is simply a polarity inversion and has no other significance.

1.8.2 Synchronizing signals

Television signals must carry the information necessary for the scanning of displays to be synchronized with the scanning that took place in the imager. This is done by adding synchronizing pulses (usually called *sync pulses*) to the horizontal and vertical blanking intervals. In order that the sync pulses do not interfere with the purpose of the blanking intervals, they go "blacker than black" so they represent the greatest signal excursion in the black direction. This also allows them to be easily separated from the rest of the video by simple clipping (passing of signals above a particular signal level). Proper clipping of the sync, called *sync separation*, requires dc restoration at the clipper.

Because the video content of the signal is constantly changing and could interfere with synchronization if it got into the sync circuits of a display or a receiver, the standard signal contains several features that facilitate clean sync separation. Providing for clipping of sync is one of them.

Another such feature is the *front porch* on the horizontal blanking interval. In Figure 1.6(e), the front porch is shown as a short inflection between the start of horizontal blanking and the beginning of the horizontal sync pulse. This makes sure that variation of picture content at the right-hand edge of the picture cannot smear over into the edge of the sync pulse and disturb the timing of the synchronization. In a signal channel that has poor high-frequency response, video signals can become slightly smeared to the right. The front porch provides a small barrier region to prevent that affecting horizontal synchronization.

There is a related problem for vertical synchronization except that it is caused by interlacing rather than video content. This gives rise to the features of the signal called *vertical serrations* and *equalizing pulses,* shown in Figure 1.6(a) and (b). In principle, vertical synchronization could be accomplished by providing a sync pulse and a front porch in the vertical interval, just like the horizontal interval. The vertical sync pulse could be wide enough that it could be separated from the horizontal pulses after clipping by simply narrowing the bandwidth (called *integration*) of the sync enough that the relatively narrow horizontal pulses are suppressed relative to the wider vertical pulse. Another clipping operation then separates the vertical pulse by itself. This is actually what is done in a TV receiver.

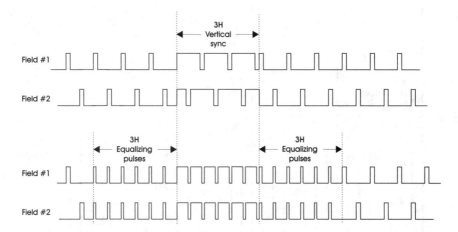

Figure 1.8 Correction of vertical sync errors with serrations and equalizing pulses

However, the wide vertical pulse would represent an interruption of horizontal synchronization. That is taken care of by making narrow slits (serrations) in the vertical pulse at horizontal sync times during the vertical pulse. The black-going edge of the serrations can be horizontal sync. At other times in the vertical interval, horizontal sync pulses could be present as normal. This way, horizontal sync continues through the vertical interval.

But, as shown by Figure 1.8, the serrations and the horizontal sync pulses cause the vertical sync to look slightly different on the odd and even fields of the interlaced scanning. That seems like a small effect, but it is great enough to cause almost 100% pairing of the interlaced lines. The solution to this problem is to make the serrations at twice-horizontal frequency and also to add twice-horizontal pulses (equalizing pulses) for a short period before and after the vertical sync. The result is that vertical sync separation by integration is greatly enhanced.

1.8.3 Spectral content of video

A monochrome video signal (or the luminance component of a color signal) covers the system bandwidth from dc to the highest frequency determined by the horizontal detail of the picture. However, the frequency spectrum of video is not continuous—it contains a fine-grained structure determined by the scanning frequencies.

This fine-grained spectral frequency structure, shown in Figure 1.9, results from the fact that video contains three periodic components, one repeating at the line rate, one at the field rate, and one at the frame rate—nominally at 15,750, 60 and 30 Hz for monochrome television. Each component can be expressed by a Fourier series, a sequence of sinusoidal terms with frequencies

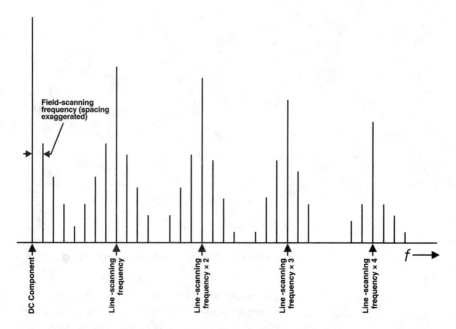

Figure 1.9 Fine-grained frequency spectrum of a video signal (field-scanning frequency components are shown with greater spacing for clarity)

that are multiples (harmonics) of the repetition rate and having amplitudes determined by the waveform of the component. The summation of the amplitudes of the terms for the three periodic components over the entire video spectrum is the frequency content of the signal. The 30 and 60 and even the 15,750 Hz harmonic components are too closely spaced to be distinguished on a frequency scale of 4.2 MHz, and they are not disclosed on a conventional amplitude-frequency graph.

The nature of these harmonic components can be visualized by considering some special cases. (For the purpose of this analysis, the vertical blanking and synchronizing pulses are ignored.)

The simplest cases are all-white pictures, or pictures consisting of vertical bars. Each scanning line is identical, and the frequency components are harmonics of the line frequency with amplitudes as determined by a Fourier analysis of the waveform, including the horizontal blanking and sync pulses. The number of harmonics is large, usually extending to the limit of the video bandpass. In a 4.2-MHz channel it would be:

$$\frac{4.2 \times 10^6}{15,750} = 266$$

Another example is a picture in which the top half is white and the bottom half black. It generates two signal components that are sequential in time and periodic at the line rate, one for the white area of the picture and one for the black. The spectrum of the signal will consist of a series of frequency components harmonically related to the line frequency and with their amplitudes determined by a Fourier analysis of the white and black waveform. In addition, however, the transitions from the black to white signals will generate another series of frequency components that are harmonically related to the field frequency, the rate at which the black-to-white transitions are repeated.

The relationship between the line and field frequency components is analogous to that between the carrier and modulating frequencies in an amplitude-modulated signal. The field frequency components cluster around the line frequency harmonics like sidebands around a carrier as shown in Figure 1.9[1,2]. The field frequency components convey the information necessary to reproduce line-to-line changes in the brightness of the image, either in time as the result of motion in the scene or in space as the result of picture content. Their number and amplitudes depend on the rate of change of brightness in the vertical direction.

In a stationary scene with typical brightness variations, most of the vertical components are contained in a band ±2 kHz on either side of each line rate harmonic for a total bandwidth of 4 kHz. This is less than half the 15,750 kHz-spacing between harmonics. If there are moving objects or unusually rapid vertical brightness variations in the image, the clusters broaden but seldom so much as to occupy more than one-half the spacing between harmonics. An exception can occur when the vertical variations in brightness are extremely rapid. Under this condition, the clusters can broaden so far that the field rate components surrounding adjacent line rate harmonics overlap. This results in aliasing. In general, however, the standard television signal is wasteful in its use of spectrum space as well as power, and half of it is unused.

In summary, the frequency spectrum of the luminance component of a color signal or of a monochrome signal consists of a series of harmonics of the line frequency, each surrounded by a cluster of frequency components separated by the field frequency. The amplitude of the line frequency components is determined by the horizontal variations in brightness while the amplitude of the field frequency components is determined by the vertical brightness variations.

1.8.3.1 Utilization of gaps in the video signal spectrum   The presence of gaps in the spectral content of luminance signals provides an important opportunity to transmit additional information without requiring more scarce bandwidth. These gaps have been utilized for the following purposes:

1. Compatible color systems. As will be described in Chapter 2, the frequency components of the color subcarrier and its sidebands, which also appear in clusters, are interleaved with the luminance frequency components. This minimizes the cross-talk between the luminance and chrominance signal components and makes it possible for the luminance channel of the receiver to utilize the full 4.2 MHz bandwidth of the video channel, even though it includes the color subcarrier. This is best accomplished by the use of a comb filter (see Section 5.11.3) in the luminance channel that removes the chrominance channel components without normally affecting the luminance.

2. Signal-to-noise improvement. A comb filter can also be used in the luminance channel of the camera to remove noise from the spectral regions between the line-scanning frequency clusters. Nearly half the noise can be removed by this technique without affecting the signal, leading to an improvement of up to 3 dB in the high-frequency signal-to-noise ratio.

1.8.4 Color video waveforms

The video waveforms shown in Figures 1.6 and 1.8 are for monochrome TV. Color TV requires the transmission of three such waveforms and that is exactly what is done in many computer systems that use the RGB color system. However, the use of separate signals for the colors is impractical for broadcasting, so they are combined by an encoding process into a *composite* color signal.

Composite color signals are designed to be viewable on monochrome displays—that is called being *compatible*. Compatibility is achieved by providing a luminance signal component that is exactly the same as a monochrome TV signal. Color information is added by encoding it on a high-frequency *subcarrier* that is superimposed on the luminance signal. Because the subcarrier frequency is high, it is not very visible on monochrome displays. There is more to it than that, which is explained in Chapter 5.

The basic sync and blanking waveforms are the same for color, but tolerances must be tightened and a color burst is added for color subcarrier synchronization.

1.8.5 Viewing video waveforms

In the operation and maintenance of a television system, it is necessary to be able to view the video waveforms at different points. An instrument for that purpose is the *waveform monitor* (WFM), which is a special-purpose oscillo-

Figure 1.10 Waveform display of an NTSC color-bar video signal

scope designed for viewing television waveforms. Figure 1.10 is a photograph of a WFM screen that shows the NTSC color signal for a color-bar pattern. The display is synchronized with the horizontal scanning and shows two lines. Vertical synchronization is also available in waveform monitors so that all the details shown by Figure 1.4 can be observed and measured.

1.9 Digital Video Signals

The concept of signal waveform that was just covered does not apply to digital video signals. Digital video being transmitted in real time is referred to as a data stream, which is a continuously-flowing sequence of digital bits. Observing it with an oscilloscope or waveform monitor, one sees a varying high-frequency signal, but the structure of an analog video signal is not visible. In order to see the structure of the digital video, its format must be known. Knowing the format, a computer or a dedicated processor can decode the data stream and produce a form of video signal. The digital video format, then, is its signal standard.

Digital data streams use the concept of a *header* to describe structure. A header is a block of data of a known format that contains information about the parameters and structure of the data that follows. It performs the functions of the sync pulses in analog video, but it actually does much more.

There will always be a header at the start of a data stream to identify it, its type, the details of the data's structure, and where the data begins. For example, a video data stream header may contain the name of the video, its resolution (pixel counts), its bits per pixel, details of any compression technique used, and other things. The header information tells the receiving system how to begin decoding the data.

There may be additional headers embedded within a data stream. For example, a video stream will have embedded headers for every frame. Thus, the characteristics of the video stream can change on a frame-by-frame basis. A header will also contain a data count so that the system can tell when it gets to the next header. Since a digital system can be as precise as needed, it can count exact numbers of bits from header to header, even if there are millions of them.

The header concept supports one of the most powerful features of a digital system—it can be programmable to as much depth as needed. In standardizing for digital video, one must standardize on the content and layout of the headers, but that content can provide for nearly infinite variability of the data itself. For example, if the horizontal and vertical pixel counts are contained in the header, then a data stream can be built for any resolution at all. The decoding system's software reads the header pixel counts and sets itself up to decode at the resolution required for the data stream. Another data stream can contain totally different numbers and the system will still work.

Carefully designed headers can provide for future options that didn't even exist when the header structure was initially set up. Therefore, a digital video system can improve in performance over time, simply by adding new software for new features. The new software is enabled when a data stream calls for it. There is no need to expect that a detailed (and limiting) standard has to last for 50 years! Digital data formats are covered in detail in Chapters 3, 6, and 7.

1.10 Analog-Digital and Digital-Analog Conversion

Modern video systems can be all-analog, partly digital, or almost-all digital. The "almost-all" is because camera imagers and display devices are still analog. Even on a computer, "real" pictures start from an imager whose output is analog, and they are viewed on a monitor that uses an analog CRT. Except for an all-analog system, every video system includes conversion to digital and back to analog. These processes are called *analog-to-digital conversion* (A–D) and *digital-to-analog conversion* (D–A). Today, each one is done by a single integrated circuit.

The basic ingredients of A–D conversion—sampling and quantizing—have already been explained. However, there are many considerations of sampling rates, filtering, digital formats, etc., that affect the performance of the conversion. These will be discussed in Chapter 3.

1.11 Summary Comparison of Analog and Digital Systems

There are many reasons for the trend to digital techniques in the design of video systems, even when the end result will be analog, such as for present-day TV broadcasting or cable systems. These will be covered in detail in Chapter 3, but here is a brief summary.

1. Digital systems can have perfect error correction so that transmission distortion, noise, or other degradation does not affect the information being transmitted. As a result, distortion does not accumulate as a system becomes larger.

2. Digital video is easily processed for delay, timing correction, signal enhancement, special effects, and storage with little, if any, quality losses.

3. Digital video can be compressed to remove inherent redundancy of information that results from the scanning process. Using these techniques, it can be transmitted with lower bandwidth requirements than analog video of equivalent subjective quality.

4. Although digital processes are often structurally more complex than their analog equivalents, they can be encapsulated into integrated circuit devices, resulting in lower-cost, higher-performance, and easier-to-build hardware.

5. The use of software-controlled (programmable) digital systems offers fantastic flexibility in signal processing. Furthermore, processing algorithms (techniques) can be changed or updated without the need to modify any hardware.

However, every comparison has its other side. As has already been explained, everything can't always be done digitally (at least not yet), so there are still analog imagers and display devices. The A–D conversion process has limitations that result in the digital signal not being an exact reproduction of the analog input. Although the digital circuits themselves may well deliver perfect reproduction, there will be an accumulating analog distortion every time the signal goes through the A–D and D–A conversions. Therefore, one must minimize the number of such conversions that cascade in a system.

1.12 The Audio Component

In the days of silent motion pictures, live music was played along with the showing of a movie so that the audience's sense of hearing was supported. Truly "silent" movies were pretty unsatisfying because an important element of realism was lacking. The same can be said of video—sound is an essential ingredient of an effective and engaging presentation. All video systems must have sound (audio).

In analog television broadcasting, audio is transmitted by a separate transmitter operating at a fixed frequency offset from the video transmitter. In computers, audio is a data stream that can be handled separately or it can be interleaved with the video data. In digital HDTV, audio is interleaved with the video. In all cases, the audio part of the system involves the same steps that were shown for video in Figure 1.1—creation, storage and transmission, and reproduction. Much of what has been said about video in this chapter also applies to audio, except that the numbers are grossly different. That makes audio a very different problem from video.

1.12.1 Audio bandwidth

Video bandwidths are in the MHz range, but audio bandwidths range only up to 20 kHz. However, the requirements for amplitude linearity and noise performance are much more demanding in an audio system than a video system. These requirements combine to make audio system design nearly as much of a challenge as video system design. In addition, audio poses another synchronization requirement—the audio must stay in sync with the video. Since the most demanding audio sync situation is when a person is talking on camera, audio-video synchronization is often referred to as *lip sync*.

Unlike video, audio cannot stand repetition of information (frames) or interruptions. Any such effects, however small, are sure to be heard.

1.12.2 Digital audio

The use of digital audio in the mass market is more advanced than the use of digital video. This is primarily because of the audio Compact Disc (CD) that has taken over the prerecorded audio field from records and tapes. The audio CD provides extremely high performance in a conservatively designed system.

Digital audio also is now widely available in personal computers. There, it has been handled in much the same way as digital video, using programmable formats, data compression, and the other digital features discussed in Section 1.7. The typical PC audio format uses less data than video, so it places a lesser demand on the capabilities of the PC and it is therefore more often used. Table 1.3 lists some typical PC audio formats.

Notice that there is a wide range of data rates from minimal speech performance up to full CD-quality digital audio. The acronym "PCM" stands for *pulse code modulation*, which is what results from simple sampling and quantization. "ADPCM" stands for *adaptive differential PCM*, which is a compression technique. ADPCM provides up to 4:1 compression, which is

Table 1.3 Digital audio formats

Mode	bpp	Stereo	Bandwidth (kHz)	Sampling (kHz)	Data Rate (kB/second)
PCM speech	8	no	5	11.02	11.0
PCM music	16	no	10	22.05	44.1
CD-quality	16	yes	18	44.1	176.4
ADPCM speech	4	no	5	11.02	5.6
ADPCM music	4	yes	10	22.05	44.1

much less than can be achieved with video compression techniques. This is because audio has far less structure and redundancy than video. The programmable nature of PCs allows any of these audio formats to be used wherever they best suit the application. The computer will play audio at whatever quality level represented in the data it receives.

In digital TV systems with audio data compression, a single audio channel may use only a small percentage of the channel data rate. Digital HDTV, for example, supports up to 5-channel audio using only about 2% of the total data rate.

1.13 Summary

This chapter has introduced some of the fundamental concepts underlying video systems, whether analog or digital. Basic system architecture was covered, and the principles of scanning standards selection, sampling, signal formats, synchronization, and data conversion have been discussed. The rest of this book covers all of these subjects in greater detail.

References

1. P. Mertz and F. Gray, "A Theory of Scanning and Its Relation to the Characteristics of the Transmitted Signal in Telephotography and Television," *Bell System Technical Journal*, Vol. 13, July 1934.

2. P. Mertz, "Television—The Scanning Process," *Proc. IRE*, Vol. 29, October 1941.

2

Color Video Fundamentals

2.1 Introduction

This chapter is devoted to colorimetry[1] and the properties of color that are the basis of color video technology.

2.2 The Properties of Color—A Summary

The term "color" as used in color video, has a dual meaning—it is both a physical property of visible light and the perception of this property by human vision. Color as a physical property can be defined and measured in objective and precise terms. Its perception is subjective, it varies with individuals, and it depends on the surrounding environment so that it must be specified less precisely on the basis of an average or "standard" observer under "standard" conditions. A major task of video engineers is to establish and define measurable criteria for color images that coincide as closely as possible with their perception.

Each picture element in a color image has three basic properties. In objective terms they are luminance, hue, and saturation. The corresponding perceptual terms are brightness, color, and purity.

The term "luminance" has approximately the same meaning in color video as in monochrome (see Sections 1.8 and 4.4). As with monochrome, a color video system must have the capability of transmitting the luminance of each

---

1 Colorimetry originally meant the measurement of color with a colorimeter, but it now has the broader meaning of the science of color.

picture element, but in addition it must transmit and reproduce its hue and color purity or saturation.

2.2.1 Hue

The most prominent characteristic of color picture elements is their hue, the sensation created by visible light that is commonly described as its color (red, yellow, orange, etc.). The term "hue" is often used synonymously with "color."

Visible light is electromagnetic radiation having a spectrum of wavelengths[2] extending from approximately 400 to 780 nm (1 nanometer = 10^{-9} meters)[3], and its hue is determined by its wavelength. The band centers of the wavelengths that produce the sensation of common hues are presented in Table 2.1.

This table is based on monochromatic light, i.e., light of a single wavelength. With the important exceptions of pure spectral colors resulting from the excitation of atoms as in a sodium vapor light or a laser, the colors that appear in nature are polychromatic, i.e., they are a mixture of many wavelengths. Each has a dominant wavelength that establishes the visual perception of its hue, but it may contain radiation components with wavelengths that extend over the entire visible spectrum. White or gray light results when the radiation at all wavelengths is present in approximately equal amounts.

Radiation with wavelengths just beyond the limits of the visible spectrum also has important applications in specialized video systems. Wavelengths in the range from approximately 780 to 25,000 nm are known as near infrared while in the range of 100 to 400 nm they are known as ultraviolet.

2.2.2 Saturation

The appearance of any color can be duplicated by a mixture of white or gray light and a pure spectral color, the dominant wavelength, in the proper proportions (see Section 2.5). The ratio of the magnitude of the energy in the spectral component to the total energy of the light defines its saturation or "purity." A pure spectral color has a saturation of 100 percent while the saturation of white or gray light is zero.

2　Wavelength rather than frequency is customarily used to specify position on the visible and neighboring regions of the spectrum. Wavelength and frequency are related by the equation: [wavelength (m)] [frequency (Hz)] = 3×10^9 m/s = velocity of light.

3　Scientists sometimes use the Angstrom unit, 10^{-10} m, rather than the nanometer for specifying wavelength.

Table 2.1 Hue vs. wavelength

Hue	Approximate wavelength (nm)
Violet	400
Blue	450
Cyan	490
Green	520
Yellow	575
Orange	590
Red	640

2.3 Color Specifications

The three basic physical properties of color radiation—luminance, hue, and saturation—can be completely specified by its spectral distribution, a plot of its radiant energy vs. wavelength. Figure 2.1 shows the spectral distributions of incident light on a reflecting surface, the reflectivity of the surface, and the reflected light, which is the product of the first two.

A graph of amplitude vs. wavelength, however, is not always the most useful format for specifying color—numerical criteria are often more convenient—and the graph does not describe the relationship between the physical properties of color and their visual perception.

Scientists have studied the nature of this relationship for the past 400 years (Sir Isaac Newton carried out basic research on this subject in the early 17th century), and they have developed numerical criteria for specifying color that are stated in technical terms but are based on the subjective perceptions of thousands of viewers. Four of these criteria are commonly used in color television:

1. Primary color mix
2. Dominant wavelength and saturation
3. Luminance and color differences
4. Coordinates on the CIE Chromaticity Diagram

All of these criteria directly or indirectly specify three independent variables. This is consistent with the trichromatic theory of color, which states that the sensation of color results from the stimulation of three sets of cones in the retina, each with a different spectral sensitivity. The perceived hue and saturation of a color is determined by the ratios of the amplitudes of the responses of the sets of cones to the color stimulation. Thus two colors will appear to have the same hue and saturation if they stimulate the same response from the retinal cones, even if the distribution of their spectral energies is different. (Their *appearance* is also affected by differences in the viewing environment, a fact well known to painters.)

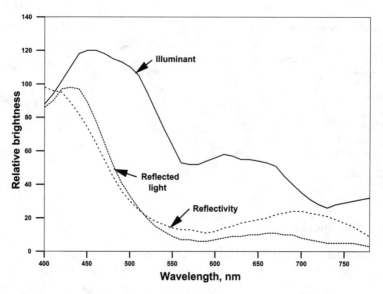

Figure 2.1 Spectral distribution of reflected light

2.4 The Primary Colors

A consequence of the trichromatic theory of color is that the hue and saturation of most colors can be duplicated by combining three *primary* colors in the proper ratio. This theory was one of the results of Sir Isaac Newton's work. It has since been developed further to meet the needs of the film and video industries.

There is wide latitude in the permissible spectral content of primary colors. The most important requirements are that they be independent, i.e., no one of the three can result from a combination of the other two, and that they be chosen so that the widest possible range of hues and saturations can be produced by suitable combinations of the three. Within these broad criteria, considerable variety is possible in their spectral characteristics.

Video colorimetry is basically different from that used in photography and painting or in the perception of color in real life. All of the latter are *subtractive* systems in which the picture or scene is illuminated by an external source of light such as sunlight, which includes components of many hues. The hue of the image is produced by the subtraction of color components by absorption, reflection, or transmission. Color television is an *additive* system since it produces hues by adding the primary color components. As a result, two types of primaries, subtractive and additive, have been defined.

Table 2.2 Subtractive primaries

Primary	Reflects or Transmits	Absorbs
Magenta	Red and blue	Green
Yellow	Red and green	Blue
Cyan	Blue and green	Red

2.4.1 Subtractive primaries

The subtractive primaries are magenta (a blueish-red), yellow, and cyan (a greenish-blue). Each absorbs one of the additive primaries and reflects or transmits the other two (see Table 2.2). Since magenta has a reddish cast and cyan is bluish, it is popularly (and erroneously) stated that the primary colors are red, yellow, and blue.

Subtractive primary colors are used in the form of inks for color printing. Because practical ink pigments do not produce perfect primary colors, laying down all three primaries does not create a good black. This problem is solved in color printing by using four inks: cyan, magenta, yellow, and black. This combination is referred to as CMYK color (the K is for blacK).

2.4.2 Additive primaries

Additive primaries (see Table 2.3) can be derived from the subtractive, for example, red can be produced by mixing magenta and yellow, the former absorbing green and the latter blue from the incident white illuminant.

The luminance of the additive primary colors, R (red), G (green), and B (blue), and the amplitude of their electrical analogues, E_R, E_G, and E_B, are the basic color television parameters, and all other color specifications are derived from them. After gamma correction (see Section 4.14), it is customary to designate the amplitude of the electrical signals as E'_R, E'_G, and E'_B. The outputs of the camera and the inputs to the kinescope are the electrical analogues of the three primary colors in the scene.

Additive primaries were seldom used until the advent of color television because it is virtually the only imaging system that creates pictures by the

Table 2.3 Additive primaries

Combinations	Hue of Combination
Red plus green	Yellow
Blue plus green	Cyan
Blue plus red	Magenta

addition of color components. The additive primaries are red, green, and blue. Other hues, including the subtractive primaries, can be produced by combining them.

It is not intuitively obvious that yellow light is produced by mixing red and green (!), but it can be confirmed by superimposing red and green spot lights on a screen.

2.5 Dominant Wavelength and Saturation

It can be shown, both by theory and experiment, that the luminance, hue, and saturation of every color in nature can be duplicated by a mixture of white or gray light and monochromatic light in the proper proportions, no matter how complex its spectral distribution. The wavelength of the monochromatic component of this mixture that establishes its perceived hue is known as the dominant wavelength.

The electrical analogues of the dominant wavelength and saturation are the basis for the NTSC, PAL, and SECAM systems that are used for the transmission of color by all present-day broadcast systems.

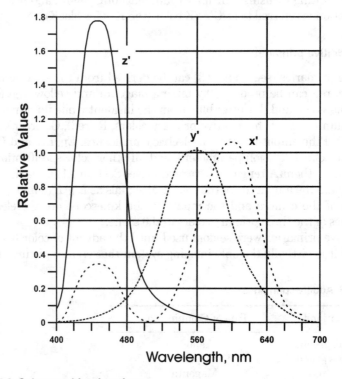

Figure 2.2 Color matching functions

2.6 Luminance and Color Difference Components

Colors can also be specified in terms of their *luminance* (Y) and *color difference* (R–Y and B–Y) components where R and B are the amplitudes of the red and blue primaries. The amplitudes of the corresponding electrical analogues are E_Y, E_R, and E_B or, after gamma correction, E'_Y, E'_R, and E'_B. The luminance component, E_Y, is given by the equation:

$$E_Y = 0.299E_R + 0.587E_G + 0.114E_B \tag{2.1}$$

The coefficients of the primary components are based on the sensitivity of the human eye to the primary colors. The magnitude of E_Y is then proportional to the perceived brightness of the scene. The electrical analogues are the basis for *component* recording and transmission, a format that is widely used for video recording and other applications with high demands for quality that do not require compatibility with broadcast transmission standards.

2.7 The CIE Chromaticity Coordinates

The CIE (*Commission Internationale de l'Eclairage*) chromaticity coordinates are based on a system for describing colors in terms of the color matching functions, x', y', and z', shown in Figure 2.2. These functions were developed on the basis of hundreds of observations, and are used to calculate the *tristimulus values*, X, Y, and Z:

$$X = \int_{380}^{780} L(\lambda)x'(\lambda)d\lambda \tag{2.2}$$

$$Y = \int_{380}^{780} L(\lambda)y'(\lambda)d\lambda \tag{2.3}$$

$$Z = \int_{380}^{780} L(\lambda)z'(\lambda)d\lambda \tag{2.4}$$

where: $L(\lambda)$ is the spectral energy density

x', y', and z' are the color matching functions.

The properties of the color matching functions are such that colors with the same tristimulus values will appear the same, even though their spectral distributions are different.

The chromaticity coordinates, x and y, are calculated from the tristimulus values:

$$x = \frac{X}{X+Y+Z} \tag{2.5}$$

$$y = \frac{Y}{X+Y+Z} \tag{2.6}$$

(Since $x + y + z = 1$ by definition, it is unnecessary to calculate z.)

The concept of chromaticity coordinates was a major milestone in the development of imaging system technology. It was developed during the late 1920s and early 1930s, and the coordinates of the colors in the visible spectrum were first calculated and published in 1931. It is an invaluable tool for the analysis and design of color television products and systems. It is also a link between the objective physical nature of color and its subjective perception.

The plot of the chromaticity coordinates of visible light is known as the CIE Chromaticity Diagram, Figure 2.3. The color matching functions and the Chromaticity Diagram are known collectively as the *CIE 1931 Standard*

Figure 2.3 The CIE Chromaticity Diagram

Table 2.4 Chromaticity coordinates of standard illuminants

Designation	Source	x	y
Illuminant A	Tungsten at 2856°K	0.4476	0.4074
Illuminant C	Daylight	0.3101	0.3516
Illuminant D$_{65}$	Daylight (revised)	0.3127	0.3290

Observer. A supplement to the 1931 paper, the *CIE 1964 Supplementary Standard Observer*, added data for viewing angles greater than 2°.

The following features of the Chromaticity Diagram should be especially noted:

1. The coordinates of spectral (monochromatic) colors are located on the horseshoe-shaped curve around the periphery of the diagram. Their wavelengths are indicated on the curve.

2. The coordinates of white light are located in the center of the diagram. The area on the graph corresponding to "white" light is rather large because there is an almost infinite variety of white hues. The three sets of coordinates for white light that have been established by the CIE are tabulated in Table 2.4. The spectral power distributions of these illuminants are shown in Figure 2.4.

   Table 2.4 includes a reference to *color temperature*, a fifth criterion for specifying the color (see Section 2.3) of illuminants. The use of color temperature as a specification results from the fact that the spectral distribution of radiant energy from an incandescent body (known as a *black body* or *Planckian* radiator) is determined solely by its temperature, usually expressed in degrees Kelvin (°K). As the temperature of a body is raised, its color changes from red to white, a fact reflected in the common terms "red hot" and "white hot."

   The spectral distribution from other illuminant sources, e.g., flourescent lights, is different from incandescent sources and the concept of *correlated* color temperature was introduced. This is the temperature of a Planckian radiator whose perceived color most nearly matches that of the source being defined. The use of color temperature as a specification has usually been limited to illuminants and has become less common as nonincandescent sources have assumed greater importance.

3. All visible colors are located within the area bounded by the spectral curve and the line (the *alychne*) joining its red and blue ends.

4. Every color whose coordinates lie on a line joining a point at the white center of the diagram to the outer spectral curve have the same hue. The dominant wavelength for the color is determined by the point at which the line intersects the spectral curve. Its saturation is indicated

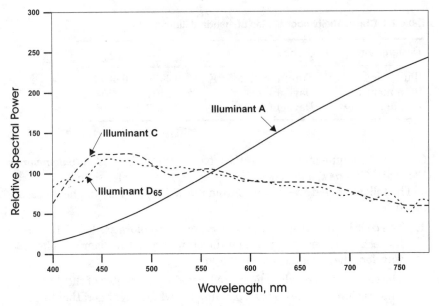

Figure 2.4 Spectral distribution of standard illuminants

on its location along this line, ranging from zero at the center to 100 percent at the spectral curve.

5. The coordinates of a primary set are located at the corners of a triangle. All colors that can be reproduced by a given set of primaries have coordinates within the triangle. Two primary sets for color display devices are shown in Figure 2.3; they are:

   - The NTSC/FCC standard primaries established in 1953.
   - The SMPTE C standard primaries established in 1982.

   These primary sets are described in Section 2.9.

6. The specification of color primaries is one of the most important applications of chromaticity coordinates (see Sections 2.8 and 2.9 below).

2.8 Camera Colorimetry

2.8.1 Camera primaries

The camera generates three primary signals that are the electrical analogues of the red, green, and blue visual components of the scene (see Chapter 8). Their colorimetry is specified by graphs of response vs. wavelength. They are determined initially by filters in an optical color splitter, but they can also be modified electrically in a camera matrix circuit by a process known as *matrixing* (see Section 8.7). With matrixing it is possible to achieve a spectral response that would be impossible with optical filters alone.

The amplitude of each primary signal is determined by the product of the spectral content of the scene and the colorimetry of the primary integrated over the entire spectrum. For example, the amplitude of the red signal, E_R, is given by Eq. (2.7):

$$E_R = R_C \int_{380}^{780} L(\lambda)R(\lambda)d\lambda \tag{2.7}$$

where
E_R = output voltage of the red channel

R_C = constant relating electrical output of the channel to luminance

λ = wavelength

$L(\lambda)$ = luminance of picture content

$R(\lambda)$ = spectral characteristic of red camera primary

The camera primaries must be compatible with the primaries of the display device, and this has led to the need for the establishment of display-device primary standards to provide the basis for the design of camera colorimetry.

The spectral content of the scene illuminant also affects the choice of primaries. If a scene is illuminated by incandescent light and the camera primaries are adjusted for daylight such as illuminant C, the reproduced image will appear reddish. Compensation for this effect is achieved by readjusting the camera matrix circuits, and a switch can be provided for these circuits with two positions, one for illuminant C and one for incandescent light.

The camera primaries recommended by the NTSC and adopted by the FCC in 1953 were based on compatibility with the NTSC recommendations for display device primaries and illuminant C. Their spectral responses are shown in Figure 2.5.

These primaries contain negative lobes that cannot be produced optically. This problem has two solutions. One is to ignore the negative lobes and the sidelobe in the red primary, as shown in the cross-hatched area in Figure 2.5. This is a compromise that produces satisfactory results but with some distortion. The other is to employ matrix circuits to create the negative lobes electrically (see Figure 2.6).

Another departure from the NTSC/FCC standards is required by industry practice, described in Section 2.9, of employing receiver primaries that deviate considerably from the original FCC specifications in order to provide greater screen brightness. This requires a corresponding modification in the camera primaries for the best results.

2.8.2 Matrix circuits

Electric matrix circuits convert an array of coefficients of independent variables into another array through a mathematical transform. They can be used to alter the spectral response of camera primaries, either to provide negative lobes that cannot be produced optically or to compensate for changes in the scene illuminant.

The operation of these circuits is illustrated by the example of the NTSC green primary shown in Figure 2.6. Optical filters cannot produce the negative values, but they can be produced electrically by a matrixing operation:

$$E_{GM} = E_G - 0.2E_B - 0.05E_R \qquad (2.8)$$

Figure 2.6 shows the close approximation of the matrixed signal to the NTSC primary.

This is a simple example involving only the addition and subtraction of linear functions. More complex circuits make use of nonlinear elements so that the matrix coefficients are dependent on the signal level. With these systems, the operator can "paint" the picture in accordance with esthetic judgments. Also, they make the precise matching of cameras possible.

2.9 Display Device Primary Standards

The colorimetry of receiver display devices is determined by the charac-

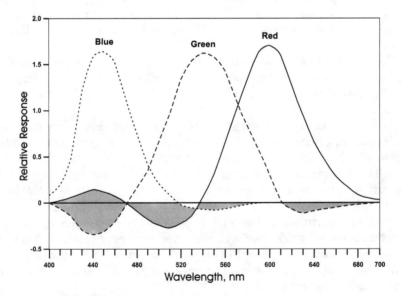

Figure 2.5 NTSC/FCC standard camera primaries

Figure 2.6 Green primaries—effect of matrixing

teristics of the phosphors used in the picture tube and is specified by their CIE coordinates.

Since the number of receivers far exceeds the number of cameras, receiver display device performance and cost are the most important factors in the selection of primary coordinates. It becomes the responsibility of manufacturers of cameras to design their colorimetry to be compatible with receiver display devices. Because of their impact on system performance and cost, the selection of display primaries has been the subject of extensive research. Standards have been established and modified from time to time in response to changes in system objectives and component capability (particularly picture tube phosphors).

The coordinates of five important display device primary sets, existing and proposed, are tabulated in Table 2.5.

2.9.1 NTSC/FCC primaries

The coordinates of display device primaries for television broadcasting, as proposed by the NTSC and adopted by the FCC in 1953 as part of its specifications for color broadcast standards, are plotted in Figure 2.3.

2.9.2 SMPTE Standard C primaries

With growing experience in receiver manufacture in the years after 1953, it was found that greater screen brightness could be obtained with primaries

Table 2.5 Chromaticity coordinates of different primary sets

	NTSC/FCC		SMPTE C		240M				274M	
					Phase 1		Appendix Phase 2			
	x	y	x	y	x	y	x	y	x	y
R	0.67	0.33	0.635	0.340	0.630	0.340	0.67	0.33	0.640	0.330
G	0.21	0.71	0.305	0.595	0.310	0.595	0.21	0.71	0.300	0.600
B	0.14	0.08	0.155	0.070	0.155	0.070	0.16	0.07	0.155	0.060
Color Gamut	Wide		Narrow		Narrow		Wide		Narrow	

that differed from the original NTSC recommendations. Customers preferred brighter pictures, and many receivers were manufactured with revised primaries, even at the expense of a small loss in color fidelity. In recognition of this reality, the SMPTE (Society of Motion Picture and Television Engineers) issued a revised primary color standard for studio monitors in 1982 that was known as SMPTE Standard C. (Receiver manufacturers were not required to adopt this standard, but it was hoped that the use of a common standard for monitors would aid in standardizing the colorimetry of broadcast signals.)

The coordinates for Standard C are plotted in Figures 2.3 and 2.7. The plots show that the added brightness is obtained at the expense of a more limited gamut of green, yellow, and red hues but with a slightly greater gamut of blues. The compromise is not serious, however, as illustrated in Figure 2.7, which compares the gamut of spectral responses of typical paints and dyes with Standard C. The color gamut of the television system (the area within the triangle) compares favorably with the color gamut of paints.[4]

2.9.3 SMPTE 240 (Appendix) standard primaries

The color primaries included in the SMPTE HDTV standard 240M essentially duplicate the SMPTE C standard with the restricted color gamut. It was hoped, however, that it would not be necessary to limit the gamut indefinitely because of the current limitations of kinescope phosphors. The 240M standard, therefore, includes an Appendix that increases the color gamut of the primaries so they very nearly duplicate the original FCC/NTSC

---

4   It should be noted, however, that the *Pointer colors*, a list of real surface colors developed by M. R. Pointer in 1980, includes a larger gamut of 576 colors. The standard C primaries will represent less than half of these. (See Section 2.8.3 for the efforts of industry to specify a standard for a wider color gamut.)

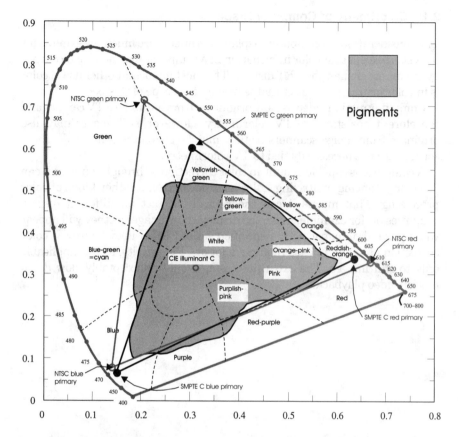

Figure 2.7 Color gamuts of dyes and paints. From K. Blair Benson (ed.), *Television Engineering Handbook,* **McGraw-Hill, New York, 1986.**

values. The adoption of the Appendix standard was planned as a Phase 2 of HDTV colorimetry. The coordinates of 240M and 274M (Appendix) are tabulated in Table 2.5.

2.9.4 SMPTE 274M standard primaries (proposed)

The SMPTE Standard 274M has been proposed for use with the Grand Alliance HDTV system (see Chapter 6), which is supposed to provide optimum performance without significant compromises. With the passage of time, however, it did not appear realistic to assume that phosphors could be developed that would combine high brightness with a wide color gamut, i.e., that would meet the SMPTE 240M (Appendix) standards. Accordingly, the primaries that were recommended for SMPTE 274M are virtually identical to the SMPTE C and 240M standards.[5]

2.10 Colorimetry of Computer Displays

The considerations of computer display colorimetry are much the same as for television except that color information in a computer may need to be output by means other than the CRT display. The most important other use of color is in color printing, a subject that is beyond the scope of this book.

Computer motion video usually originates from a standard TV camera and therefore it has standard TV colorimetry. However, still images may also originate from image scanners or they may be computer generated—these sources are not necessarily tied to TV colorimetry.

A computer is capable of built-in color correction—through software it can do color balancing, matrixing, color gamma control, or other kinds of color processing. That means it can theoretically correct for different display primaries or, for example, make the display show images as they will appear when printed on paper with a specified set of CMYK inks. However, color correction of digital images requires a lot of CPU activity, so it is applicable only to still images. The extra processing would unacceptably slow down motion video playback.

---

5   Grand Alliance Expert Group on Scanning Formats/Compression, *Report on Colorimetry*, July 1995.

3

Introduction to
Digital Technology

3.1 Digital

Digital means "comprised of digits or numbers." *Digital video* then is video represented by a series of numbers. This chapter explains how images that contain smoothly varying brightness, hue, and saturation values can be conveyed by a sequence of numbers and why it is so valuable to do that.

3.2 Number Systems

To most people, a number is one or more of the arabic numerals: 0, 1, 2, 3, 4, 5, 6, 7, 8, or 9. Each numeral in this system is a *digit* that can have one of ten possible values. No more than ten values are represented by a single digit. It is a *base-10* number system. Of course, more than ten values are obtained by combining multiple digits, such as in the number "25." The "2" represents a value of 20 (2 multiplied by the base 10), since it is in the second position to the left. Similarly, in the number "325," the "3" represents 300—the 3 is multiplied by the base squared. By combining enough digits, where each one to the left is multiplied by a successively higher power of the base, any *integral* value can be represented.

There is nothing magic about the base-10 for a number system. Any number can be used as the base—some common bases and their names are:

base-2	binary
base-3	ternary
base-8	octal
base-10	decimal
base-16	hexadecimal

An analog system allows an infinite number of values to be possible and valid; it can be thought of as a number system of base infinity. In such a system, one cannot distinguish between valid values and errors because all values are valid. In a system of lower base, however, circuit values that are not exactly equal to an integral digit value are clearly in error and are prohibited. In electrical transmission of numerical values, the process of *quantization* is used to force all output to be exactly integral values.

3.2.1 Binary number systems

Quantization effectively eliminates transmission errors that are less than the reciprocal of the base. That means the lower the base value, the more robust is the transmission capability of the system. So the base-2 (binary) system is the most tolerant of transmission errors. However, as the base is lowered it takes more digits to reproduce useful numbers and this makes any processing or transmission more complex.

In the early days of television and computing, different choices were made regarding the tradeoff between robustness and complexity based on differing needs. In television, the inherent redundancy in the scanning process and the characteristics of human vision provided considerable tolerance for errors, so the simplicity of an analog system was an easy choice. However, in computers, the requirement for complete elimination of errors at almost any cost shifted the tradeoff to a binary number system even though it was much more complex.

Today, with solid-state integrated circuits, the complexity of binary circuits has moved inside IC chips and most system designers now think it is easier to build binary hardware than analog hardware. The complexity issues of binary systems have been taken over by IC designers so that system designers now work at a much higher level and don't have to worry about the details of fundamental binary processes. In video systems, the choice between binary or analog is now just a matter of whether there are ICs available to perform a process at appropriate speeds for the video performance required. Of course, this is oversimplifying things, but the conclusion that nearly all video systems are moving to binary technology is still correct.

The developmental energies of the computer and the IC industries has primarily been directed to binary number systems and little hardware is available for anything other than base-2 systems. For this reason, the word *digital* has become synonymous with *binary*. It is used that way throughout this book.

3.2.2 Binary numbers

A binary digit, called a *bit*, can have only two values: 0 or 1. It takes many more than two levels to make a useful digital video system, so multiple bits per pixel value must be used. A typical value is 8 bits per pixel—that gives

Decimal value = 183

Figure 3.1 Binary numbers

256 levels for a pixel value. Computer people call a group of 8 bits a *byte*. Binary numbers are represented as a series of digits (0 or 1). For example, the number 5 would be 0000 0101 in binary. Further explanation of this is in Figure 3.1.

Working directly with binary numbers quickly becomes tedious, so there are several techniques to simplify the process. One that is widely used is *hexadecimal (hex)* notation that represents each 4-bit group as a single hexadecimal digit. Hex digits have 16 possible values, which is accomplished by using the numerals 0–9 and the letters A–F for values above 9. This is shown in Table 3.1. Because hex numbers that contain no digit values above 9 can become confused with decimal numbers, it is common to put the suffix "H" or the prefix "0x" on hex values. Some examples:

Decimal	Binary	Hex	Hex-Decimal
0	0000 0000	00H	$(0*16 + 0)$
5	0000 0101	05H	$(0*16 + 5)$
31	0001 1111	1FH	$(1*16 + 15)$
255	1111 1111	FFH	$(15*16 + 15)$
2550	0000 1001 1111 0110	09F6H	$(0\times4096 + 9\times256 + 15\times16 + 6)$

Notice that the binary values above are shown in groups of 4 bits. This helps in the process of conversion between binary and hex.

A digital audio system usually requires more than 8 bits per sample—16 bits per sample (2 bytes) are often used. This gives 65536 levels (2^{16}).

Table 3.1 Decimal, binary, and hexadecimal values

Decimal	Binary	Hexadecimal
0	0000	0
1	0001	1
2	0010	2
3	0011	3
4	0100	4
5	0101	5
6	0110	6
7	0111	7
8	1000	8
9	1001	9
10	1010	A
11	1011	B
12	1100	C
13	1101	D
14	1110	E
15	1111	F

3.3 Analog-to-Digital Conversion

Conversion from analog to digital (ADC) may be required at several points in a system, depending on how much of the system is digital and what parts they are. An all-digital system would perform ADC at the imager and everything beyond that would be digital. Another system might use digital technology only for recording. In that case, ADC (and DAC—see Section 3.9) would be inside the recorder(s). Of course, if video is being generated from scratch by a computer, there is no ADC.

Section 1.5 explained how a video signal from an imager is sampled by the scanning process. However, the resulting video signal is still a series of analog values. To convert completely to a digital data stream, the analog values must be sampled and *quantized*. That subject was introduced in Chapter 1. But even that is not the whole story because there are many ways to translate sampled and quantized values into groups of bits— a process called *encoding*. This section expands on sampling, quantizing, and encoding as parts of the process of *analog-to-digital conversion* or *digitization*.

Figure 3.2 shows functional diagrams of A-to-D converters for composite and component analog signals. The first element (optional) is an anti-aliasing filter. This is a low-pass filter with a cut-off below the Nyquist limit (see Section 1.5.1.1), i.e., below one-half the sampling rate. This eliminates the signal frequency components that could cause aliasing. It is desirable to have the sampling rate high enough so that there is a reasonable separation between the nominal cut-off frequency of the filter and the Nyquist frequency. This makes it possible to use a filter with a gradual cut-off without the

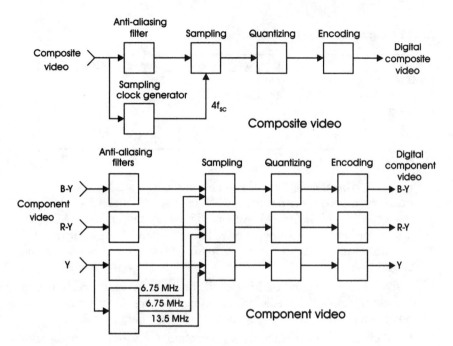

Figure 3.2 Analog-to-digital converters

excessive phase shift of sharp cut-off filters. In the component format, the cut-off of the filter in the color difference channels is approximately one-half of that in the luminance channel in accordance with their reduced bandwidth. The anti-aliasing function in the D-to-A converter is followed by the sampling, quantizing, processing, and encoding functions.

Detail specifications for the ADC process depend on where it is occurring in a system. That will determine what is the analog input format and what digital output format will be needed. The resulting digital data stream must conform to a format standard so that other systems will know what to look for in the data. It may be a proprietary format that applies to a specific situation, or it may be a public standard that will be recognized worldwide. This is one of the major concerns of national and international standardizing bodies, including the SMPTE, IEEE, EBU (European Broadcasting Union), and ITU (International Telecommunications Union).

3.3.1 Horizontal sampling rate

The horizontal sampling rate or its reciprocal, the sampling interval, is a fundamental specification. It should meet a number of conditions:

1. It should meet or exceed the Nyquist limit to avoid degrading the horizontal resolution and to minimize aliasing (see Section 1.5). This re-

Table 3.2 Digital television bit rates and bandwidth

System	Bit rate, Mb/second	Bandwidth, MHz
NTSC ($4f_{sc}$, 8 bpp)	114.5	58
PAL ($4f_{sc}$, 8 bpp)	141.9	72
Component		
Luminance	108	54
Color difference	54	27

quires it to be at least twice the upper limit of the analog signal frequency components or a rate of 8.5 MHz (2 × 4.25 MHz) for NTSC systems and 10.0 MHz for PAL systems. These are minimum rates, and it is standard practice to employ rates that are somewhat higher than the minimum to permit the use of an anti-aliasing filter with a gradual cut-off.

2. It should be an integral multiple of the line rate. With this relationship, the sampling points on adjacent scanning lines are directly above each other. This is illustrated in Figure 3.3 for component signals.

3. For digitizing composite signals that contain a subcarrier, the sampling rate should be an integral multiple of the subcarrier frequency. This is not an absolute requirement, but signal processing and decoding are greatly simplified when the sampling rate meets this criterion.

4. For component signals, it should be the same for 525-line 30-frame systems and 625-line 25-frame systems.

In response to these requirements, the television industry has achieved a degree of standardization in sampling rates as shown in Table 3.3. (The bandwidth figures in the table assume simple binary transmission. With more sophisticated modulation, bandwidths can be less—see Section 3.7.2.)

The sampling rate of 13.5 MHz was chosen for component signals because it is an integral multiple of both the NTSC and PAL line rates. It is not, however, an integral multiple of either the NTSC or PAL subcarrier frequency. For composite signals, sampling rates of $3f_{sc}$ or $4f_{sc}$ can be used. $3f_{sc}$ does not exceed the Nyquist frequency by a comfortable margin, and therefore a sampling frequency of $4f_{sc}$ is the most commonly used and is an SMPTE standard (244M).

In the computer industry, the choice of sampling rate depends not only on the input signal format as described in the list above, but also on the desired output display format. If a composite signal is being sampled, there is a need to decode the composite format because computers generally use a component format to describe the color value of each pixel. There are many variations of computer formats as explained in Chapter 7.

Sampling rate considerations for HDTV are discussed in Chapter 6.

Table 3.3(a) Sampling frequencies—composite signals

Standards:	NTSC		PAL
		SMPTE 244M	
	3fsc	4fsc	4fsc
Bandwidth, MHz	4.2	4.2	5.5
Subcarrier (fsc), MHz	3.58	3.58	4.43
Sampling frequency, MHz	10.74	14.318	17.72
Samples per total line	682	910	1134
Samples per active line	576	768	939
Bit rate, Mbits/s (8-bit words)	85.9	114.5	141.8

Table 3.3(b) Sampling frequencies—component signals

Standards:	525 Lines, 59.94 Fields ITU BT.601-4 4-2-2 SMPTE 125M	625 Lines, 50 Fields ITU BT.601-4 4-2-2
Luminance channel		
Bandwidth, MHz	5.5	5.5
Sampling frequency, MHz	13.5	13.5
Samples per total line	858	864
Samples per active line	720	720
Bit rate, Mbits/s	108.0	108.0
Color difference channels		
Bandwidth, MHz	2.2	2.2
Sampling frequency, MHz	6.75	6.75
Samples per total line	429	432
Samples per active line	355	358
Bit rate, Mbits/s (8-bit words)	54.0	54.0

3.3.2 Location of sampling points

The locations of the sampling points on individual lines for the ITU BT.601-4 component format are shown in Figure 3.3. The luminance and two color difference sampling pulses are synchronized so that the color difference points are co-sited with alternate luminance points.

In the formats for composite signals in Table 3.3(a), the sampling points have a specified relationship with the phase of the burst. For both PAL and NTSC, there are four sampling points for each cycle of the burst. For PAL, the sampling points are at the 0°, 90°, 180°, and 270° points of the burst waveform. For NTSC, the first sample is located at 57°, the I axis (see Section 3.14.1). The others are located at 147° (the Q axis), 237°, and 327° (see Figure 3.4). This is known as I-Q axis sampling.

The sampling rate and phase must be synchronized with the line and subcarrier frequencies and phases in order to maintain these precise loca-

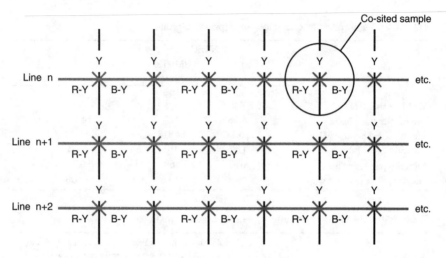

Figure 3.3 Location of sampling points, component signals

tions. For composite signals the synchronizing element can be the subcarrier burst. This system element that generates the sampling frequency is called the *sampling clock*.

3.3.3 Quantizing

The next step in the A-D process is to quantize the pulse stream resulting from the sampling process.

3.3.3.1 Quantized signal levels Quantizing is depicted in Figure 3.5. The signal amplitude range is divided into discrete intervals, and a fixed or quantum level is established for each interval. This level is assigned to

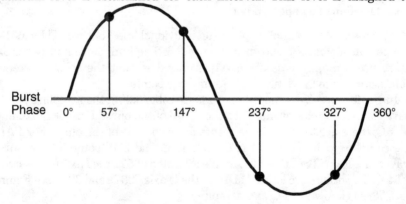

Figure 3.4 Location of sampling points, composite signals

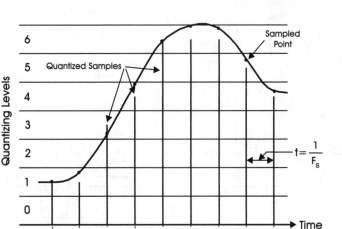

Figure 3.5 Quantizing

samples having an analog value that falls within the interval. The difference between the quantum level and the analog signal at the sampling point is the quantizing error.

Most video quantization systems are based on 8-bit words with 2^8 or 256 discrete values. (2^{10} or 1024 levels are sometimes used for pre-gamma signals in cameras.) Not all of the levels may be used for the video range because of the need to use some of the values for data or control signals.

Figure 3.6 shows typical utilization of the 256 8-bit quantized levels for composite and component signals. The entire composite signal, including sync pulses, is transmitted within the quantized range of levels 4 to 200.

In component systems, sync pulses fall outside the quantized range and are not transmitted. They are unnecessary because line and field synchronization is provided by digital codes. The luminance signal occupies levels 16 to 235. The color difference signals can be either positive or negative, and the zero black level is placed at the center of the quantized range because there are no negative numbers in the binary number system used for encoding quantized signals. Note that neither range includes the binary words 0000 0000 (0) or 1111 1111 (255), and they are available for synchronizing.

3.3.3.2 Quantizing errors   If the sampling rate is harmonically related to a signal frequency component, the result of quantizing errors is distortion of the waveform. In the more unlikely event that the relationship is random,

Figure 3.6 Utilization of quantizing levels for composite and component video

the result of quantizing errors is noise. The peak-to-peak signal-to-rms quantizing error ratio in dB is given by Eq. (3.1):

$$\frac{S}{Q_e} = 10.8 + 6.02n + 10\log\left(\frac{F_s}{F_v}\right) \tag{3.1}$$

where: S = peak-to-peak signal

Q_e = rms quantizing error

n = the number of quantum levels as a power of 2

F_s = the sampling rate

F_v = the video bandwidth

For $n = 8$, $F_s = 14.4$ MHz, and $F_v = 4.2$ MHz, $S/Q_e = 64.3$ dB.

The nature of the noise depends on the magnitude of n. If n is 6 or greater, the noise will be random as from thermal effects. If n is less than 6, it will appear as false contours on sharp edges.

Equation (3.1) assumes that the original analog signal is noise free. With a noise-free input signal, the sampled waveform will be distorted when reconverted to an analog signal. For example, a diagonal line in the image will be reproduced as a stairstep with discontinuities equal to the quantizing intervals. If there is noise in the analog signal, however, it will partially or totally obscure the discontinuities. If the rms value of the noise exceeds one-third the quantizing interval, the discontinuities will disappear in the noise. The ratio, S/Q_e (in dB), under these conditions is given by Eq. (3.2):

$$\frac{S}{Q_e} = 6.02n \tag{3.2}$$

When $n = 8$, $S/Q_e = 48.2$ dB.

In some cases it may be desirable to trade off signal distortion for noise by deliberately introducing noise. This is known as *dithering*, an intentional rapid and random variation of some parameter of a signal that eliminates the visibility of quantizing discontinuities.

3.3.4 Encoding

The final step in digital-to-analog conversion is to *encode* the quantized levels of the signal samples. The encoding and processing of digital signals has been the subject of intensive mathematical analysis. Hundreds of papers have been published on the subject as well as a number of excellent textbooks. A complete description of this subject is beyond the scope of this chapter. Rather, the purpose here is to describe the basics and define the principal terms of digital signal encoding as a foundation for a more extensive study of the subject.

3.3.4.1 Pure binary encoding When the binary digits directly represent values multiplied by powers of 2, the encoding is called *pure binary*. The range of 8-bit pure binary encoding goes from 0 to 255 (256 levels). This is satisfactory to represent composite video waveforms, but it is not suitable to represent zero-center (bipolar) waveforms such as audio or color-difference signals. In that case, analog zero has to be offset to the center of the quantizing range (128). This is known as *offset binary* encoding, but it is not satisfactory because the offset values tend to interfere when the signal is processed.

3.3.4.2 Two's complement encoding A better strategy for bipolar waveforms is what is known as *two's complement* binary encoding, illustrated in Figure 3.7. In two's complement, the lower half of the pure binary range (when MSB = 0) is defined as negative numbers. That is accomplished by offsetting the analog signal so that it fits into the normal range of binary values (0–255 for 8 bits) with zero at the center, digitizing, and then inverting the MSB of the resulting digital values. Since zero remains at digital zero,

Figure 3.7 Two's complement encoding

this system of encoding does not interfere with processes such as addition (mixing) that depend on that.

The largest positive decimal number that can be represented in the two's complement code is $2^{(n-1)}-1$, where n is the number of bits. The absolute value of the largest negative number that can be represented is $2^{(n-1)}$. For 8-bit values, the range is from -128 to $+127$. The two's complement code for a positive number equal to or less than $2^{(n-1)}$ is the same as its pure binary code. The two's complement code for a smaller negative number, $-X$, is equal to the pure binary code for the positive number, $2^n-|X|$. The first digit in this code for positive numbers is always 0 while for negative numbers it is 1.

More sophisticated coding is often employed as a part of an error management system. Some examples of that are discussed in Section 3.8.

3.4 Serial and Parallel Formats

The bits in a digital code can be transmitted or recorded either serially over a single path, or in parallel with a separate path for each numbered bit in the code. Thus, the transmission (or recording) of an 8-bit byte by parallel transmission requires eight separate communication paths. The use of parallel transmission simplifies the terminal equipment, but the necessity for a multiconductor transmission path makes it impractical except for short distances such as the interconnections within a single equipment or within a studio. Serial transmission, therefore, is more commonly used for medium and long distance transmission.

In parallel transmission, a separate circuit is provided for each of 8 (or 10) bits. In serial transmission, the 8 (or 10) bits in each code, which are produced simultaneously, are stored in a shift register and read out sequentially to a single circuit.

The minimum bandwidth requirement of a baseband digital channel is nominally one-half the bit rate. An 8-bit parallel system would require eight separate channels, each with one-eighth the bandwidth of an equivalent serial channel. Each mode has advantages and disadvantages.

Video bit streams are usually created in a parallel format, and the cost and complexity of serialization equipment are avoided with parallel transmission. Thus, parallel systems are cheaper and simpler for local signal distribution. On the other hand, multiconductor cable is more expensive and bulky, and parallel transmission is more costly for long distance transmission. Matching the delay times of parallel paths may be difficult in parallel systems, and there is the possibility of cross talk between paths. Matrix switching between multiple inputs and outputs is difficult. Signal processing equipment is more costly and complex for the parallel mode. As a result, serial transmission is widely used where distances are great or for complex signal processing. This has led to the development of inexpensive serialization integrated circuits and the cost of serialization today is minimal.

In the absence of data compression, the bandwidth requirements of either serial or parallel transmission, 54 MHz or higher for 8-bit composite or luminance signals, are a serious handicap to the use of digital long distance transmission. The bit rates for 8-bit signals do not fall neatly into one of the steps in the standard 1.54 Mb/s T-1 communications hierarchy, which has levels at 44.7 and 274.2 Mb/s.

On the other hand, video compression technology is rapidly advancing and bandwidth problems will soon be reduced for digital video transmission at all levels. For example, the compression developed for HDTV delivers high-resolution video at 19 Mb/s or less. By the use of advanced modulation methods (see Section 3.7.2) this data rate can be transmitted in a bandwidth of 6 MHz or less. In fact, many systems are being designed today for *multichannel* serial transmission of compressed digital video.

Digital video transmission on metallic coaxial cable is limited to short distances unless repeater amplifiers are used as in CATV (see Chapter 16). The normal bandwidth of C-band satellite and microwave channels is 36 MHz (see Chapter 17).

Fiber optics is the best prospect for increased bandwidth of wired transmission. Fiber-optic circuits usually employ digital transmission and have circuit capacities upwards of 200 Mb/s.

Although bandwidth receives the major emphasis in digital transmission system design, envelope delay distortion must be kept within reasonable limits to avoid interference between adjacent bits in the bit stream. Digital signals are considerably more tolerant to noise and amplitude linearity than analog signals.

3.5 Binary Arithmetic

A branch of mathematics known as Boolean algebra (named for the English mathematician, George Boole), deals with the relationships between and manipulation of binary numbers. It includes laws on deductive logic that are used for the analysis of complex relationships between many interacting components, for example in the design of machine control systems.

3.5.1 Binary addition

In principle, the addition of pure binary numbers is similar to that of decimal numbers. Starting with the LSB column, the binary digits in each column are added. If the sum of the digits, n, in a column is 0 or 1, that number is entered as the total for that column. If n exceeds 1 and is an even number, 0 is entered as the total and $n/2$ is carried into the next column to the left. If n is an odd number, 1 is entered in the total column and $(n-1)/2$ is carried into the next column to the left. Two examples are:

	Binary number	Decimal equivalent	Binary number	Decimal equivalent
Carryover	111		1	
Binary A	1101	13	101	5
Binary B	1111	15	111	7
Binary C	0011	3	Total 1100	12
Total	11111	31		

Offset binary numbers can be added in the same fashion, but it must be remembered that the sum will include the sum of the offsets.

Addition of two's complement binary numbers is often necessary. Addition of two's complement numbers can be carried out with the same rules as with pure binary numbers.

	Binary number	Decimal equivalent
Carryover	101	
Binary A	0111	7
Binary B	0100	4
Binary C	1011	−5
Total	0110	6

When adding positive numbers, whether using a pure binary or a two's complement code, care must be taken that the sum does not exceed the range available for the bit number of the system, i.e., 2^n for pure binary code and 2^{n-1} for two's complement. If the overload results from negative numbers in the two's complement format, some bits may actually be deleted. Some systems have overflow control circuits to avoid this effect.

3.5.2 Binary multiplication

Binary multiplication is not as simple as addition, but it is essential to many signal processing functions such as fading and mixing. Multiplication can be accomplished by repeatedly adding the bits in each multiplicand column in a shift register in accordance with the value in the multiplier. The sum for each column is weighted in accordance with its significance, ranging from the LSB to the MSB.

Binary notation	Decimal conversion
[101]	[4+0+1]=5
×[111]	[4+2+1]=7
=[100011]	5×7=[32+3+1]=35

This procedure requires a tremendous number of operations, but it is accomplished quickly and accurately with integrated circuits.

3.5.3 Boolean logic

An important feature of computers is the ability to perform *logical* operations on or between binary values. A simple application of logic is found in the situation where a group of bits—say a byte—is used to represent eight individual yes/no flags instead of a numerical value. This is often done in file header structures to define attributes of the file's data. Using the logic operations, the bits can be individually set, reset, or tested.

There are four basic logic operations: NOT, OR, AND, and exclusive-OR. Their behavior can be described by means of a *truth table*, which shows all the possible outcomes of the operation. Figure 3.8 shows truth tables for each of the four basic logic operations. The truth tables show the outcome of the operation as a matrix (shaded area) based on the possible values (0 and 1) of the 1-bit operands A and B. (The NOT operation has only one operand, A.)

Figure 3.8 Truth tables for the basic logic operators: NOT, OR, AND, and XOR

3.5.3.1 NOT operation This operation simply reverses the value of a bit, regardless of what it was.

3.5.3.2 OR operation This function outputs a 1 when either or both of the input values is 1. It outputs 0 when both inputs are 0.

3.5.3.3 AND operation This function outputs a 1 when both inputs are 1. If either input is 0, the output is 0.

3.5.3.4 Exclusive-OR This operation, often called XOR, is the same as OR except that it outputs 0 when both inputs are 1.

3.5.4 Bit-flag example

Returning to the example of a byte that contains 8 bit-flags, the logic operators are used as follows to control the flags:

The OR operator is used to set a bit flag. In logical operations, the flag bits are identified by their decimal values as indicated in Figure 3.1. Thus, to set the second flag from the right (0000 0010—whose value is 2), the flags byte is ORed with a byte whose value is 2.

The NOT and AND operators are used to clear a flag. In this case, a byte must be created that has all bits 1 except the one to be cleared. This is done by using the NOT operator on the flag value and ANDing the result with the flags byte.

The XOR operator has the property of reversing the value of a bit. Thus, if the flags byte is XORed with 2, the 2 bit will be flipped.

Various combinations of these operations can be used to modify several flags at once.

3.6 Digital Signal Processing

Digital video systems have the capability to perform signal conversion and processing functions that would be difficult or impossible with analog signals. After an A-to-D conversion, the signal consists simply of a series of numbers that can be manipulated with little or no loss of picture quality. This capability is enhanced by the ease with which the bit signals can be stored in memories for rapid retrieval as needed.

Digital processing can be done in special-purpose hardware that performs video processes on a signal in real time. This is the form of most digital devices intended to be embedded in video systems. However, it is also possible to do many processes in general-purpose, programmable, hardware (a computer) under control of software. This has exciting prospects because of the extreme flexibility that is available, but many video processing functions are still too complex to be handled by software in real time. That is changing every day as the power of computing hardware continues to advance, but most digital video systems today use a combination of hardware and software processing. This subject is discussed further in Chapter 7.

Four important signal processing functions that can be performed with digital signals are time-base error correction (Section 3.6.1), standards conversion, e.g., PAL to NTSC (Section 3.6.2), postproduction editing (Chapter 11), and bandwidth reduction (Section 3.11).

3.6.1 Time-base error correction

Maintaining a stable time base for video signals and accurately preserving their original time relationships is a problem with video tape recorders because of timing instabilities introduced by the mechanical systems involved. The time-base errors at the output of uncorrected helical scan recorders with their long recording paths on an elastic tape medium can be particularly severe.

An adequate degree of stability can be achieved for normal television viewing without time-base correction with consumer-type recorders that use the "color under" analog technique (see Chapter 10). However, more precise correction is required for professional recorders because of the necessity of performing editing and other postproduction functions on their outputs (see Chapter 11). These tasks require fully stabilized signals.

A number of methods have been developed for providing precise time-base correction (TBC) with the signal in analog form, using a variable electrical delay line as the basic element. Timing stability better than ± 2 nanoseconds has been achieved; that is approximately ± 2.5° at the NTSC color subcarrier frequency. This is a satisfactory tolerance, but the range of correction required for helical-scan recorders is so great that long and expensive delay lines, covering several microseconds or more are required.

Time-base correction can be performed more cheaply and easily with digital techniques. The availability of low-cost, wide-range correctors has made helical-scan recorders standard in professional recording (see Chapter 9). The improvement in stability is quite impressive—timing errors of $10\,\mu$sec or greater can be reduced to a few nanoseconds.

The principle of operation of digital time-base correctors is straightforward. The signal from the replay process is converted to a digital bit stream, which is read into sequential locations of a memory in real time. The stored data is then read out at a rate that is precisely synchronized with the system's clock. In effect, the memory acts as a buffer that absorbs the timing errors of the input signal.

3.6.2 Standards conversion

Prior to digital television, standards conversion was accomplished by the brute force methods of kinescope recording or of focussing a camera operating at the new standards at an image on a kinescope operating at the old. The picture quality of analog conversion systems was marginal at best. Standards-converted images suffered not only from the degradations that are inherent in the conversion of line and frame rates, but much worse was the cumulative degradation of the signal-to-noise ratio, gray scale, and definition that are unavoidable in cascaded analog systems.

The use of the digital format does not necessarily solve the inherent problems of rate conversions, but it eliminates or greatly reduces the cumulative quality degradation of analog systems.

Digital standards converters employ the same principle as time-base correctors—storage at the old standards and read-out at the new. The process is considerably more complex, however, because the number of lines, fields, frames, and sampling rates must be changed without destroying time relationships of the events in the image. Also, it is difficult to convert the burst and subcarrier from one standard to another, for example from NTSC to PAL, and the first step in the conversion process is to decode the composite signal to its luminance and color difference components. That eliminates the incoming subcarrier frequency. The line, field, and frame rates are then converted to the new standards, and the signal components at the new standards are re-encoded to the new composite format.

The difference between the line period for 25-Hz, 625-line systems and 30-Hz, 525-line systems (64.0 vs. 63.6 μsec) is so small that it is usually ignored. The change in the number of lines per frame from 525 to 625 and the number of frames per second, however, must be taken into account in the conversion process. Repeating or deleting scanning lines or frames is not a satisfactory solution. Repeating or deleting lines will cause discontinuities in the image and repeating or deleting fields or frames will cause an appearance of jerky motion. The solution is to generate new signals by interpolation between lines and frames that approximate what the signals would have been had they been originally produced at the new standards. Digital circuits easily handle the arithmetic required for interpolation.

3.6.3 Postproduction editing

The art of postproduction film editing—the production of an edited master from a combination of short film or tape segments and special effects—was developed over a period of decades by the motion picture industry, and its practitioners are highly skilled artisans. The television industry tried for years to duplicate the complex postproduction functions performed routinely in film studios, but these efforts were only modestly successful with analog systems.

Digital formats can produce unlimited multiple generations of recorded material without significant loss of quality, incorporate precise computer control for switching and effects, and insert computer-generated program material. This has made digital video an extraordinarily powerful production medium. With the improvements that have occurred in the performance of video cameras, video production techniques are now practical for the production of programs for use on television.

The digitized analog signals used for editing can either be composite such as NTSC or PAL, or in component form, usually R, G, and B or Y, R–Y, and B–Y. Composite digital signals have the same basic problems of luminance-chrominance cross talk as analog, and it is desirable, therefore, to perform complex editing functions with component signals. Video editing systems are described in Chapter 11.

3.7 Digital Video Transmission

Transmission is the sending of video data from one point to another (point-to-point) or from one point to many others (broadcasting). Digital point-to-point transmission is widely used for connecting signal sources to signal-assembly centers via land lines, microwave links, and satellite links. Digital broadcasting of video is emerging, with the first example being the multichannel satellite services. However, with the digital standardization of HDTV, digital broadcasting is the wave of the future. This section discusses some of the special considerations of digital transmission of video, including packetization, modulation, error control, and synchronization.

3.7.1 Packetization

A digital bit stream must have a structure to facilitate synchronization and identification of the data. This is commonly done by dividing the data up into blocks called *packets*. Each packet begins with a header that identifies the packet and contains a unique pattern for synchronization (see Section 3.7.4), followed by a specified amount of signal data. Packets may be of any size, either fixed or variable (variable-size packets must have a size parameter in their headers), and there are arguments pro and con for each type.

Packetization allows interleaving of different types of data in a bit stream, such as video, audio, auxiliary data, or even multiple programs. The bit stream capacity (data rate) can be flexibly allocated to the different data types by varying the number and/or size of the packets for each data type.

Error handling also requires blocking of the data and it is convenient to choose a packet architecture that will support the error handling methods used in the system. Error considerations usually lead to rather small packet sizes, such as a few hundred bytes.

A detailed example of fixed-length packetization for a digital video system is the Transport layer of the Grand Alliance HDTV system (see Section 6.6).

3.7.2 Digital modulation modes

Most transmission media are inherently analog systems. To efficiently handle digital data, the data must be converted to a format that is compatible with the characteristics of the transmission medium. This process is called *modulation*. An important objective of the choice of modulation method is to achieve an appropriate match to the bandwidth and noise performance of the transmission channel.

These two parameters can actually be traded for each other—for example, if better noise performance is available, the bandwidth for a given data rate can be reduced. There are two measures for this tradeoff:

1. The bandwidth requirement is expressed in terms of the bit rate that can be transmitted in a bandwidth of 1 hertz. This factor is sometimes called the *transmission speed*.

2. The noise performance of the system is expressed as the signal-to-noise ratio required for a specified bit error rate (BER). A BER of 10^{-4} (1 error in 10,000 bits) is often used. Most systems require BER performance better than this; they achieve it by using *error protection* techniques (see Section 3.8). Error protection involves adding additional bits to the data (called *overhead* bits) and therefore the net data rate of the system becomes lower than the data rate of the raw channel.

Based on its bandwidth, a transmission channel is capable of transmitting a certain number of independent events per second. These events are called *symbols* and, depending on the modulation method, each symbol may represent one or more digital bits. When there is one bit transmitted by each symbol, the symbol has only two valid values—just like the bit it represents. A symbol that represents two bits has four valid values (2^2), and so on. Because multiple bits per symbol expand the number of valid values of a symbol, such modulation methods require higher channel S/N ratio, but at the same time they are transmitting more data per second. This is how the tradeoff between bandwidth and S/N works—bandwidth can be reduced, but the necessary signal-to-noise-ratio increases.

All digital modulation modes establish conditions of amplitude, frequency, or phase (or a combination of these) to represent each of the values of a symbol. There is a wide variety of possibilities, and a summary comparison of the speed vs. bandwidth relationship and the power requirements of four modes that span the range of possibilities is shown in Table 3.4. They are:

1. on-off keying or OOK, which has one bit per symbol, represented by the two conditions of carrier and no-carrier.

2. quaternary phase-shift keying (QPSK), which transmits two bits per symbol using phase modulation of a constant-amplitude carrier with four possible phases spaced by 90°.

3. frequency-shift keying (FSK) transmits one bit per symbol using frequency shift between two frequencies.

4. quadrature amplitude modulation (QAM) is a combination of amplitude and phase modulation of the carrier that transmits four bits per symbol (16 values of the symbol).

The table also shows a *constellation diagram* for each of the modes; this diagram is a polar diagram showing the amplitude and phase of each of the valid carrier states. Among the four modes in the table, QAM can transmit the highest bit rate in a fixed bandwidth, but the power level must be high to achieve a satisfactory BER. Accordingly, it is often used in microwave systems

Table 3.4 Performance of digital signal transmission modes

	OOK	QPSK	FSK	QAM
Constellation Diagram				
bits/symbol	1	2	1	4
Transmission Speed, bits/Hz	0.8	1.9	0.8	3.1
Required E_b/E_a for BER=10^{-4}	12.5	9.9	11.8	13.4

Key:
E_b = energy per bit
E_a = white noise spectral density
OOK = on-off keying
QPSK = quaternary phase-shift keying
FSK = frequency-shift keying
QAM = quadrature amplitude modulation
SOURCES: J. D. Oetting, "A Comparison of Modulation Techniques for Digital Radio," *IEEE Transactions on Communications*, 23(12), 1979; Alladin Javed, "Signal Transmission Modes," in A. F. Inglis (ed.), *Electronic Communications Handbook*, McGraw-Hill, New York, 1988, Chap. 11.

where adequate power is available. QPSK presents an attractive compromise between power requirements and transmission speed, and it is frequently used in satellite system. The references cited in Table 3.4 describe other digital transmission modes. Modulation modes for magnetic recording are discussed in Section 9.12.3.

3.7.2.1 HDTV modulation modes The Grand Alliance HDTV transmission system described in Chapter 6 uses multilevel vestigial sideband (VSB) amplitude modulation, referred to as 8-VSB or 16-VSB (the numbers give the number of amplitude levels involved). For example, the 8-VSB system transmits 3 bits per symbol using its 8 levels and the 16-VSB system sends 4 bits per symbol.

3.7.2.2 Trellis coding The 8-VSB system actually sends only 2 bits of data per symbol; the data is expanded to 3 bits per symbol for transmission by means of a *trellis-coding* algorithm, an error protection strategy that can improve channel performance without increasing bandwidth. (The name comes from a diagram of this process that looks like a garden trellis.) In this approach, there are only four possible data states but eight possible symbol states, so not all combinations in the symbol are valid. The effect is to increase the differences between adjacent symbol states, which makes it easier to detect valid states.

At the receiver, detection is facilitated by the use of a *soft-decision decoder* (sometimes called a *Viterbi*[1] decoder), which operates on the channel signal before quantization (noise removal) to decide what a symbol value must be based on the context of adjacent symbols. Since the trellis coding has invalidated some symbol states or sequences, the decoder can correct many single-symbol errors.

The digital HDTV system with 8-VSB trellis-coded modulation achieves a data rate of 19.29 Mb/sec in a 6-MHz channel, which is a transmission speed of 3.2 bits/Hz. It has a noise threshold of 14.9 dB. These performance numbers are not directly comparable to those in Table 3.4 because they describe the complete system including the error protection, whereas Table 3.4 shows raw modulated-channel performance at 10^{-4} BER.

The 16-VSB HDTV system does not use trellis coding, so it achieves twice the data rate of the 8-VSB system. This is intended primarily for cable service, where the necessary higher S/N ratio can be obtained. The noise threshold for 16-VSB is 28.3 dB. HDTV modulation is described further in Section 6.6.3.

3.7.3 Component and composite signals

As described in Section 3.2, either component or composite analog color signals can be converted to a digital format. Of course, digitizing a composite signal cannot remove signal impairments that already exist due to the

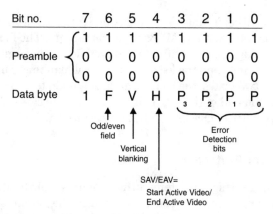

Figure 3.9 Serial synchronization pattern TRS-ID

composite format. The limitations of composite color signals are especially serious for postproduction editing techniques that require extensive use of multigeneration recording. Component signals are supportive of the advantages of the digital format—lower noise, lower distortion, greater flexibility for editing, and freedom from artifacts from luminance-subcarrier cross talk. Accordingly, the component and digital formats are frequently used in combination, particularly in recording (see Chapter 9).

In many system applications, however, it is necessary or desirable to transmit or record the signal in a composite form such as NTSC or PAL. As a result, industry standards have been established for the frequency and location of sampling points for both composite and component signals (see Section 3.3 and Table 3.2).

3.7.4 Synchronization

The digital bit stream from the A-to-D converter must not only convey analog signal information but also digital words to synchronize it with the D-to-A converter at the output of the system or with other processes occurring within the system. This is accomplished by the use of digital code words, which make it possible for the D-to-A converter to identify positively the following:

1. In serial systems, the beginning of each word.
2. In component systems, the location of the beginnings of lines, the active regions of lines, fields, and frames.

The code words must have values that the decoder will not confuse with pixel values. Words 0000 0000 and 1111 1111, which are outside the quantizing range for pixels (see Section 3.3.3), meet this requirement.

Figure 3.9 shows a four-word synchronization pattern, known as TRS-ID (timing reference and identification signal), that is used for serial transmission. It is repeated for each line at the end of the horizontal sync pulse and synchronizes word beginnings, lines, fields, and frames. The first three words, the TRS, define the LSBs and MSBs in the bit stream. The fourth word, the ID, contains bit flags that identify odd and even fields, the beginning and end of the active line time, and the location of vertical blanking. Since each bit in the ID has only two states, the identification is limited to two choices such as the beginning and end of the active line time. In some cases, the ID has bits that are used for parity checking and error correction.

3.8 Digital Error Protection

A major advantage of digital systems is their ability, within limits, to recognize and correct bit errors that are caused by noise in the system. If the number of errors exceeds the limits, an additional technique, error concealment, can be applied that reduces the visible effect of the errors. This technique is particularly used in video recorders and transmission systems.

3.8.1 Visual effect of bit errors

The visual effect of a single-bit error depends greatly on the significance of the bit. In an 8-bit PCM system, an error in the least significant bit will be 1 part in 256 with a barely noticeable effect. An error in the most significant bit, on the other hand, will cause a major disturbance in the signal. This range in sensitivities is considered in some error-correction systems, but it is not much of an issue any more because modern error-correction systems correct almost all single-bit errors. Furthermore, systems that involve data compression may make all bits equally significant and a single-bit error can be highly visible. This means that error correction is more important in compressed-data systems.

Typical transmission systems can have *burst errors*, where bit errors occur in a group of adjacent bits in the bit stream. These are more difficult to correct directly and most systems use a strategy such as *interleaving* (see Section 3.8.3.3) to spread burst errors out into multiple error-correction blocks.

Figure 3.10 Error-correction system

3.8.2 Error-correction system functions

Figure 3.10 is a functional diagram showing the essential elements of error-correction systems. The redundancy code is a bit or word added to the digital bit stream to make it possible to detect bit errors with a high degree of probability.

The retransmission request function is applicable only to systems that have two-way transmission and that can tolerate random delays in the data transmission. The latter requirement usually precludes the use of retransmission for audio or video systems because it would interrupt the smooth reproduction of the audio or video.

Since bits have only two states, the correction of a bit error is straightforward once it has been detected—the bit value is simply reversed.

Error concealment, applicable primarily to digital recorders, is a last resort process. It is used when error correction is ineffective, e.g., when a large number of bits are in error. Its principle is to reconstruct the missing or erroneous bits by interpolation of spatially and temporally adjacent bits. However, proper application of error concealment requires that the existence of an error be known, so a means for error detection is still required.

3.8.3 Encoding and error detection

A variety of redundancy codes have been developed to make it possible to identify bit errors with a high degree of certainty and probability. The simplest code is a single 1 or 0 bit, known as a *parity* bit, added to the end of each word as required to make the sum of the bits an even number. This code is illustrated for 2-bit words:

Signal words	Parity bit	Sum
0 0	0	0
0 1	1	2
1 0	1	2
1 1	0	2

If the sum of the bits in a word plus its parity bit after transmission or recording is an even number (0 is considered to be an even number), it is probable that none of the bits is in error, although there is a possibility that more than one are incorrect.

This simple parity check identifies single errors, but it does not identify the offending bit so that it can be corrected. It is particularly useful in situations where errors are most likely to occur one at a time, as in an electronic memory. It is not effective by itself for errors occurring in groups (known as *burst* errors).

Error correction can be accomplished in parity systems by considering blocks of data as a two-dimensional matrix. For example, an 8-byte data block can be considered to be an 8×8 matrix of 64 bits. By assigning a parity bit to each row and column of this matrix—resulting in 2 bytes of parity overhead—single bit errors can be located by looking at the intersection of the row and column that show errors. The offending bit can then be reversed to correct the error. This use of a two-dimensional error protection block is known as a *crossword* or *product code*.

The simple row and column parity scheme described above works only for one error per block. However, more sophisticated codes have been developed that can operate with higher error rates. Two of the most common ones used in digital video are the *Hamming*[2] code and the *Reed-Solomon*[3] code.

3.8.3.1 Hamming error protection code To generalize on the product code example above, a certain number of check bits are created from the data so that the pattern of failed check bits can be used to identify the data bits in error. The data bits of a block are transmitted unchanged and they are followed by the specified number of parity bits. This is known as a *systematic* code. In a Hamming code, the matrix for the generation of the check bits is designed so that the pattern of failed and unfailed check bits is an address to the failed data bits. In practice, Hamming codes are easy to implement in hardware and they can readily detect and correct single-bit errors per block. Additional errors can sometimes be detected, but they cannot be corrected.

A special form of the Hamming code is the *cyclic redundancy check* code (CRC), which is often used in data transmission. However, for digital video and audio, Hamming codes have been superceded by the much more powerful Reed-Solomon codes.

3.8.3.2 Reed-Solomon error protection codes One of the most widely-used codes in digital audio and video is Reed-Solomon. It is used in CD audio, CD-ROM, digital video recorders, and digital HDTV. It is a block coding scheme that can correct bursts of errors up to a limit governed by the amount of error correction overhead that is chosen at design time. Although the processing is quite complex, it has been implemented in integrated circuits and is now easily included in system hardware.

3.8.3.3 Reed-Solomon as a product code By defining a macro-block of data that contains a number of R-S blocks within it, a two-dimensional form of error correction is achieved that has much more power than a single R-S implementation. This is often with the data from an entire TV frame or with the data from a tape track segment in a magnetic recorder. R-S coding is applied to the incoming data after which the entire macro-block of data is stored in a memory arranged as a 2D matrix. The data is then read out from the memory by reading the memory matrix perpendicular to the way it was read in and a second R-S process is applied. For example, the data could be read into rows of the memory matrix and then it would be read out by columns. The first R-S process is called *outer* error coding and the second process is *inner* coding. At the receiving end, inner decoding is done first, the memory matrix transformation is done, and then the outer decoding is done to restore the original data. This scheme allows correction of large burst errors such as from tape dropouts.

3.8.3.4 Data interleaving Although Reed-Solomon coding can handle multiple bursts of errors in one data block, transmission channels can sometimes have larger burst errors that would overload correction in a single block. This problem can be overcome after R-S encoding by *interleaving* the data between blocks so that the data for each block is not transmitted in normal sequence. Instead, it gets spread out over the whole interleave pattern. At the receiving end, the interleaving is removed before R-S decoding.

For example, if eight blocks were interleaved, then one-eighth of the data would be sent from each block in sequence, followed by the next eighth of each block, and so on. The result is that a single burst error in the channel cannot cause more than one-eighth of a block to be lost unless the channel burst gets longer than a whole data block. This process can be extended to deal with almost any size of channel interruption, but of course, the amount of processing and memory required increases with larger interleave patterns. In the inner-outer type of R-S coding, interleaving is done between the outer and inner coding.

3.8.4 Effectiveness of error correction

Figure 3.11(a) illustrates the performance of an error-correction system as the bit error rate is increased. In a system with no error correction, the output bit error rate increases with the input (assuming that no additional errors

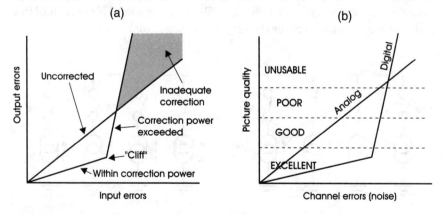

Figure 3.11 Error correction with high error rates

are introduced). Error correction is effective for input error rates up to a critical value. Above this level, sometimes called the "cliff," the error correction is not only ineffective but the output BER is actually worse than without correction.

In comparing digital and analog systems, the cliff effect is sometimes presented as a disadvantage to digital systems. However, in a properly designed digital system, the cliff would not be reached until the transmission characteristics are so bad that analog transmission in that channel would also be unsatisfactory. This is illustrated in Figure 3.11(b).

3.9 Digital-to-Analog Conversion

Digital video signals must ultimately be reconverted to an analog format to be displayed. Figure 3.12 illustrates the principle of one type of digital-to-analog converter. A circuit is clocked so that each bit in a word, from the LSB to the MSB, controls a current flow that is proportional to its position in the word. The summation of these currents is proportional to the quantized level in the analog signal. The analog signal is then recovered by passing the output through a low-pass filter that removes the sampling frequency components.

An obvious problem with this system is the necessity of controlling the current in the MSB path with a high degree of accuracy. With an 8-bit code, the error in the MSB path must be substantially less than 1 part in 256 or it will overwhelm the effect of the LSB.

An alternative conversion system causes a constant current to flow into an integrator for a time interval that is determined by the position of the bit in the word.

The output of these conversion systems is a reconstruction of the pulse train that was created by the sampling process in the A-to-D converter. It is necessary to pass it through a low-pass filter having a cut-off at one-half the pulse rate in order to recover the original analog signal.

(a)   (b)

Figure 3.12 Digital-to-analog converter

3.10 Video Bandwidth Reduction

When digital video transmission or storage is viewed in terms of raw bit rates and analog bandwidth requirements, the numbers are so large that many tasks appear to be impractical. However, this view overlooks the potential of digital video and audio data to be *compressed*—that is, to be reduced. Because of the large amount of redundancy that exists in scanned video data, compression factors up to 100:1 or more are possible in some cases. This is the reason that all new video transmission systems will soon be digital.

It is an apparent paradox that digital formats, which inherently require more bandwidth than analog, can be used to reduce the bandwidth required for video signals. The technologies for reducing the bandwidth of video signals by digital data compression are developing rapidly; and it has been estimated that as many as 40 companies worldwide are engaged in this endeavor. Digital video compression can accommodate the growing requirements of computers, satellites, cable systems, and HDTV.

For the past several years, the video technical community has been engaged in two major development programs—HDTV and video compression—both directed toward more efficient use of spectrum space but with different objectives. The major purpose of HDTV development is to improve television picture quality with a minimum increase in bandwidth requirements. The major purpose of video compression development is to reduce data requirements with a minimum loss of picture quality.

These developments have now come together in the Grand Alliance HDTV system that is expected to be adopted in the United States. Through MPEG video compression (see Section 3.12.2) and a creative digital transmission system (see Chapter 6), HDTV video and multichannel high-fidelity sound are transmitted over the same 6 MHz channels that now handle NTSC analog video.

Communications theory[4,5] shows that under the conditions encountered in practical video systems the rate at which information can be transmitted in a communications channel is proportional to its bandwidth. The name of the game in bandwidth reduction thus is to reduce the amount of information needed to send video pictures.

If video signals were random current or voltage variations, bandwidth reduction would not be possible without a major loss of picture quality. Fortunately this is not the case. Video signals are highly structured, and they are repetitive in time and correlated in space. Further, the rate of information transfer varies widely from frame to frame and in different areas of the frame. Areas of an image with rapid motion and a large amount of detail require more bandwidth than non-moving areas with little detail. The role of bandwidth reduction systems is to make use of these properties to reduce this rate with a minimum loss of picture quality. It should be emphasized, however, that any bandwidth reduction system will result in some loss in picture quality—for example, loss of resolution at the edges of moving objects—and an objective is to make the defects least visible.

3.11 Digital Data Compression Technologies

High-quality video compression requires extremely complex processing and effective techniques and hardware are still being developed. A compression method is called an *algorithm*, which is a description of a mathematical process that performs the compression or the decompression required to restore the compressed signal to its original form. Many algorithms already exist for different purposes and different quality levels. Because of the tremendous diversity of existing algorithms and the vast potential that still exists for new development, progress toward worldwide compressed video standards has been slow.

3.11.1 Algorithms

As with other complex digital processing functions, the algorithm is the basic tool for specifying the mathematical operations that are to be performed for compression. Figure 3.13[6] is a diagram of some of the different techniques of compression that may be combined into an algorithm. Each of these techniques is potentially applicable to digital video at the pixel or line (spatial), or frame (temporal) level.

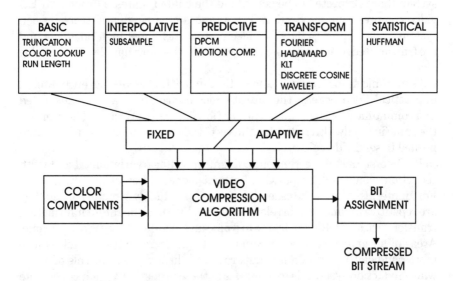

Figure 3.13 Video compression techniques

3.11.2 Fixed and adaptive techniques

Figure 3.13 illustrates five common techniques for video compression:
Basic
Interpolative
Predictive
Transform
Statistical

This is by no means a complete list of techniques, but it is illustrative of the range of possibilities.

The techniques have two modes, "fixed," and "adaptive." In the fixed mode, their application is independent of image content. In the adaptive mode, their application depends on the content—for example it may operate differently for images with a great deal of detail or motion. The resulting algorithm processes the incoming digital signals continuously, and the output is a compressed bit stream.

3.11.3 Truncation

Truncation is a brute force technique that reduces the bit rate by dropping one or more of the least significant bits (LSBs). In the fixed mode, truncation has the obvious problem that it lacks the ability to distinguish between the brightness levels represented by the LSBs, and the variations in image brightness will be discontinuous. This can be avoided by an adaptive transform that drops the LSBs only to the extent that the amount of picture detail permits.

The bit stream with variable word length is fed into a buffer after which the missing LSBs are restored. A potential problem with this system is that the buffer can become overloaded if the picture contains an unusual amount of motion or detail.

3.11.4 Subsampling and interpolation

Subsampling and interpolation is a widely used bandwidth reduction technique, and it can be used with samples that have not been digitized. Figure 3.14 illustrates one example of the technique.[7] The sampling frequency is reduced by one-half (subsampling) and the samples on adjacent lines are offset by one-half the sampling interval. The values of the samples that are skipped as the result of subsampling are interpolated from the values of adjacent sampled elements, both horizontal and vertical. Two possible interpolation equations for the example in Figure 3.14 are:

$$i = \frac{b+c+d+e}{4} \tag{3.3}$$

or

$$i = \frac{c+d}{2} + \frac{b-e}{2} \tag{3.4}$$

Equation (3.3) is the arithmetic average of all the adjacent sample values. This is sometimes called "A" interpolation. Equation (3.4) is the average value of the adjacent points on the same line plus the change in value of points on an adjacent line—a function of the high-frequency components in the signal. The choice between these formulas is largely subjective.

3.11.5 Prediction

Spatial prediction techniques are based on the fact that the difference in signal levels between adjacent pixels is usually small. If this is the case, the difference can be represented by a smaller number of bits per word, say 4 instead of 8. This technique is called *differential pulse code modulation* (DPCM).

A DPCM system is overloaded when there is an occasional large difference between adjacent pixels as at a very sharp edge. This is called *slope overload* and results in a smeared edge.

DPCM also has the problem that the amplitude of each pixel signal is the sum of the amplitude of the adjacent pixel signal and the difference. If an error creeps into one of the pixels, it will be repeated until the level is reset. As the result of these problems, DPCM is not widely used by itself, but it is a useful technique when combined with others.

Prediction can also be done temporally by examining adjacent frames and finding out what parts of a frame already exist in the preceding frame. These parts of the new frame do not have to be transmitted again. The technique is called *motion compensation* (see Section 3.12.2).

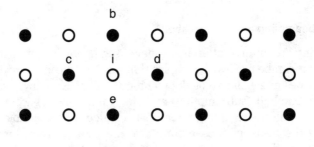

Figure 3.14 Spatial subsampling

3.11.6 Transformation

Video compression can be accomplished by a transformation of the values of a group of pixels into another set, which can be transmitted with less data. After transmission, an inverse transform is performed that recovers the original values.

Transforms have been the subject of extended investigation, their objective being to be effective and to operate speedily in real time. As a simple example, assume the values A, B, C, and D are transformed to the values W, X, Y, Z by the equations:

$$W = A$$
$$X = B - A$$
$$Y = C - A$$
$$Z = D - A$$

W, X, Y, and Z are transmitted. Since X, Y, and Z are value differences, they are usually smaller than the untransformed originals, B, C, and D and do not require as many bits per word.

After transmission, A, B, C, and D are recovered by a *reverse transform*:

$$A = W$$
$$B = W + X$$
$$C = W + Y$$
$$D = W + Z$$

Adaptive transformation is another technique. The amount of information that must be transmitted for a particular portion of the image is proportional to the fineness of detail in that portion. A portion of the picture with little detail can be transmitted with very few bits, and this provides extra time for transmitting portions with high detail. A buffer is used to restore the original spatio-temporal relationship. If the entire picture has high detail, the buffer may become overloaded and the rate of information transfer reduced by increasing the quantum intervals. This, of course, reduces the image quality by introducing artifacts.

Transforms have been the subject of extended mathematical analysis, and many are in use. Some of the better known ones are the *Hadamard* transform, the *Fourier* transform (not to be confused with the Fourier transform that converts from the frequency domain to the time domain), and the *discrete cosine transform* (DCT).

3.11.6.1 The discrete cosine transform   The DCT is a version of the Fourier transform that takes account of the fact that the input is a limited set of samples of continuous waveforms (thus, the word "discrete"). A Fourier transform produces both sine and cosine terms in its output in order to specify the phase of the frequency components. However, by performing the discrete transform on a block of samples and its mirror image, the sine terms can be made to cancel, leaving only the cosine terms. This is actually a simplification and it explains why the word "cosine" is in the name of the transform.

The DCT is typically done on 8 × 8 blocks of pixels, so there are 64 total samples to be processed. The output of the transform is 64 new values representing coefficients for the spatial frequency components in the block. So far, there is no compression because there were 64 values going in and the same number of values came out. The zero-frequency coefficient is referred to as the DC coefficient; it is the average of all the pixels in the block. The remaining coefficients are for finite spatial frequencies in the block. The beauty of this transform is that it defines most of the block's information in the lower spatial frequencies and many of the higher-frequency coefficients come out to be zero. This is because adjacent pixel values are most often similar. Although the transform itself does not provide any compression, the transformed values are amenable to substantial compression by subsequent processes as described below.

In most algorithms, the DCT coefficients are quantized according to an adaptive table and then they are ordered in ascending spatial frequency (see Section 3.12.1), which causes most of the zero values to be grouped. This facilitates further compression by run-length or statistical techniques.

Because the DCT equations involve transcendental functions, which can only be approximated in a real system, this transform is inherently lossy. However, it lends itself to making the losses invisible in most images and is therefore a highly effective compression technique.

3.11.7 Statistical techniques

Statistical techniques can be used for bandwidth reduction by making use of the fact that some image values are less frequently encountered than others. Shorter words are used to define the more frequently encountered values with a reduction in the total number of bits. This technique, often called *Huffman coding*[8], requires the transmission of a code table so that the length of each word can be identified at the receiving end.

3.11.8 Cascaded compression

To achieve a desired degree of compression it may be necessary to combine more than one technique and to compress the signal in a number of successive steps. For example, a compression of 32 to 1 could be achieved by 4:1, 4:1, and 2:1 compressions performed successively.

3.12 JPEG and MPEG Compression

The *International Electrotechnical Commission* (IEC) some years ago recognized the need for digital video and audio compression standards and formed the *Motion Picture Expert Group* (MPEG) to develop standards in this important technology area. The work of MPEG has led to the MPEG-1 and MPEG-2 standards, which are becoming the basis for most widely-distributed forms of compressed video. Additional MPEG standards are under development.

MPEG was actually preceded by another working group of the IEC, the *Joint Photographic Expert Group* (JPEG) for standardization of still-image compression techniques. This was important because still-image compression is widely applicable to the fields of desktop publishing, graphic arts, facsimile, medical imaging, and others. JPEG developed a single standard[9] that is configurable and applicable to this wide range of end uses. The standard supports a variety of algorithms and defines how these are combined to address different applications.

The idea of configurability is significant because it does not lock the industry into a single algorithm; in fact, it even allows for new algorithms to be added in the future. Combining this with the flexibility of software-controlled processing allows continued growth of compression system performance. The JPEG still-image compression concepts form the basis of MPEG motion video compression.

3.12.1 JPEG Still-Image Compression

The JPEG standard supports both lossless and lossy compression of images of practically any size and bits per pixel structure. It is parameterizable so the user can make tradeoffs of degree of compression vs. quality. It has four modes of operation:

1. Sequential—encode in the order that the image was scanned
2. Progressive—multipass encoding so that a coarse image is transmitted rapidly, followed by repeated passes at progressively higher resolutions
3. Lossless—exact reproduction is guaranteed (all other modes are lossy)
4. Hierarchical—the image is encoded at multiple resolutions

Figure 3.15 shows the JPEG architecture for the lossy compression mode. Basic compression uses the DCT algorithm, encoding 8 × 8 blocks of pixels at a time (see Section 3.11.6.1). The coefficients resulting from the DCT are quantized according to a quantization table that is supplied by the application. This is one place where the algorithm is configurable—the quantization table becomes part of the compressed data stream. The next step, called *zig-zag ordering*, arranges the DCT coefficients serially so that zero-value coefficients tend to be together at the end of each block. This facilitates the next step,

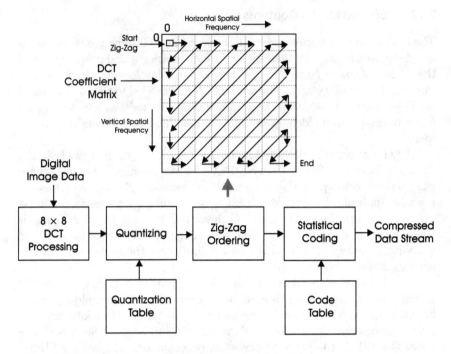

Figure 3.15 JPEG coding architecture—sequential, lossy

which is statictical coding. Huffman coding is typically used here and again, the coding table is configurable.

With most high-resolution color images, JPEG can compress 10:1 or so with excellent results. For example, a 16- or 24-bpp image can be transmitted at less than 2-bpp and reproduced with no loss of resolution. In some cases, compression up to 30:1 is achievable. The amount that can be used depends on the image content and the viewing conditions for the final decompressed image.

The JPEG standard is usable for motion video compression by performing the algorithm separately for each frame of the motion sequence. This is called *motion-JPEG* and has been applied extensively in situations that don't need the maximum compression. However, it fails to exploit the opportunity for further compression that is inherent in the redundancy between frames of a motion sequence. The MPEG algorithms were designed to do that.

3.12.2 MPEG Architecture

The technique for exploiting frame-to-frame redundancy is called *motion compensation* and it is done to reduce the frame data before it gets coded by techniques similar to JPEG.

3.12.2.1 Motion Compensation The basic idea is to find the parts of each frame that have moved or changed from the preceding frame and code only the changes, which are called *residuals*. Each frame is then built by adding the decoded residuals to the prediction based on the previous frame. This is a form of adaptive prediction, where the previous frame predicts the next frame except for places where something has changed or moved. These frames are called *P-frames*.

The example in Figure 3.16 shows how frames are divided into blocks for the purpose of motion compensation (the number of blocks is reduced to simplify the figure). The frames are taken from a sequence of a crow flying over a city. The camera tracked the crow, so the city background is moving to the left. The previous frame is searched for each block of the current frame (this is called *block matching*). If a match is found, the location of the block (called a *motion vector*) in the previous frame is coded. A trial predicted frame is then built by moving blocks from the previous frame using the motion vectors. This process is shown in the figure for one block. The predicted frame is subtracted from the actual current frame to make the residual image, which is then coded for transmission. At the receiving end, the process is reversed: the predicted frame is built from the previous frame, the residual image is decoded and added to the predicted frame.

Searching the entire previous frame for a block from the new frame requires a fantastic amount of computing. Since the probability of finding the block drops as one searches further away from the block's position in the new frame, the search can be restricted to a small region surrounding each new block. This is defined by an encoder parameter called the *search range*. If the search range is too small, the encoder will be overloaded for even small amounts of motion from one frame to the next. Greater search range allows more motion to be handled, but it makes the encoder more expensive.

Transmitting a long motion sequence by means of frame prediction has two problems:

1. If an error occurs in one frame, it will tend to be propagated to future frames until a change obliterates it. This makes errors more visible.

2. In many cases video must be editable, meaning that it should be possible to cut any piece from a video sequence and use it in another sequence. With predictive coding, a video sequence must always begin at the first frame.

Both problems are solved by periodically sending a full frame that is not a prediction of the previous one (similar to a JPEG image). If this is done at a moderate rate (say every 10th frame), it doesn't add much data to the stream. Thus, in a 30 frames/second system, edits could be started at any of 3 points in every second. This is not frame-accurate editing, but it is sufficient for many purposes. Such full frames are called *intracoded* (I-frames).

There is a third strategy for frame coding in MPEG, which uses *bidirectional interpolation* between two frames. These are called *B-frames* and they

Figure 3.16 Motion compensation encoding

I = intracoded
B = bidirectional predicted
P = forward predicted

Figure 3.17 A typical MPEG frame sequence

provide the most compression, but they also require the most processing during compression and the most complex decompression strategy. Figure 3.17 shows how the three frame types might be combined in a data stream and how they depend on each other. The exact frame sequence is configurable.

3.12.3 MPEG-1 and MPEG-2

The first MPEG standard was MPEG-1[10], which is designed for data rates up to 1.5 Mbits/second. This supports the single-speed CD-ROM rate used in computers and the T-1 data communication channel. MPEG-1 performance is often described as similar to a VHS videotape recording. It is satisfactory for many entertainment applications or for partial-screen display on computers, but it is not as good as NTSC or PAL broadcast television.

MPEG-2 is designed for higher data rates, up to 15 Mbits/second, and higher-quality performance, including HDTV (see Chapter 6). It supports full-screen display on both computers and television receivers.

3.13 Summary

It is safe to say that digital video technology has reached the point where it is the preferred approach for all new system designs. In most cases, digital video performance vs. price has surpassed analog systems. Satellite broadcasting is already digital, and with HDTV, terrestial broadcasting and cable systems will be digital. Digital video technology allows integration with computer data networks and it will be an essential ingredient to the developing Information Superhighway.

References

1. A. J. Viterbi, "Error Bounds for Convolutional Codes and an Asymptotically Optimum Decoding Algorithm," *IEEE Trans. Information Theory*, Vol. IT-13, pp. 260–269, April, 1967.

2. R. W. Hamming, "Error Detecting and Error Detecting Codes," *Bell Syst. Tech. J.*, 26:147–160, 1950.

3. I. S. Reed and G. Solomon, "Polynomial Codes Over Certain Finite Fields," *J. Soc. Ind. Appl. Math*, 8:300–304, 1960.

4. R. V. L. Hartley, "Transmission of Information," *Bell Syst. Tech. J.*,Vol. 7, July 1928.

5. C. E. Shannon, "A Mathematical Theory of Communication," *Bell Syst. Tech. J.*, Vol. 27, July–October 1948.

6. Arch Luther, *Digital Video in the PC Environment*, McGraw-Hill, New York, 1991.

7. R. Kishimoto, N. Sakuri, and A. Ishikura, "Bit-Rate Reduction in the Transmission of High-Definition Signals," *Jour. SMPTE*, 96(2), February 1987.

8. D. A. Huffman, "A Method for the Construction of Minimum Redundancy Codes," *Proc. IRE*, 40, pp. 1098–1101, 1952.

9. ISO/IEC 10918-1. Digital Compression and Coding of Continuous-Tone Still Images, 1993.

10. ISO/IEC 11172. Coding of Moving Pictures and Associated Audio for Digital Storage Media at up to about 1.5 Mbits/s, 1993.

Chapter

4

Elements of Image Quality

4.1 Overview

The technical performance of a video system is measured by the quality of the images it produces. An image is judged to be of high quality if it is a near-perfect replication of the original scene, which may be a natural scene viewed by a camera or an artificial image created by an artist or a computer. Deliberate distortions of the image may be introduced for esthetic purposes—for example, by defocusing the background areas of the scene—but the basic technical objective is to duplicate the appearance of the original as closely as possible.

The use of the word "appearance" in the preceding sentence indicates that the evaluation of video images is ultimately subjective. The video engineer, however, cannot be satisfied with exclusively subjective criteria for image quality. The design and operation of television systems require that they be defined in objective and measurable terms. An important engineering task, then, is to develop objective picture quality yardsticks that coincide with the subjective perceptions.

Television systems are designed to optimize the presentation of natural scenes while computer displays are designed to best present artificial images, graphics, or text. However, there is more and more need to show both image classes on all systems. The basic criteria for computer display of "artificial" images are covered in Chapters 3 and 7.

This chapter will concentrate on the criteria for display of natural images, whether on television or computer. It begins with a brief summary of the most important quality criteria for monochrome video and the luminance component of color television. That is followed by the application of these criteria to practical operating systems that employ NTSC broadcast and HDTV production standards (SMPTE 240M). The implications of the criteria to the display of natural images on computers is also treated. The criteria for the chrominance components of color television are described in Chapter 2.

4.1.1 Television and film

The complexity of establishing objective quality criteria for television images is increased by the necessity of comparing television and photographic systems. Photography preceded television, and film quality has historically provided a basis for defining television image quality. A frequently stated goal for the design of broadcast television systems is to try to duplicate the quality of 16 mm motion picture film. Similarly, the goal of HDTV (high-definition television) systems is often described as the duplication of the quality of wide-screen theatrical film.

Comparison of the quality of television and film images is difficult and cannot be rigorous. There are inherent differences between the television and photographic processes that make it almost impossible to duplicate precisely the appearance of a film image with a television system or vice versa. The television engineer nevertheless must establish practical image quality standards that are equivalent to those used for film, even though the comparison cannot be exact.

The challenge of establishing objective image quality criteria for television and of comparing the quality of inherently dissimilar film and television systems has stimulated an enormous amount of research by the photographic and television industries over the past decades. Practical definitions of the most significant measures of television image quality, together with their specifications and measurement procedures, have emerged from this research. A comprehensive listing of the criteria for evaluating the quality of television images, together with performance standards for both color and monochrome, is contained in a joint publication of the Electronic Industries Association (EIA), and the Telecommunications Association (TIA): "EIA/TIA STANDARD—Electrical Performance for Television Transmission Systems, EIA/TIA-250-C." Although this document is intended to set standards for transmission systems alone, it provides a useful basis for evaluating complete system performance. Since performance degradations may be cumulative, the standards for the transmission system are more rigorous than for the complete system.

4.1.2 Production, transmission, and display standards

Video systems have three components—production, transmission, and display. Each of these has a different set of constraints that affect image quality.

The NTSC television system was designed to operate within the constraints of a 4 MHz bandwidth and to have compatibility with monochrome video standards. It is fundamentally a transmission standard (see the EIA/TIA standards noted above), but of necessity it became a *de facto* production standard as well. This results in reduced performance in applications such as studio program production where limited bandwidth and monochrome compatibility are not required. This problem has been recognized and efforts are now underway to develop production standards for high-performance systems that employ NTSC scanning standards but that are not limited by transmission system constraints or a requirement for compatibility. Component systems (see Section 5.11) are an example.

In HDTV, the differing requirements of production and transmission were recognized early and separate standards have been proposed or established for each.

The SMPTE 240M is a widely used HDTV production standard. It originally specified an analog signal, but a digital version has now been developed. The 240M standard is used to define the performance of HDTV systems in this chapter.

A transmission standard for HDTV broadcasting has been recommended by the Advisory Committee on Advanced Television Service, an industry committee that was established to provide expert advice to the FCC. This specifies a digital system with the parameters listed in Table 6.1. Note that with digital bandwidth compression it is possible to transmit an HDTV signal in a 6 MHz bandwidth.

4.2 Basic Image Quality Criteria

Image quality criteria can be divided into two categories—basic criteria that apply to all imaging systems, and image defects, some of which are unique to television images. Basic criteria include image definition, limiting resolution, gray scale, signal-to-noise ratio, and colorimetry.

4.2.1 Image definition

Definition is the degree of the "in-focus" appearance of the image. It is primarily measured by the sharpness of the transitions between dark and light areas of the picture. Although it is related to the resolution of fine detail in the image and is sometimes confused with it, image definition describes a different aspect of picture quality.

4.2.2 Limiting resolution

Limiting resolution indicates the maximum number of alternate light and dark lines that can be distinguished per unit distance in the image. This is a measure of reproduction of fine detail.

4.2.3 Gray scale

The gray scale specifies the image brightness as a function of scene brightness. It is sometimes called the *transfer characteristic*. It can be specified by three parameters: highlight brightness, the brightness of the brightest areas of the image; contrast ratio, the ratio of the brightness of the brightest areas of the image to the darkest; and gamma, the slope of the transfer characteristic of image brightness as a function of scene brightness, both expressed logarithmically.

4.2.4 Signal-to-noise ratio

In the broadest sense, "noise" is any unwanted signal, but in this usage it usually means random or thermal noise of a type that produces "snow" in a television image.

4.2.5 Summary

Definition, gray scale, and signal-to-noise ratio are fundamental criteria that apply to all imaging systems. They are also interrelated. For example, the perceived definition of an image is strongly influenced by its gray scale. For analytic purposes, however, it is usual to consider them separately. These basic quality criteria are described in detail in Sections 4.5 to 4.16.

4.3 Image Defects

In addition to the basic criteria described above, image *defects* must be considered in evaluating television picture quality. Image defects can be divided into three categories:

1. Defects that are inherent in the system such as flicker and aliasing.
2. Defects such as lag and smear that result from deficiencies in the production of the video signal.
3. Spurious signals that are introduced by recording or transmission systems such as hum and other forms of electrical interference, cochannel and adjacent channel interference, receiver-generated interference, and "ghosts."

Sections 4.3.1 to 4.3.8 summarize the nature of these defects. Section 6.16 contains more detailed descriptions.

4.3.1 Flicker

Flicker (see Section 4.16.4) results when the frame repetition rate is not high enough to cause the eye to perceive it as a continuous image. In an interlaced system, the visibility of large area flicker is determined by the field rate and of small area flicker by the frame rate.

4.3.2 Aliasing

Aliasing (see Section 4.8) is the production of spurious signals as the result of sampling in space or time.

4.3.3 Lag

Lag is a measure of the rate of change of the video signal at a fixed point on the raster when the scene changes. Ideally, the signal voltage at any point would change instantaneously with a change in scene; if it does not, the result is a loss of resolution of moving objects or in extreme cases a trailing tail.

4.3.4 Geometric distortion

Distortion of image geometry, i.e., of the shape or size of objects, results from defects in the camera's optical system or nonlinearities in the display device or the scanning process.

4.3.5 Hum

The term "hum" is borrowed from audio engineering and it describes interference caused by spurious power source voltages that have been introduced in the system. In its most common form it results in a narrow horizontal "hum bar" that moves vertically through the picture at a rate equal to the difference between the field rate and power source frequency.

4.3.6 Cochannel interference

Cochannel interference occurs when two television stations operating on the same frequency are received at the same location. The most common effect is "venetian blinds," alternate black and white horizontal bars across the picture that result from the "beat" between the two carriers.

4.3.7 Receiver-generated interference

A number of spurious signals can be generated by receiving systems. Two of these, local oscillator radiation and i-f images, result from the use of the superheterodyne principle in receivers. A third, intermodulation interference, may result when strong signals are received from two stations having carriers separated by the i-f frequency of the receiver. These problems are particularly troublesome for UHF stations, but they are mitigated by the FCC's policy of channel assignments that reduces the possibility of receiver-generated interference from the use of the superheterodyne principle. Another potential source of interference is intrachannel beat frequencies between the sound and picture carriers. The problems of receiver-generated interference are discussed in Chapter 14.

4.3.8 "Ghosts"

A "ghost" is a duplicate of the main image, slightly displaced from it, usually to the right, and much fainter. It is caused by multipath transmission, i.e., the signal radiated from the transmitter arrives at the receiver over two or more paths, usually a direct path and a reflection from a building or other large object. In spite of the enormous velocity of radio waves, the path-length differences between direct and reflected waves that are often encountered in practice are sufficient to produce noticeable displacements of the ghost image. Radio waves travel 63,000 feet during the time interval of one scanning line, and a 1000-foot path difference would produce a displacement of 1/63 of the scanning line length or slightly more than 1/3 inch on a picture tube with a 20-inch width.

4.4 Characteristics of Human Vision

The human eye is the final link in the perception of television images, and a consideration of its capabilities is critical in the design of video systems and the establishment of objective performance criteria. To state the obvious, there is no advantage in designing a system with image quality that exceeds the ability of the eye to perceive it.

The characteristics of human vision are exceedingly complex, and a detailed discussion of this subject is beyond the scope of this volume. The most significant characteristics for television systems design are the perceptions of brightness and contrast, the visibility of sharpness and detail, and flicker. These properties are described in this chapter. The perception of color was described in Chapter 2.

4.4.1 Brightness and contrast perception

Brightness is a subjective term that indicates the magnitude of the visual sensation produced by a source of light. *Contrast* is the ratio of brightnesses between the darkest and lightest sections of the image. For example, if the darkest area of an image has a brightness of one one-hundredth of the brightest area, the contrast ratio is 100:1.

Luminance is the objective equivalent of the subjective sensation of brightness and is a measure of the *luminous intensity* of the surface of an extended light source. The luminous intensity is equal to the *luminous flux* or *lumens* emitted per unit solid angle. It is calculated by multiplying the energy or

Figure 4.1 The luminosity function

radiated flux expressed as a function of wavelength by a weighting factor, the *luminosity function*, that expresses the sensitivity of the eye to radiation of different wavelengths. Luminance is specified in *lamberts*, or *foot lamberts*. Figure 4.1 shows the luminosity function as standardized by the CIE (*Committee Internationale del'Eclairage*) in 1924.

Care should be exercised to avoid confusing *luminance* and *illuminance*. The former is a measure of the brightness of an area of an image. The latter is a measure of the illumination of a scene, for example a television studio set, by an external light source. Illuminance is specified in *foot candles*.

The visibility of an object in an image depends on the brightness contrast between different areas. In common with most physical sensations, the magnitude of this perception tends to be proportional to the brightness ratio rather than to the absolute brightness difference. This is expressed in *Weber's law*:

The increase in stimulus necessary to produce an increase in sensation of any of our senses is not an absolute quantity but depends on the proportion that the increase bears to the immediately preceding stimulus.

The perception of contrast between two areas also depends on the sharpness of the boundary. If the boundary is sharp, a much smaller brightness difference can be perceived.

4.4.2 Detail and image sharpness perception

One of the most significant characteristics of the eye for establishing the design parameters of a television system is its ability to discern detail and the sharpness of edges in an image. This attribute determines the eye's perception of two of the basic quality criteria for television images—definition and signal-to-noise ratio. It also has a direct effect on the bandwidth requirements for the system.

The ability to discern detail is known as *visual acuity*, which affects the perception of limiting resolution. The eye's visual acuity can be described by specifying the smallest angular separation at which individual lines can be distinguished. Visual acuity varies widely, as much as from 0.5′ to 5′ (minutes of arc), depending on the contrast ratio and the keenness of the individual's vision. An acuity of 1.7′ is often assumed in the design of television systems.

The angular separation between lines can be converted to television lines after specifying the *viewing ratio*, the ratio of viewing distance to picture height, and this is a more common method for specifying this parameter.

With 480 active scanning lines (as in 525-line broadcast systems) and a visual acuity of 1.7′, individual scanning lines can be distinguished at a viewing ratio as great as 4.4. At a viewing ratio of 4 (the minimum usually assumed for broadcast systems), individual scanning lines can be faintly distinguished by a typical viewer. A viewing ratio of 3 or less is usually assumed in designing HDTV, while a range of viewing ratios from 4 to 8 is assumed for broadcast television. Desktop computer displays operate at viewing ratios as low as 1:1.

The difference in the eye's acuity for detail and edge sharpness at the viewing ratios assumed for HDTV, broadcast television systems, and computer displays establishes some of the most important differences in their design requirements. HDTV pictures, observed at a low viewing ratio, must have higher definition to give the same perception of sharpness. Broadcast quality pictures seen from a low viewing ratio will appear fuzzy. On the other hand, an HDTV picture will appear no sharper than a broadcast quality picture at greater viewing ratios. The practical result is that HDTV requires a larger screen at normal viewing distances to take advantage of its superior definition and signal-to-noise ratio. Conversely, a large screen image will appear "soft" and noisy with broadcast quality definition and signal-to-noise ratio (see Section 4.13).

4.5 Image Definition

The sharpness of an image is expressed by its *image definition* parameter.

4.5.1 Image definition defined

The dictionary meaning of image definition, the distinctness of the outlines in the image, is also satisfactory as a technical definition. A less precise definition is the degree to which the image appears to be "in focus." Stated in still other terms, the definition of an image is a measure of the sharpness of the transitions or edges between its dark and light areas. In a high-definition system, these edges must be very sharp.

While the perceived definition of an image is determined primarily by the width of the edges between light and dark areas, there are second-order effects that depend on the *shape* of the edge transition. As compared with film, the slope of the transition curve for a television system is steeper in the center of the brightness scale, and there is often an overshoot. This tends to give the television image a "harder" appearance that is sometimes described pejoratively by the film industry as "the television look." The perceived definition of an image is also affected by the gray scale and the signal-to-noise ratio. These factors are treated with greater detail in later sections.

4.5.2 Criteria of image definition

Although the perceived definition of an image is determined by the sharpness of its edges, edge width is not a satisfactory working criterion for image definition. It is difficult to measure and equally difficult to handle analytically. As a result, television image definition is usually specified by one of two indirect criteria—aperture response and limiting resolution. The measurement of aperture response is more complex, but it is a more complete criterion and is indispensable for system analysis. Limiting resolution is easy to measure and is particularly useful for routine maintenance.

4.6 Aperture Response

Aperture response is a universal criterion for specifying picture definition and other aspects of imaging system performance. It can be used for film images, camera lenses, television camera imagers, video amplifiers and other bandwidth limiting components, the scanning process, receiver picture tubes, and for the human eye. It is used frequently in this volume for describing the performance of television system components.

4.6.1 Definition of aperture response

The term "aperture response" is defined in Figure 4.2. Assume that a pattern of black and white lines of varying widths is scanned by a narrow light beam (the aperture), and the peak-to-peak variation in the reflected light from the black and white lines is measured. On lines that are much wider than the diameter of the spot, these variations will be of full amplitude. As the width of the lines is decreased so that the scanning spot always overlaps a portion of a black and white line, the amplitude of the variations will decrease. When the width of the lines is twice the diameter of the spot, the variations disappear.

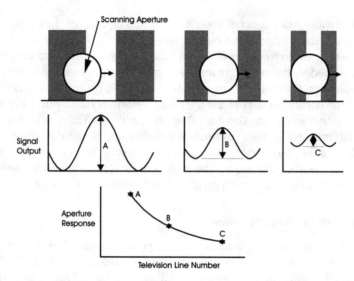

Figure 4.2 Definition of aperture response

The width of these lines is specified by its reciprocal, the number of alternate black and white lines (counting both black and white lines) that can be fitted into the vertical dimension of the picture. This parameter is known as the *television line number* (TVL).

The aperture response of a component or system is a graph of the peak-to-peak amplitude of its response, or of the variations in reflected or transmitted light, as a function of the television line number.

Aperture response can be specified either by the response to a square-wave pattern, i.e., alternate dark and light bars, known as the *contrast transfer function* (CTF), or of the response to a mathematically defined pattern in which the cross-sectional darkness of the bars varies sinusoidally, known as the *modulation transfer function* (MTF). The CTF is physically measurable, but the MTF is more useful for analytic purposes. The conversion can be made by means of Eq. (4.1):

$$\text{MTF}_n = \frac{\pi}{4}\left(\frac{\text{CTF}_{3n}}{3} - \frac{\text{CTF}_{5n}}{5} + \frac{\text{CTF}_{7n}}{7} \cdots \right) \tag{4.1}$$

where: MTF_n = the modulation transfer function (aperture response to a sinusoidal pattern) at line number, n

CTF_n = the contrast transfer function (aperture response to a square wave pattern) at line number, n

The aperture response curves, MTF_n and CTF_n, for a representative charge-coupled device imager are shown in Figure 4.3.

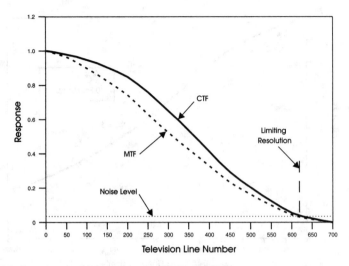

Figure 4.3 Aperture response, showing MTF and CTF curves

4.6.2 Aperture response of components in series

The aperture response of a system having several components in series can be calculated by multiplying the MTFs of the individual components, just as with the frequency response of cascaded electrical components. The ability to calculate the cascaded aperture response of a number of components in this manner is a very useful characteristic of aperture response as a criterion of performance.

4.6.3 Limiting resolution and aperture response

Limiting resolution is determined by visual acuity (see Section 4.4.2) and also by the line number at which the aperture response curve disappears into the noise (see Figure 4.3).

4.6.4 Aperture response of the human eye

Figure 4.4 shows the aperture response, MTF, of a typical human eye at viewing ratios of 3, 4, and 8. This response function should be considered as nominal rather than precise because the visual acuity of individuals varies widely.

4.6.5 Aperture response and image definition

The correlation of aperture response with definition or edge sharpness is not intuitively obvious, but it can be demonstrated by calculating the output of a system of known aperture response with a step function input, i.e., to an input signal that makes an infinitely steep transition between two levels. The

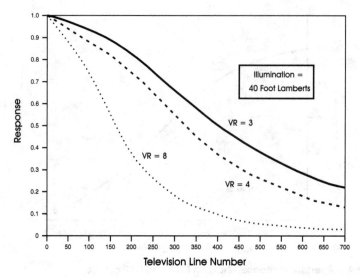

Figure 4.4 Aperture response of the human eye

sharpness of this transition as a function of the system aperture response is calculated by an *inverse Fourier transform*:

$$e(x) = \int g(n) e^{jn} dx \tag{4.2}$$

where: $e(x)$ = the amplitude response of the component or system as a function of its position on the scanning line, i.e., along the x axis of the raster.

$g(n)$ = the aperture response of the system as a function of the television line number, $n = 2\omega x$.

With the aid of this equation, the correlation between aperture response and definition can be verified.

4.6.6 Equivalent line number, N_e

An aperture response curve is the most complete indication of the definition of a system, but it is not as convenient or easily compared as a single number. In an effort to define a single number that would specify the definition of a system to a reasonable degree of approximation, Schade[1,2] developed the concept of equivalent line number or N_e. Although N_e is not widely used as a criterion of picture quality, it is very useful for system analysis.

N_e is defined in Figure 4.5. It is the line number that defines a rectangle having the same area as the area under the aperture response squared curve. (It is equivalent to the *noise bandwidth* in the electrical analog.)

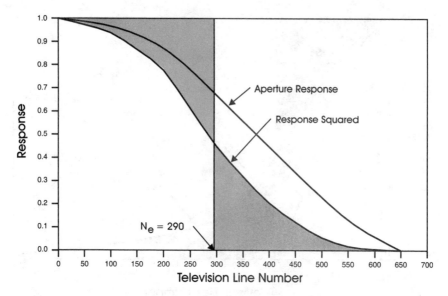

Figure 4.5 Definition of N_e

The validity of N_e as a measure of picture definition can be demonstrated by calculating the edge configurations of systems having the same N_e but differently shaped response curves. Figures 4.6 and 4.7 compare two imaging systems with the same N_e but with aperture response curves typical of television and film system components. Figure 4.6 shows the square of their aperture responses while Figure 4.7 shows their responses to a step function input. Note that the resolution of the film system is much higher but its response is lower at lower line numbers. This causes minor differences in the shape of the edges, but the edge width is approximately the same, thus confirming the validity of N_e as a measure of image definition.

The edge width is given by Eq. (4.3):

$$\text{Edge width} = \frac{\text{picture height}}{N_e} \tag{4.3}$$

Within reasonable limits of inequality, the approximate perception of definition of images that have unequal horizontal and vertical definition is given by Eq. (4.4):

$$N_e = \sqrt{N_{e(\text{vert})} N_{e(\text{horiz})}} \tag{4.4}$$

Because of limitations imposed by the scanning process and the improved performance of modern cameras, however, the inequality often exceeds "reasonable limits" and the horizontal definition is noticeably greater.

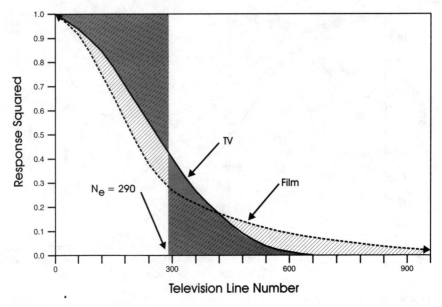

Figure 4.6 TV and film MTF compared

Figure 4.7 TV and film edge sharpness compared

4.7 Sampling and Aliasing

Sampling is a fundamental technique that is pervasive in video systems. It was introduced in Section 1.3. Applications of sampling in analog television include the following:

Sampling process	Parameter
Scanning lines	Vertical detail
Frames	Image motion
Shadow mask kinescope	Horizontal detail
CCD imagers	Horizontal detail

Sampling is an essential ingredient of digital television systems (see Chapter 3) and in these systems the image is sampled horizontally as well as vertically. In a related process, the amplitude of the signal is *quantized*, i.e., it is represented by discrete levels rather than by a continuous function. Thus digital systems sample the image in both dimensions and quantize the signal amplitude. Sampling and signal amplitude quantizing as employed in digital systems are described in Chapter 3.

4.7.1 Sampling aperture response (MTF)

Sampling a signal potentially gives rise to two distortions. The first is aliasing, as described above, for sampling rates below the Nyquist criterion. The second is a reduction in the MTF for higher frequency components of a spatial or temporal signal. The $MTF_{(H,V)}$ of the sampling process for the horizontal (H) or vertical (V) components of a spatial signal can be calculated by means of Eq. (4.5). This equation is strictly valid only for image frequency components below the Nyquist limit; higher frequencies are subject to aliasing (see Sections 4.7.4 to 4.7.7 below).

$$MTF_{(H,V)} = \frac{\sin x}{x} \qquad\qquad (4.5)$$

$$x = \pi \, \frac{W_{(H,\,V)} \, f_{(H,\,V)}}{F_{(H,\,V)}}$$

where: $F_{(H,V)}$ = spatial sampling frequency in the horizontal or vertical dimension

$W_{(H,V)}$ = (sample width)/(sample spacing) e.g., (scanning line width)/(scanning line spacing)

$f_{(H,V)}$ = spatial frequency component of sampled image

Examination of Eq. (4.5) shows that the MTF is determined by the sample width and is independent of the sampling frequency provided that frequency exceeds the Nyquist frequency.

4.7.2 Scanning line aliasing

Vertical brightness variations are sampled by the horizontal scanning lines, and if the dimensions of the vertical detail are less than the scanning line spacing, aliasing will occur. The result is a third and spurious set of lines having a line number equal to the difference between the number of scanning lines in the raster and the line number of the vertical detail. With typical subject matter, this produces an objectionable moire pattern in the image.

This form of aliasing is particularly visible with subject matter that has a strong geometric pattern, e.g., a herringbone suit. Motion in the picture also enhances the effect, and a geometric pattern may appear to twinkle. It is most noticeable in sharp high-definition images, and it was not a serious problem in early television pictures because early cameras had limited resolution and their fine detail did not have sufficient amplitude to interact visibly with the scanning lines. With the improved definition of current broadcast systems, it has become a significant problem, and it is a major consideration in HDTV system design.

Figure 4.8 shows the MTF of the vertical scanning sampling process (without vertical aperture correction) and its relation to aliasing as calculated by Eq. (4.5) for an NTSC and representative HDTV signal that meets the SMPTE HDTV production Standard 240M.

Figure 4.8 Scanning MTF curves for NTSC television and HDTV. Parameters: NTSC—525 scanning lines, 480 active lines, $W_V = 0.9$; HDTV—1125 scanning lines, 947 active lines, $W_V = 0.9$

Figure 4.9 The EIA resolution chart

Aliasing for the NTSC signal begins at 240 TV lines, one-half the number of active scannning lines. The MTF is high for spatial components above 240 TV lines, and, if scanning were the only element determining the MTF, aliasing would be severe. In practice, the system MTF at these higher line numbers is substantially reduced by the aperture responses of an optical filter, the camera lens, the camera imager, and the display device. The vertical MTF for a complete NTSC system, which includes these elements (as described in Section 4.9), is shown in Figure 4.15. Aliasing is present but at a much lower amplitude than would be produced by the scanning pattern alone because of the attenuation of the other elements. System vertical aliasing can be seen as twinkling in the horizontally-oriented line patterns of the EIA resolution chart, Figure 4.9.

Vertical aliasing for the HDTV system begins at 474 TV lines, one-half the number of active scanning lines. The increase in the beginning of aliasing from 240 TV lines in NTSC systems to 474 is one of the major factors in the improved picture quality of HDTV systems.

4.7.3 Frame aliasing

Frame aliasing occurs when a cycle of a repetitive horizontal pattern moves vertically more than one scanning line during the frame interval. For example, if a fine-grained repetitive pattern moves *upward* 1.5 lines during the frame interval, the visual effect will be the same as though it had moved *downward* 0.5 lines. This is the cause of the motion picture phenomenon in which wagon wheel spokes appear to be turning backward.

Frame aliasing is an inherent sampling defect, and can only be reduced or eliminated by increasing the frame repetition rate.

4.7.4 CCD aliasing

CCD signals are generated by rows and columns of discrete photosensitive elements (see Chapter 8) and hence are the result of a sampling process in both the horizontal and vertical dimensions. This causes aliasing in both dimensions as described more fully in Chapter 8.

4.8 Limiting Resolution

4.8.1 Limiting resolution as a measure of system performance

Limiting resolution has been used historically to specify the performance of optical instruments and film systems. It indicates the maximum number of alternate light and dark lines (or points of light) that can be distinguished per unit distance in the image. Following the lead of the film industry, limiting resolution was initially used to specify the performance of television systems. It has, however, two serious weaknesses as a performance criterion for television systems that reduce its value for this purpose.

The first difficulty is that limiting resolution defines only one point on the aperture response curve —the line number at which the curve falls below the system noise level. This is not a problem when comparing two systems with aperture response curves of very nearly the same shape—as for two types of film—but it can be very misleading if the shapes are different—as with film and television. Image definition, as measured by edge sharpness, is determined by the aperture response over the whole range of line numbers, not just at a single point. Figure 4.6 shows examples of film and television systems with almost equivalent definition, although the limiting resolution of the film is much higher.

A second problem arises when limiting resolution is used to evaluate the performance of systems that employ sampling (see Section 4.7). The relationship of image detail to the sampling points is random, and the limiting resolution is less than the number of sampling points. More seriously, the Nyquist criterion predicts that aliasing will occur for image detail components that have line numbers more than one-half the sampling rate. The limiting resolution may be an ambiguous and irrelevant figure if the Nyquist number is less than the limiting resolution, and significant aliasing occurs before the limit of resolution is reached.

In spite of its flaws as a criterion of image definition, however, limiting resolution plays a useful role in the design and maintenance of television systems, and it continues to be widely used. It is easy to measure, and the result is a single, easily understood number. The Electronic Industries Association (EIA) standard resolution test chart, Figure 4.9, has been widely used for evaluating system performance.

4.8.2 Vertical resolution and the Kell factor

Because of the random relationship of image detail and scanning lines, the vertical limiting resolution is less than the number of scanning lines. The ratio between the observed limiting resolution and the number of scanning lines is known as the *Kell factor*. This must be evaluated empirically rather than analytically, and various values have been determined by different observers in the range from 0.53 to 0.85, with 0.70 being a commonly accepted value. All of these values are above the Nyquist limit of 0.5 and are, therefore, in the region where aliasing occurs.

In television's early years, the aperture response of cameras and other system components at line numbers above the Nyquist limit of the scanning process was low, and the aliasing was small enough to be of minor significance. Limiting resolution under these conditions was an important specification, and it was frequently quoted.

With cameras of recent design, there has been a major increase in the aperture response in the region above the Nyquist limit, and aliasing is now a significant factor in system design. This is illustrated in Figure 4.15, which shows the locations of limiting resolution calculated with a Kell factor of 0.7 and the onset of aliasing in the vertical MTF of a representative NTSC television system. The limiting resolution of 336 lines is well above the Nyquist limit, and it is relatively insignificant as a measure of system performance since aliasing begins at lower line numbers. In consideration of this fact, the vertical resolution of cameras is seldom specified.

4.8.3 Horizontal resolution and bandwidth

Aliasing in the horizontal dimension occurs with CCD imagers because there is sampling along each scanning line. The number of horizontal samples per line is much higher than the number of scanning lines, however, typically over 1100 lines with spatial offset imagers (see Chapter 8) and the Nyquist limit is then over 550 lines. At these line numbers, the aperture response of the camera and other system components is typically low enough that the aliasing amplitude is relatively insignificant. Under these conditions, the horizontal limiting resolution of the system components rather than the Nyquist limit determines system performance.

The horizontal limiting resolution of an *analog* television system—the number of black and white vertical lines that can be distinguished in a dimension equal to the picture height—may be determined by its electrical bandwidth or by the aperture response of the system components. In most cases the bandwidth is limiting, and each cycle can reproduce a black and white line-pair, i.e., two television lines. This leads to Eq. (4.6) for the limiting horizontal resolution of analog systems:

$$R_H = 2 \frac{C_H \, BW}{AR \, NL \, FR} \tag{4.6}$$

where: R_H =   the horizontal resolution

C_H =   the fraction of time in each scanning line devoted to the transmission of picture information after subtracting the time required for horizontal blanking

BW =   the system bandwidth

AR =   the aspect ratio

NL =   the total number of scanning lines per frame

FR =   the frame rate per second

The horizontal limiting resolution, R_H, of NTSC broadcast systems (U.S. and Japan), PAL broadcast systems (Europe), and the analog HDTV system defined by the SMPTE Standard 240M are compared in Table 4.1.

High definition requires a high limiting resolution, and the 4.2 MHz bandwidth limitation of standard broadcast channels is a fundamental problem for HDTV. A tremendous amount of engineering research has been devoted to the development of systems which can acheve high resolution and definition with a limited bandwidth. These efforts were only marginally successful with analog signals, and it now appears that most HDTV signals will employ digital formats. These formats can use compression techniques

Table 4.1 Horizontal resolution comparison

	NTSC	PAL	HDTV (SMPTE 240M)
BW (MHz)	4.2	5.0	30
NL	525	625	1125
C_H	0.85	0.80	0.87
AR	1.33 (4/3)	1.33 (4/3)	1.78 (16/9)
FR	29.97	25	30
R_H (TV lines)	352	409	869

that make it possible to equal the performance of wide-band analog systems in much narrower bandwidths. The bandwidth requirements for digital HDTV systems are described in Section 3.7.2.2.

4.8.4 Picture elements (pixels)

A television raster can be described as a grid of picture elements or pixels. A pixel is defined basically as the smallest element of a picture whose properties can be described independently of adjacent elements. The height of each pixel is usually considered to be the spacing between scanning lines (or alternatively the spacing divided by the Kell factor) and its width is the dimension of the narrowest object that can be reproduced. For analog signals, this is one-half cycle of the highest video frequency in the transmission bandwidth. The number of pixels in a frame is a rough indication of its information content.

The number of pixels in a frame is:

$$NP = 2 \frac{C \; KF \; BW}{FR} \tag{4.7}$$

where: C = the fraction of time occupied by active scanning (after subtracting vertical and horizontal blanking periods)

KF = the Kell factor (optional)

BW = the bandwidth of the system in Hertz

FR = the frame rate

In the broadcast and HDTV systems described in the preceding sections, the broadcast system has 152,000 equivalent pixels while the HDTV system has 1,162,000.

4.9 NTSC Television System Aperture Response

This section describes the aperture response of the functional components of a representative NTSC television system and their cascaded response in a complete system.

The components can be divided into two categories: the electro-optical components that generate and display the images, and the processing and transmission components that generate and process the video signal and transmit it from camera to receiver. The principal system components are:

Electro-optical components
    Camera lens
    Optical filter
    Camera imager

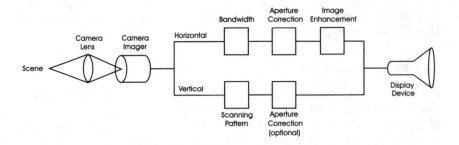

Figure 4.10 Aperture response determining components

Display device
Processing and transmission components
    Video bandwidth
    Aperture correction
    Image enhancement
    Vertical scanning
    Vertical aperture correction (optional)
    Figure 4.10 shows the relationship of these components in the complete system.

4.9.1 Electro-optical components

Aperture response curves for typical broadcast-quality electro-optical components and their overall response when cascaded are shown in Figure 4.11. The performance of commercial products varies widely, and these curves are representative and not intended to describe specific models.

4.9.1.1 Camera lens   A high quality, fixed-focus lens such as that used as an example in Figure 4.11 has a very high response out to 300 lines, the approximate upper limit of NTSC systems, and it has only a minor effect on the performance of these systems. A zoom lens makes some compromises in aperture response in order to provide the variable focus feature, and the performance of these lenses must be considered when used in HDTV systems.

4.9.1.2 Optical filter   CCDs, currently the most common type of camera imager, suffer from aliasing, an inherent potential result of sampling. Once introduced, aliasing cannot be eliminated from a signal by electrical filtering. It can, in theory, be removed by filtering the spatial frequency components above the Nyquist limit from the optical image. This could be done in a brute force fashion by defocusing the lens, but this would produce an unacceptable sacrifice of picture definition. An ideal solution would be to employ an optical

Figure 4.11 Electro-optical components MTF—Broadcast

prefilter with unity aperture response below the Nyquist limit established by the scanning pattern or the video bandwidth and zero response above it. Such a filter does not exist, but this response can be very roughly approximated by practical filters.

The response of a typical optical filter is shown in Figure 4.11; it is a characteristic sin x/x curve with a zero ($x = \pi$) at 1200 TV lines (14.8 MHz).

4.9.1.3 Camera imager   A wide range of camera imager aperture responses is available from commercial cameras as described in Chapter 8. Figure 4.11 shows the response for a camera with a 768(h) × 494(v) CCD and spatial offset.

4.9.1.4 Display device   The display device aperture response curve shown in Figure 4.11 is for a shadow-mask color kinescope (see Chapter 5) with 763 apertures per picture width and a television line number of 574 (763/1.33). Aliasing, in this case the interference pattern between the apertures and horizontal detail, begins to appear at about 287 (574/2) lines. The horizontal limiting resolution, unless it is limited by other factors, is 400 lines (574 lines times a Kell factor of 0.7), and aliasing rather than resolution limits the performance of the device.

4.9.1.5 Cumulative response, electro-optical components The cumulative horizontal aperture response of the electro-optical components shown in Figure 4.11 is approximately 0.5 at a television line number of 340, the upper limit established for the U.S. broadcast system by the video passband. The response can be improved further by the use of image enhancement techniques, see Section 4.9.2, to give better results within the bandwidth, aliasing, and scanning line limitations of the system.

The composite response of the broadcast quality components described above is marginal or unsatisfactory for HDTV systems. This dictates use of higher-definition camera transducers and display devices.

4.9.2 Horizontal channel processing and transmission

4.9.2.1 Video channel bandwidth The video bandwidth of an analog system establishes a fundamental limit on its resolution and definition. The resolution is given by Eq. (4.6). The aperture response of the video channel, which is also its amplitude response, is one of the factors that determines the system's definition.

In analog systems, the amplitude response drops to zero at TV line numbers above the limit set by the video bandwidth as given by Eq. (4.6).

4.9.2.2 Video channel aperture response The amplitude response of the video channel can be converted to aperture response by translating electrical frequencies to television line numbers (TVL) in accordance with Eq. (4.8):

$$\text{TVL} = 2\,\frac{C_H\,f}{AR\;NL\;FR} \tag{4.8}$$

where: C_H = the fraction of time in each scanning line devoted to the transmission of picture information (after subtracting the time required for horizontal blanking)

f = the frequency

AR = the aspect ratio

NL = the total number of scanning lines per frame

FR = the frame rate per second

Controlling the amplitude response of the video channel (it is important that this be accomplished without introducing phase distortion) is an effective way of increasing the aperture response of the system and hence improving its definition. A number of techniques with the generic title of *image enhancement* have been developed to accomplish this.

Aperture correction is a commonly used method. It boosts the amplitude response of the video channel to the higher frequency components in the signal (taking care that this does not introduce unwanted phase shifts) to compensate for the fall-off in the response of the electro-optical components.

Another method (once called "contours out of green") is to create a waveform by passing the green signal through a linear-phase high-pass circuit that produces spikes at each rapid change in the signal. These spikes are then combined with the luminance signal in a manner that sharpens its edges.

Figure 4.12 shows the aperture response of a video channel with a flat response out to a bandwidth limit of 4.2 MHz and with a response enhanced to compensate for the reduced MTF of the electro-optical components at higher frequencies.

4.9.3 Vertical processing and transmission

4.9.3.1 Scanning   The aperture response of the scanning pattern, which affects the sharpness of horizontal edges and the reproduction of vertical detail, is not a straightforward single-valued function, but is the complex result of the size of the scanning aperture and the sampling process.

The major criteria of scanning are its aperture response or MTF, its

Figure 4.12 Video amplitude response—broadcast

limiting resolution, and aliasing. These are described in Sections 4.7.3 and 4.7.4.

4.9.3.2 Image enhancement   Image enhancement by aperture correction or equivalent techniques is more difficult in the vertical dimension than in the horizontal because the correction cannot be accomplished along a single line. Vertical image enhancement circuits use delay lines that make it possible to compare the video signals on adjacent lines. The image enhancer accentuates the differences in signal levels between corresponding points on adjacent lines as would occur on a horizontal edge, thus sharpening its appearance. (The same technique can be used for horizontal image enhancement by the use of delay lines that delay the signal by one picture element.)

4.10 HDTV System Aperture Response

4.10.1 Electro-optical components

HDTV research has emphasized the problem of obtaining high definition performance within a limited, e.g., 4.2 MHz, bandwidth, but improving the aperture response of the electro-optical components is equally important. Figure 4.13 shows the MTF curves for HDTV electro-optical components.

4.10.1.1 Lens   The aperture response of a high quality, fixed focus lens is adequate for use in HDTV systems. The response of zoom lenses is compromised somewhat to achieve the variable focus feature, and it must be considered in HDTV systems.

4.10.1.2 Optical filter   The optical filter for HDTV cameras should be designed to have a higher cut-off frequency than for NTSC cameras. Aliasing begins at higher line numbers; this permits a higher cut-off, and a higher cut-off is needed to avoid needless attenuation of the higher frequency components in the optical image. The filter response in Figure 4.13 has its first zero at 2000 lines (25 MHz) as compared with 1200 lines in the filter assumed for the NTSC example in Figure 4.11.

4.10.1.3 Camera imager   The aperture response of the camera imager used in an HDTV system must be substantially greater than for an NTSC broadcast system, and it must be designed for a wider aspect ratio, usually 16:9 rather than 4:3. If the imager is a CCD, the size of the photosensitve elements must be reduced to improve the aperture response and their number increased to reduce aliasing. A typical HDTV CCD has a 1920(h) ×

1036(v) element matrix as compared with 768(h) × 494(v) for NTSC. Figure 4.13 shows the aperture response of a 1920 × 1036 CCD.

4.10.1.4 Display device   To obtain the full visual benefit of HDTV, it is necessary to use larger display devices. They may be either direct view or projection. Projection displays can be larger, but at the present time do not equal direct view in brightness or aperture response. For example, the specified aperture response of current projection displays varies from 5 percent to 35 percent at 300 lines. This compares with a response in excess of 80 percent for direct view kinescopes. The maximum practical size for direct view devices employing kinescopes is 35–40 inches.

Direct view tubes may employ dot triads as in the original shadow mask tubes or color-stripe triads as for trinitrons (see Chapter 12).

One trinitron design has a picture diagonal of 38.5 inches with 1861 vertical stripe triads. The aperture response or MTF of this kinescope design is shown in Figure 4.13.

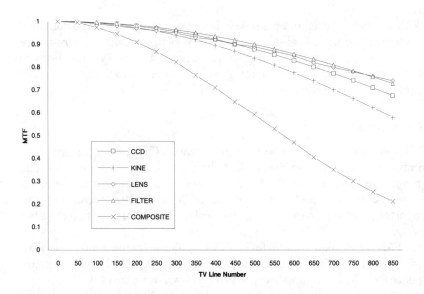

Figure 4.13 Electro-optical component MTF—HDTV

4.10.1.5 Cumulative response, electro-optical components The cumulative response of the other electro-optical components of the HDTV system described above is also shown in Figure 4.13.

4.10.2 Video channel

HDTV systems generally employ digital rather than analog formats for the processing and transmission of signals. The signals are generated in an analog format but are then converted to a digital format.

The first step in the conversion process is to sample the analog waveform. The sampling rate must be at least twice the highest frequency component in the analog signal, the Nyquist frequency, to avoid aliasing. The sampling rate for the luminance component of the SMPTE 240M Standard signal is 74.25 Mbs.

Most HDTV systems use some form of video compression (see Chapter 6) which makes it possible to transmit the signal at a substantially lower rate.

The aperture response of the sampling process is determined by the ratio of the width of the samples to their spacing. If the width is one-half the spacing, the aperture response exceeds 0.9 at frequencies just below the Nyquist limit. Thus, the aperture response of the sampling process can be maintained almost at unity out to the Nyquist limit.

Image enhancement can be introduced in the digital as well as the analog domain. One technique is to generate a detail signal by digital signal processing. A composite signal is then generated by adding the primary and detail signals.

With image enhancement possible in both the analog and digital domains, the system designer has great flexibility in selecting an aperture response curve that is optimum for the application. Figure 4.16 shows the aperture response of a video channel with image enhancement sufficient to maintain an overall system response at unity out to 400 lines.

4.10.3 Scanning

HDTV systems have more scanning lines than NTSC, partly for improved definition but also to move the onset of aliasing to higher line numbers. The high aperture response of HDTV electro-optical components would cause severe aliasing above the Nyquist limit, and a large number of scanning lines is important, not only to increase the vertical definition but also to reduce the aliasing.

The SMPTE Standard 240M for HDTV has 1035 active scanning lines with aliasing occurring for TV line numbers greater than 517. The aperture response for a 517 scanning-line pattern is shown in Figure 4.8.

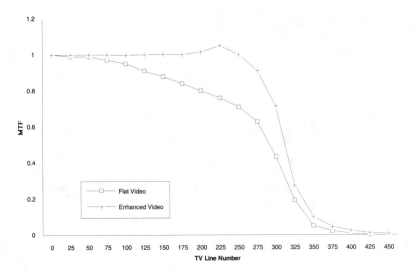

Figure 4.14 Horizontal definition—broadcast

4.11 Image Definition—Broadcast Systems

The next two sections describe the image definition as measured by the modulation transfer function or MTF (also the aperture response) of NTSC broadcast and HDTV systems. N_e is used to specify the definition by a single number.

4.11.1 Horizontal definition

The horizontal definition can be calculated by cascading the aperture response of the electro-optical components (see Figure 4.13) with the response of the video channel. The results of this operation are shown in Figure 4.14.

The N_e of the system is 206 with a flat video response and 286 with an enhanced response. The N_e of the enhanced system could be increased further by greater enhancement but a high N_e in combination with a sharp frequency cut-off requires very careful control of the phase/frequency response to minimize overshoots on the picture edges.

The maximum resolution is limited by the 4.2 MHz video bandwidth to 355 TV lines.

The CCD imager assumed in the system has 640 columns of photosensitive elements, and aliasing of the horizontal signal will begin at 240 (640/1.33/2) lines unless spatial offset (see Chapter 8) is used in the CCD. This would approximately double the TV line number at the onset of aliasing to 480.

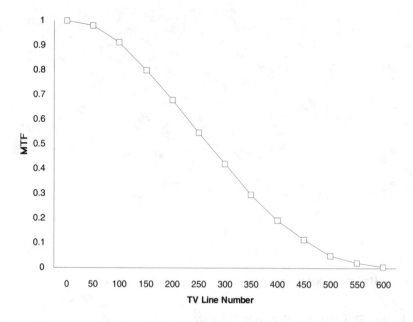

Figure 4.15 Vertical definition—broadcast

4.11.2 Vertical definition

The vertical definition can be calculated by cascading the aperture response of the electro-optical components (see Figure 4.13) with the response of the scanning pattern (see Figure 4.8). The results of this calculation for the system in this example are shown in Figure 4.15.

The vertical N_e is 194. This could be increased to more than 225 lines by the use of vertical aperture correction as described in Section 4.9.3.2. With the development of solid-state delay lines, vertical aperture correction has become practical.

Aliasing begins at one-half the number of active scanning lines or 240 lines. The limiting resolution as calculated with a Kell factor of 0.7 is 336, but because of aliasing it has little significance.

4.12 Image Definition—HDTV Systems

The determination of the image definition of HDTV systems is not as straightforward as for NTSC systems because they employ digital compression to reduce the bandwidth requirements while retaining the perceived definition of conventional wider band analog systems. One technique is to transmit the stationary portions of the image in high definition at a slower

Figure 4.16 Horizontal definition—HDTV

rate while transmitting the moving portions with less detail. This method takes advantage of the eye's reduced acuity to detail for moving objects.

An analog system based on the SMPTE production Standard 240M is assumed for the purpose of comparing the performance of NTSC and HDTV systems in this chapter.

4.12.1 Horizontal definition

The cascaded aperture response of the HDTV electro-optical components and the video channel is shown in Figure 4.16. The responses with two video channels are shown, one with flat response out to the limit established by a 30 MHz bandwidth and one with aperture correction or other image enhancement that maintains a unity aperture response for the system out to 400 TV lines.

The N_e of the system with flat video response is 375 TV lines and of the enhanced system, 539 lines. The limiting resolution as established by the 30 MHz bandwidth is 808 lines.

4.12.2 Vertical definition

In the absence of vertical aperture correction, the vertical aperture response would be that of the electro-optical components (see Figure 4.13) cascaded

Figure 4.17 Vertical definition—HDTV

with the response of the scanning pattern (see Figure 4.8) which in the SMPTE 240M standards has 1035 active scanning lines. This is shown in Figure 4.17.

The vertical N_e of this system would be about 350 lines and this could be increased to more than 400 lines with vertical aperture correction. Now that inexpensive memory circuits, a prerequisite for vertical aperture correction, are available, it is expected that the use of vertical aperture correction will become common in future HDTV systems.

The high aperture response of the components at 400 lines and above indicates the importance of increasing the number of scanning lines in HDTV systems to limit aliasing.

4.13 Visual Perception of Broadcast and HDTV Images

Figure 4.18 shows the combined aperture responses of the eye and of NTSC and HDTV television systems at viewing ratios of 3 and 8. They are calculated by cascading the response of a normal eye as shown in Figure 4.4 with horizontal aperture response of NTSC and HDTV systems as shown in Figures 4.14 and 4.16.

Figure 4.18 illustrates the following:

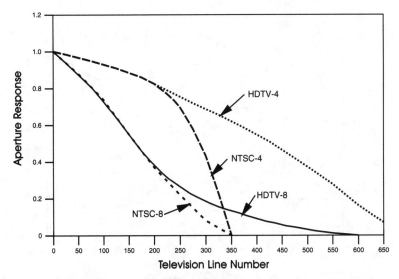

Figure 4.18 Visual perception of NTSC and HDTV images

1. At a viewing ratio of 8, there is very little difference in the perception or appearance of NTSC and HDTV systems. A smaller viewing ratio is necessary to take advantage of the higher definition of HDTV images.

2. As the viewing ratio is reduced below 8, the difference between HDTV and NTSC images begins to become apparent. The difference is striking at a viewing ratio of 3.

3. At viewing ratios of 8 and above, the eye rather than the television system becomes the major limitation in the perceived definition of the image. Since the eye is also the limitation when viewing the scene directly, these images will appear relatively sharp.

4.14 Image Gray Scale

The importance of gray scale as a measure of image quality can be demonstrated by turning on the room lights while photographic slides are being projected on a screen. With the lights out, the images appear crisp and in focus. When the room lights are turned on, thus increasing the ambient lighting on the screen, the images appear washed out and fuzzy. The addition of a fixed amount of ambient illumination to the slide image over a large area of the screen not only distorts the image's gray scale but also its apparent definition.

In color systems, as described in Chapter 2, careful control of the *transfer function* (the parameter that specifies gray scale, see below), of the three primary color signals is required to avoid color distortion.

Figure 4.19 Television transfer function

Unlike image definition, improved gray scale reproduction requires no changes in bandwidth, numbers of scanning lines, or other transmission standards. It is nevertheless, as important to the production of high quality HDTV pictures as improved definition. A better gray scale will also improve the quality of standard broadcast pictures with no change in transmission standards.

4.14.1 Gray scale transfer function

The gray scale of an image is specified by its transfer function, a plot of image brightness as a function of scene brightness. Figure 4.19 is the transfer function of a typical system, including camera, video channel, and display device. It is the television equivalent of the H & D (Hurter and Driffield) curves that specify film density as a function of exposure for photographic negatives. Because the eye perceives brightness differences in terms of ratios rather than numerical differences, a logarithmic scale is used for both scene and image brightness.

The transfer function can also be used to specify the input-output performance of television system components, including camera imageries, video amplifiers, and display devices. The system gray scale function is obtained by cascading the transfer functions of its components.

As with most imaging systems, the television transfer function is an "S" shaped curve with a linear section in the middle that slopes off at both ends. The upper limit of the curve is called the *white field brightness*, and is usually established by the limitations of the display device, ordinarily a kinescope. The lower limit is called the *unexcited field brightness* and is the brightness of the kinescope when it is turned off, primarily due to reflection of ambient light.

It is often convenient to specify the gray scale of a television image by numerical parameters rather than graphically. The three most commonly used parameters, as shown in Figure 4.19, are as follows:

1. Highlight Brightness: The white field brightness, i.e., the brightness of the brightest areas of the image.

2. Contrast Ratio: The brightness ratio of the brightest and darkest areas of the image. For an imaging system, it is the ratio of the white field and unexcited field brightnesses.

3. Gamma: The slope of the transfer function curve. Gamma is a term borrowed from photography where it is used to specify the slope of the H & D curve at a single point in its linear portion where the slope is the greatest. Its usage is slightly different in television where it can specify the slope of the transfer function at any point.

4.14.2 Brightness and contrast ratio

The brightness and contrast ratio of television images are determined primarily by the characteristics of the kinescope (or other viewing device) and the ambient lighting. Great progress has been made in recent years in increasing both the brightness and contrast ratio by increasing the white field brightness and reducing the unexcited field brightness of kinescopes.

Between 1962 and 1982 the white field brightness of typical color tubes was increased from 10 to 65 foot lamberts. This was achieved by utilizing metalized backing on the face-plate, improved phosphors, higher beam voltage, and improved electron gun design. The greater brightness of the phosphors was achieved by deviating somewhat from the NTSC primary color specifications (see Chapter 2). This has not been a serious concern for broadcast systems, but it has been questioned for HDTV systems that strive for perfection.

Most of the unexcited or dark field brightness results from reflection of ambient light from the picture tube face. Its effect was originally reduced by placing a neutral density filter in front of the kinescope. Light from the kinescope passed through the filter once, while light reflected from the kinescope passed through it twice. If the filter transmitted one-half the light, the white field brightness would be reduced by one-half but the unexcited field brightness would be reduced by three-quarters and the contrast ratio doubled.

The unexcited field brightness of modern color kinescopes has been reduced by the *black matrix/black surround* and other techniques that substantially reduce the light reflected by the phosphors. The results can be seen by observing the kinescope of a color receiver of recent design when the

Figure 4.20 Effect of ambient light

set is turned off. By contrast with older sets in which the kinescope was whitish gray, new sets appear black.

A high contrast ratio is a technical ideal and, with some exceptions, an aesthetic one as well. Improved kinescopes have made it possible to meet the desired contrast ratio standards for high quality broadcast TV, provided the ambient lighting is not excessive. Optimum results for HDTV require control of ambient lighting by creating a theater-like environment for home viewing.

Because of the eye's logarithmic response to brightness differences, the reflected ambient light from the kinescope screen, which establishes the unexcited field brightness, can have a greater effect on the contrast ratio than the highlight or white field brightness. A relatively small numerical difference in reflected light can have a major effect on the contrast ratio.

This effect is illustrated in Figure 4.20, which shows the transfer function under three different conditions. All assume a white field brightness of 50 foot lamberts, but with unexcited field brightnesses of 1, 2, and 4 foot lamberts as the result of differences in ambient lighting. The resulting contrast ratios are 50, 25, and 12.5. The gamma is also reduced as the ambient light is increased. The gamma of the image with low ambient lighting is 0.94, and it decreases to 0.76 as the lighting is increased.

The combination of a lower contrast ratio and a reduced gamma with high reflected light results in a softer, less sharp picture, the opposite of the effect desired for HDTV and undesirable for broadcast television.

Table 4.2 shows the approximate ranges of contrast ratios that are typical of or desired for operating systems. In practice, it depends to a great extent on the ambient lighting on the screen, a parameter over which the system designer often has no control.

Table 4.2 System brightness and contrast ratio (brightness values in foot lamberts)

System	Desired Brightness	Typical Brightness	Desired Contrast	Typical Contrast
Motion picture theater	25	10–12	150:1	50–100:1
HDTV (studio)	150	60	300:1	100–150:1
HDTV (home display)	60	40	150:1	10–30:1
Broadcast TV (studio)	150	100	200:1	100–150:1
Broadcast TV (525 home)	60	40	80:1	10–30:1

4.14.3 Gamma and picture quality

An image gamma of unity would indicate that it is precisely reproducing the gray scale of the scene. This is the technical ideal, although deliberate distortions may be introduced for aesthetic reasons.

If gamma is greater than unity, the image will appear sharper but the scene contrast range that can be reproduced is reduced. If gamma is increased sufficiently, the image will be a silhouette with only blacks and whites. (This was a problem with many early kinescope recordings, which had such a high gamma that most of the image areas were either black or white with very little shades of gray between.)

Reducing the gamma has a tendency to make the image appear soft and "washed out."

Selecting the television picture gamma that is most pleasing is an aesthetic judgment. For color images, a gamma of 1.2 to 1.5 is generally accepted as optimum.

4.14.4 Gammas of television images and components

The gamma of the image produced by a complete television system can be calculated by multiplying the gammas of its individual series components, specifically the camera imager, the display device, and any nonlinear video amplifiers that are added to the system for *gamma correction*.

With the exception of vidicons, which have a gamma of about 0.7, most camera imagers have a gamma of unity.

Kinescopes, which are almost universally used as display devices, have gammas from 2.2 to 2.4

Gamma correction circuits usually have a gamma of less than unity to compensate for the high gammas of kinescopes. For example, if it is desired to have a system gamma of 1.2 with a unity-gamma camera imager and a kinescope gamma of 2.2, the gamma correction circuit should have a gamma of 1.2/2.2 or 0.55.

The FCC Rules for NTSC Color Transmissions specify gamma by saying: "gamma-corrected red, green, and blue voltages suitable for a color picture tube having a transfer gradient (gamma exponent) of 2.2 associated with each primary color." "Suitable for" is not a precise specification, and no tolerance was set on the value of gamma.

The SMPTE Standard 240M for HDTV production also assumes a display device gamma of 2.2 but it specifies the signal gamma more explicitly and in greater detail; the specified transfer function is linear at low brightness levels up to 2.28 percent of reference white and logarithmic with a gamma of 0.45 thereafter. The specification is:

$$V_C = 4\,L_C \quad \text{for } L_C < 0.0228 \tag{4.9}$$

$$V_c = 1.1115 L_C^{(0.45)} - 0.1115 \ \text{for } L_C > 0.0228 \tag{4.10}$$

where: L_C = light input

V_C = camera output voltage

(both are normalized to equal unity at reference white)

This more explicit definition was adopted in anticipation of the availability of technology for achieving compatibility between the transfer functions of a variety of camera and display devices.

4.14.5 Gamma correction and system noise

Gamma correction circuits with a gamma of less than unity "stretch" the black portion of the signal, which causes the noise in the black areas of the picture to be amplified more than noise in the white. The effect on the perceived signal-to-noise in the image will depend on the location of the gamma correction circuit with respect to the noise source.

Noise generated in the dark areas of the picture ahead of the gamma corrector, e.g., in the imager preamplifier, will be amplified more than the picture signal. Depending on the picture content, this may reduce the overall signal-to-noise ratio by several decibels. The adverse effect of this phenomenon is exacerbated by the fact that noise is more visible in the dark areas.

Conversely, dark area noise generated after gamma correction, as in the transmission link, will be amplified no more than the signal, and its amplitude will be reduced by the high gamma of the kinescope.

4.15 Signal-to-Noise Ratio

4.15.1 Sources and properties of noise in television systems

Random electrical disturbances, which are usually thermal in origin, produce "snow" in the picture, and are the most common type of noise in video signals. The relationship between temperature and noise is close and *noise temperature* (see Chapter 17) is a frequently used specification for system components.

The most significant sources of random noise are components where low signal levels are encountered. They include camera imagers, camera preamplifiers, video tape recorders, cable circuits, microwave and satellite circuits, and broadcast receivers. The noise generated by these elements establishes basic limitations on system performance and is fundamental in establishing its design parameters.

Most random noise is "white," i.e., the noise energy is uniformly distributed through the frequency spectrum as with white light. An important exception is the noise generated in microwave and satellite links that employ frequency modulation for analog signals. These system components have the characteristic that the noise power generated per hertz increases as the square of the frequency (see Chapter 17).

The nature and amplitude of the noise generated by CCD imagers is described in Chapter 8.

4.15.2 Definition of signal-to-noise ratio

The signal-to-noise ratio is the basic specification for random noise. It is the power ratio, usually expressed in decibels, of the peak-to-peak signal voltage or current, from black level to reference white but not including the sync pulses, to the rms value of the noise.

4.15.3 Unweighted and weighted noise

Noise power can be specified either as *unweighted* or *weighted*. Unweighted noise is measured by an instrument with uniform frequency response. Weighted noise, a measure of its visibility, is measured with an instrument having a frequency response, the weighting function, that simulates the cascaded aperture response of the eye (see Figure 4.4) and the system components following the noise measurement point.

The variables in the calculation of the weighting function include the viewing ratio, the assumed response of the eye, the screen brightness, the width of the scanning lines, and the aperture response of the kinescope and other system components following the point of measurement. There are differences between the visibility of noise in monochrome and color images because of interaction of the high frequency color subcarrier and noise

Figure 4.21 EIA noise weighting factor

components in this frequency range. The weighting factor, therefore, cannot be both universal and precise.

The industry has found it important, however, to establish a standard weighting function that approximates practical situations so that systems can be specified and compared on a uniform basis. The Electronic Industries Association has published a widely used standard for broadcast transmission systems, which is part of EIA/TIA-250-C and is shown in Figure 4.21. It is based on a somewhat more pessimistic assumption as to the aperture response of the eye than is indicated by Figure 4.4.

The difference between the weighted and unweighted signal-to-noise ratio is called the *weighting factor*. Using the weighting function shown in Figure 4.21, it is about 8 dB for white noise and 12 dB for FM noise. It is larger for FM noise because most of the noise is concentrated at the upper end of the transmission spectrum where its visibility is less and the weighting function is lowest.

At present, a standard industry weighting factor for HDTV does not exist, but it has been variously estimated as being 3 to 5 dB less than that for broadcast service. This would indicate that the unweighted signal-to-noise ratio for HDTV should be 3 to 5 dB higher for the same weighted value. The weighting factor is smaller because a lower viewing ratio is assumed and fine-grained high-frequency noise is more visible.

4.15.4 Signal-to-noise ratio in digital systems

Noise power is additive in analog systems, and the signal-to-noise ratio for the complete system is inevitably lower than that of the worst stage. Similarly,

Table 4.3 Signal-to-noise ratio standards (dB)

Service	Weighted	Unweighted (white)	Unweighted (FM)
HDTV	55	50	47
Broadcast (good)	53	45	42
Broadcast (fair)	43	35	32
Broadcast (poor)	37	37	26

with multiple generation recordings, the signal-to-noise of the final generation is lower than any individual generation.

With digital transmission or recording (see Chapter 3), noise causes *bit errors*, and so long as the *bit error rate* (BER) is low enough, correction circuits can restore the original bit stream and there is virtually no degradation of the signal-to-noise ratio. If the noise power is increased sufficiently, however, the bit error rate becomes so large that error correction is not effective and the signal is lost.

In summary, an analog signal suffers *graceful degradation* as it passes through successive stages or recording generations. Its quality gets steadily worse, but it remains useful as noise is added. By contrast, the signal-to-noise ratio in digital systems suffers little reduction as the channel noise increases to a critical level, but at that level it suddenly becomes unusable (also see Chapter 3).

4.15.5 Signal-to-noise ratio standards

A number of industry and governmental standards have been established for the video signal-to-noise ratio. Some are for a complete system from camera to receiver while others cover only a portion of the system, e.g., a microwave link. Some are for white noise and others for FM noise. Some are for weighted and others for unweighted noise. When using published standards, it is important to determine exactly what is being specified and the basis of the specification.

Table 4.3 shows one commonly accepted set of signal-to-noise standards for broadcast systems. It also shows possible goals for HDTV system standards. The HDTV goals are very demanding, possibly not always obtainable, but would result in superb noise-free pictures. The broadcast standards are based on a study by TASO (Television Allocations Study Organization) for the FCC. Although the report is a number of years old, its conclusions are still reasonably valid.

Since noise is additive for analog transmission, a practical standard must take the number of system stages into account. As an example, the EIA weighted signal-to-noise standards for analog transmission circuits are:

Circuit	S/N
Short haul	67
Medium haul	60
Satellite	56
Long haul	54
End-to-end system	54

4.16 Image Defects (Continued from Section 4.3)

The remainder of this chapter expands the description of television image defects that were defined in Section 4.3. The magnitude of these defects, together with the system's performance with respect to the basic criteria of definition, gray scale reproduction, and signal-to-noise ratio, determine the quality of the images that it produces.

4.16.1 Hum

Hum bars were a major problem in the design of early television receivers. To alleviate the problem, field rates equal to this frequency (60 Hz in the United States, 50 Hz in Europe) were chosen. Spurious patterns resulting from hum were then stationary and less noticeable. Small deviations from this frequency became necessary as the industry developed. Networking brought signals into areas where the frequency of the power source was not necessarily synchronized with that at the program source. And NTSC color standards specified the use of a field rate, 59.94 Hz, a submultiple of the subcarrier frequency, that was slightly below the nominal 60 Hz power source frequency.

The effect of these small deviations, which would cause the hum pattern to move slowly up or down on the raster, were eliminated by better receiver designs that greatly reduced the cross talk of the power source into the video signal. The design of modern receivers has now been improved to the extent that hum should not be a problem even with large differences between power source and field frequencies.

4.16.2 Cochannel interference

The FCC originally specified a minimum desired-to-undesired carrier ratio of 20:1, 26 dB, to reduce cochannel interference to an unobjectionable level. Subsequently it was discovered that the visibility of the interference (venetian blinds) could be greatly reduced by an offset carrier technique in which the carrier frequencies of adjacent cochannel stations were offset by ±1/2 the line rate or ±7,625 Hz, for a total difference of 15,250 Hz. The interference bars then have a width equal to that of the scanning lines and are less visible by about 20 dB than the wide bars that are produced when the carriers are only slightly separated in frequency.

This offset was applicable to two stations, but if a third station was added, two of the three would be on the same frequency. A compromise was adopted in which the carriers are separated by ±10 kHz. This results in an improvement of about 12 dB in the signal-to-noise ratio and offset carriers are now specified by the FCC.

The use of digital transmission also shows great promise as a means of reducing the minimum acceptable desired-to-undesired signal ratio. Field testing is now underway (1994) under the auspices of the FCC Advisory Committee on Advanced Television Service to determine the amount of reduction in cochannel interference that can be expected from digital transmission.

4.16.3 Geometric distortion

Geometric distortion of the final image may originate either in the camera or the display device. The distortion produced by the display device is usually much greater, in part because it is a mass-produced product with wider tolerances and in part because the very wide deflection angle required by most picture tubes makes scanning linearity more difficult.

The geometric distortion produced by a camera chain can be measured by the EIA linearity test chart shown in Figure 4.22, sometimes called a *ball chart*. The inner diameter of each circular ring is 1 percent of picture height while the outer diameter is 2 percent. Linearity is measured by superimposing an image of the chart as produced by the camera with an electronically generated grating pattern in which the intersections of the vertical and horizontal lines should coincide with the centers of the circles. A well-designed camera will maintain the intersections within the 1 percent circles (±0.5%) in a circle in the center of the raster having a diameter equal to picture height and within the 2 percent circles (±1.0%) in the remainder of the raster.

Before the advent of color, competitive cost pressures caused manufacturers to design receivers with marginal scanning linearity and geometric distortion. Color, however, imposed a requirement for the registration of the three colors that could not be compromised. Eliminating registration errors also led to improved linearity and reduced geometric distortion. Typical geometric distortion specifications for production receivers are as follows:

Aspect ratio	5%
Overscan	5–10%
V and H linearity	5–8%
Parallelogram	90±1° at each corner

The vertical and horizontal linearity of 5–8 percent is to be compared with the typical 1 percent tolerance for cameras (see above).

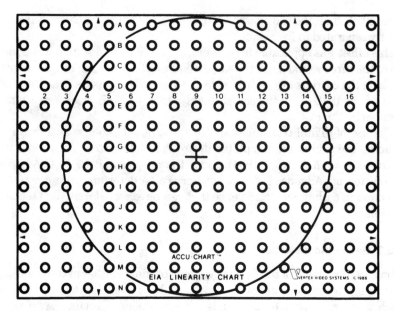

Figure 4.22 EIA linearity chart

4.16.4 Flicker

Flicker is an inherent defect that results from the frame rate that is used in the display of video signals. It can be minimized or eliminated by the use of a sufficiently high repetition rate. Both video and motion pictures depend on the retentivity of the eye to merge a rapid sequence of images into a single continuous one. If the repetition rate of the images is too low, the eye will fail to merge them, and flicker results. The lowest repetition rate at which a continuous image is perceived is called the *critical fusion frequency*.

Two types of flicker are encountered in video systems that employ interlaced scanning. *Large area flicker* occurs at the field rate, 60 fields per second in the NTSC system and 50 fields per second for PAL. *Small area flicker*, described in Chapter 1, affects only vertical detail. It occurs at the frame rate, 30 per second for NTSC and 25 for PAL.

Large area flicker involves the entire image area and is the most troublesome. The magnitude of the flicker effect depends on two parameters, the field repetition rate and the illumination of the retina by the image. Flicker can be reduced and ultimately made unnoticeable by increasing the field repetition rate or reducing the retinal illumination.

The visibility of flicker is extremely sensitive to the field rate and the retinal illumination. At the 60 field/second rate employed by NTSC systems it is seldom a problem, but it can be a problem at the 50 field/second rate. One study showed that the retinal illumination can be as much as three times as great at 60 fields as compared with 50 fields before flicker is observed.

Table 4.4 Flicker-brightness tolerance

	Field/Second	Relative Brightness (%)
NTSC TV	60	100
PAL TV	50	29
Theater movies	48	20
Home movies	48	20

The retinal illumination is determined by the image brightness and the size of the pupil's aperture. The latter, in turn, depends on the brightness of the area around the image, the *surround brightness*, as well as the image itself. In a darkened motion picture theater, the surround brightness is low, the pupil aperture is relatively large as the eye automatically adjusts to the low light level, and the retinal illumination is high in relation to the screen brightness. In a typical home environment, the surround brightness is higher, the pupil aperture is smaller, and the retinal illumination is lower in relation to the image brightness. As a result, the image brightness can be greater without causing visible flicker in a home TV set than with motion pictures in a theater. The difference is increased still further by the use of the lower field rate of 48 per second in motion picture systems.

The image brightnesses that can be tolerated without flicker under typical viewing conditions, relative to a 60 field per second NTSC system, have been estimated to be as shown in Table 4.4.

Large area flicker has not been a problem with 60 field/second NTSC systems with the image and surround brightnesses that are normally encountered. Looking ahead to future HDTV systems, it is a potential problem. In an effort to increase the contrast ratio, viewing conditions with a lower surround brightness may be chosen and the white field brightness increased. Both of these conditions would increase the vulnerability to flicker. Also, the use of a wide screen and lower viewing ratio places the edges of the screen on the periphery of vision where flicker is more noticeable (presumably one of nature's natural protective functions to make the individual aware of threatening movements out of the line of vision).

Small area flicker affects only vertical detail, i.e., horizontal edges, and it occurs at a rate of 30 or 25 Hz in systems employing interlaced scanning. It has not been a problem with broadcast systems because the vertical definition is not great enough to produce sufficiently sharp edges. The high numbers of scanning lines proposed for HDTV systems significantly reduce this problem, and the use of sequential scanning would eliminate it. It is doubtful, however, that small area flicker is serious enough to warrant the increase in bandwidth.

The situation with computer displays is quite different. They are viewed at close distances and the subject matter typically is small alphanumeric characters. Under these conditions, small area flicker can be extremely annoying. Accordingly, most computer displays employ sequential scanning.

4.16.5 Camera defects

In addition to the image defects described above that are the result of system imperfections, there are others including lag, smear, blooming, and comet tails that are characteristic of camera performance (see Chapter 8).

4.17 Comparison of Film and Television Picture Quality

4.17.1 Overview

A comparison of the quality of film and television images is complex and controversial, and it has been the subject of an extensive literature. The complexity of the comparison is illustrated by two recent papers[3,4], Reference 3 written from the point of view of the film industry and the more recent Reference 4 written from the perspective of the television industry.

There are a number of reasons for this complexity:

1. The inherent dissimilarity of photographic and television images as described earlier in this chapter, with the result that images produced by the two media will never appear exactly the same. (A somewhat analogous situation arises in comparing pipe organs and electronic organs. Although it is possible in theory to duplicate the sound of a pipe organ by electronic synthesis, the duplication is never exact. On the other hand, there are sounds and effects produced by an electronic organ that cannot be made by a pipe organ. The two instruments simply are different, but this does not imply that one is "better" than the other.)

2. The number of criteria that must be considered in the comparison and the problem of weighing their relative importance.

3. The wide range in the specifications of different film types and electronic imagers so that no single comparison is valid for all of them.

4. The rapid changes in the technologies of both media so that a comparison at any particular point in time is a moving target. The changes have been particularly significant in recent years, and there have been major improvements in the performance of both film and TV cameras since the first edition of this volume was written.

5. The basis of the ultimate comparison must be aesthetic rather than technical. This means that it will be subjective and subject to individual preferences.

In spite of these problems, it is useful to make this comparison, although one must recognize its inevitable limitations. The remainder of this section discusses the quantifiable differences.

4.17.2 Production and display systems

It is necessary to begin the comparison by specifying whether it is production or display systems that are being compared. Four combinations are possible:

Production	Display	Designation
Film	Film	F–F
Film	Television	F–T
Television	Television	T–T
Television	Film	T–F

The difficulties of comparing film and television display systems are the greatest and probably of the least importance. This comparison would have to consider not only the properties of the medium but also the ambiance of the display area. Film is almost invariably displayed in a darkened theater or room—in fact it is necessary to achieve a satisfactory contrast ratio. By comparison, satisfactory results can be achieved with television under conditions of ordinary room lighting. Film displayed in a darkened room may have a better gray scale than television in a well-lit room but this is not an indication of the relative performance of the media but rather of the manner in which they are used.

This is one reason for limiting the discussion in this volume to systems with television displays, i.e., F–T and T–T. They are the most important in the television industry, and this limitation greatly simplifies the comparison.

4.17.3 Image-quality determining components

Table 4.5 lists the components that determine the image quality in each of the four systems. Subsequent sections compare their individual and cascaded performance in the F–T and T–T systems.

4.17.4 Component performance

Representative performances of the components of the F–T and T–T systems listed in Figure 4.4 are summarized in Table 4.6 and described below.

4.17.4.1 Definition (MTF)   The telecine in the F–T system reproduces an image of the scene on negative film rather than the scene itself. This would appear to give the T–T system an advantage in its MTF because there are fewer aperture limiting components. This advantage has been minimized by significant recent improvements in film performance and can be offset by employing a telecine with a higher MTF than a live camera. The higher telecine performance is possible because it can be designed for maximum

Table 4.5 Film and television system components

F–F	F–T	T–T	T–F
Production components			
Camera lens	Camera lens	TV camera	TV camera
Film negative	Film negative	Lens	Lens
Film print	Telecine	Imager	Imager
	Lens	Scanning	Scanning
	Imager	Sig. processing	Sig. processing
	Scanning	Bandwidth limit	TV/film transfer
	Sig. processing	VTR	Film print
	Bandwidth limit		
	VTR		
Display components			
Projection lens	Kinescope	Kinescope	Projection lens
Viewing ambience			
Darkened room	Room lighting	Room lighting	Darkened room

MTF without the necessity of trade-offs for sensitivity, variable focal length lenses, and other constraints that limit the MTF of live cameras.

The MTF of the F–T systems at 400 and 600 lines tabulated in Table 4.6 are as reported in Reference 1. To demonstrate the potential of the film medium, the telecine employed an experimental high-definition storage tube and optimized optics with a fixed focus lens. Similar results could probably be obtained with high quality CCD imagers (see Chapter 7). The MTF of the TV systems was as described in Section 4.9 for NTSC and 4.10 for HDTV (for HDTV it was very close to the performance of the HDTV camera as reported in Reference 1).

The performance of the NTSC TV system was compared with two film formats, 16-mm and wide-screen Super 16. The film type was Eastman EXR 7245 (the sensitivity of the faster type ECN 7292 was also shown).

The performance of the HDTV system was compared with 35-mm type EXR 5245.

This comparison of T–T and F–T systems on the basis of selected points on their MTF curves shows an advantage for film, which is small if the ultimate display medium is television. This advantage can easily be overcome by the use of aperture correction.

There is a subtle but important difference when the complete MTF curve is considered. The film MTF has a more gentle cut-off as the limiting resolution is approached, and this produces a different edge shape as shown

Table 4.6 A comparison of TV and film image quality

Parameter	NTSC Standards FILM (TV Playback) 16mm	S-16	TV (CCD Imager) (818H × 513v)[1]
Definition (MTF)			
400 lines	0.60	0.69	0.59^2
600 lines	0.30	0.40	0.28^2
Resolution (TVL)			
Horizontal	650	675	700
4.2 MHz BW	352	352	352
Vertical	315^3	315^3	315^3
Aliasing region			
Frames	>24 fps	>24 fps	>30 fps
Scanning (vertical)	>226 lines	>226 lines	>226 lines
Pixels (horizontal)	None	None	>560 lines[1]
S/N (unweighted)	$42\,dB^4$	$44\,dB^4$	$50\,dB^5$
Sensitivity (see Chapter 7)			
Exposure index	50–320	50–320	575

1 Include effect of spatial offset.
2 Before aperture correction.
3 Based on Kell factor of 0.7. Value is ambiguous because of aliasing.
4 Noise equivalent for film granularity.
5 Based on 51 dB VTR S/N.

in Figure 4.7. This difference is reduced but not eliminated when the film is reproduced by a television system, and it probably accounts for much of the different visual perceptions described as the "film look" and the "television look."

Although a comparison with F–F systems is beyond the scope of this study, it should be noted that the F–F system has at least two apertures in series after the negative film (positive film and projection lens), and a projected image may have an MTF that is no higher or possibly even lower than that of an F–T system.

Table 4.6   A comparison of TV and film image quality (continued)

	HDTV 240M Standards	
	FILM	TV
	(35 mm film)	(CCD imager)
Parameter	(TV playback)	(1920H ×1036v)[1]
Definition (horizontal MTF)		
400 lines	0.80	0.78
600 lines	0.65	0.59
Resolution (TVL)		
Horizontal	>1000	>1000
30 MHz bandwidth	869	869
Vertical	724[3]	724[3]
Aliasing region		
Frames	>24 fps	>30 fps
Scanning (vertical)	>517 lines	>517 lines
CCD (horizontal)	None[4]	>490 lines
S/N (unweighted)	50 dB[5]	54 dB
Sensitivity (see Chapter 7)		
Exposure index	50–320	575

1 Includes experimental high-definition telecine.
2 Before aperture correction.
3 Based on Kell factor of 0.7. Value is ambiguous because of aliasing.
4 No scanning processes: film camera uses storage tube telecine.
5 Noise equivalent for film granularity.

4.17.4.2 Resolution   The horizontal limiting resolution in each case is determined by the bandwidth of the system, 4.2 MHz for NTSC and 30 MHz for
HDTV, and neither medium has an advantage.

The ambiguities in the use of vertical resolution as a measure of performance because the scanning lines are a form of sampling were described in
Section 4.8.2.

4.17.4.3 Aliasing   Frame aliasing, e.g., wagon wheels appearing to turn
backward, is determined by the frame rate—or more precisely, by one-half
the field rate. The telecine image is derived from film that was photographed
at a frame rate of 24. The TV image has a frame rate of 30 and is superior in

this respect (although audiences have become accustomed to this effect and seem to expect it).

Scanning line aliasing begins at one-half the number of active lines, which is the same for both the F–T and T–T systems.

CCD aliasing basically begins at one-half the number of horizontal sensors per picture height. By the use of spatial offset, however (see Section 8.4.4), the line number at which aliasing begins can be almost doubled. A factor of 1.9 was used in Table 4.6.

4.17.4.4 Signal-to noise ratio The high S/N ratio of the latest CCD imagers has made film granularity a major noise source in systems that use film. This accounts for the better S/N performance for the T–T systems.

4.17.4.5 Sensitivity The latest CCDs have a sensitivity approximately equal to that of very fast film. If the better performance of slower film is desired, CCDs would have the advantage in sensitivity.

4.17.5 Picture quality summary

Recent improvements in imager performance give television production a slight edge over film production in the tangible criteria for picture quality, although each has pluses and minuses. Film has a small advantage in MTF and television production systems have a small advantage in frame aliasing, sensitivity, and S/N. Film production systems that employ storage tubes in the telecine do not produce aliasing of a type characteristic of CCDs.

These comparisons do not take into account subtle and difficult-to-quantify differences described as the "film look" and "television look." As a result of these differences, film and TV-produced programs will never look exactly alike, but neither is inherently "better" than the other. A comparative evaluation must be made by personal preference rather than tangible, objective measurements.

Further improvements can be expected in the picture quality of both television and film production systems for television display, but the quality produced by both is now approaching the limit imposed by the human eye. The picture quality available from both systems is so nearly alike that other factors will probably be preeminent in the choice.

References

1. Otto H. Schade, Jr., "Electro-Optical Characteristics of Television Systems," *The RCA Review*, Vol. 9, March, June, September, December, 1948.

2. Otto H. Schade, Jr., *Image Quality, A Comparison of Photographic and Television Systems*, Princeton, N.J., RCA Laboratories, 1975.

3. Glenn Kennel, LeRoy DeMarsh, and John Norris, "A Comparison of Color Negative Films and HDTV Cameras for Television Program Production," *SMPTE J.*, May 1991.

4. L. Thorpe, F. Nagumo, and K. Ishikawa, "A Comparison Between HD Hyper-HAD Cameras and Color Film for Television Program Production," *SMPTE J.*, June 1994.

Analog Video Systems

5.1 Introduction

This chapter describes the analog video systems (as distinguished from digital systems; see Chapter 6) that are used for the transmission and storage of color video signals. They are based on the colorimetry of television as described in Chapter 2.

Analog systems can be classified as *composite* or *component,* depending on the format of their signals. Composite formats multiplex the luminance and chrominance components into a single signal while component formats employ a separate signal for each component.

Composite formats are used for broadcasting. Component formats are widely used for recording in complex program production systems where the highest quality must be achieved. They are also used for satellite transmission.

The principal composite analog formats are NTSC, PAL, and SECAM. The configurations and signal formats of NTSC and PAL composite systems and of the principal component systems are described in this chapter.

5.2 Standardization

The objective of most analog video systems is mass distribution of programs and information. This may happen by broadcasting or other means of distribution such as cable transmission or video tape. In any case, the equipment at the receiving end must know the signal format before successful reception is possible. That means the signal format must be standardized—precisely defined so that any manufacturer around the world can build receiving equipment. Because this is a multicompany and multinational problem, various organizations have been formed specifically for this purpose.

For example, in the United States, standards for broadcasting are set by the Federal Communications Commission (FCC). However, the FCC does not develop standards, they only endorse them. Other industry bodies are relied upon to propose standards to the FCC. These are organizations such as the Society of Motion Picture and Television Engineers (SMPTE), the Institute of Electronic and Electrical Engineers (IEEE), the Electronic Industries Association (EIA), and special ad-hoc groups that are formed from time to time. One of the latter class is the National Television Systems Committee (NTSC), which was formed in the 1950s to develop color television standards that were proposed and later adopted by the FCC. Similar approaches have been taken by other countries.

However, the television industry is worldwide and it would not be practical for every country to have a different standard. That would require receiver manufacturers to have different products for every country and that would increase the cost of equipment for everyone. Therefore, another class of standards groups exists with a worldwide purview. These include the International Standards Organization (ISO), the European Broadcasting Union (EBU), the Consultative Committee on International Radio (CCIR), and the Consultative Committee on Telephone and Telegraph (CCITT). Like the FCC, these organizations usually do not develop standards but rather they rely on others to make proposals. However, they will in some cases form special committees to work in specific areas, such as the JPEG and MPEG groups of the ISO (see Setion. 3.12).

The standards resulting from these activities are *voluntary*; the standardizing organizations do not have authority to enforce their standards. However, most manufacturers comply with them because of the important commercial advantages of worldwide standards.

5.3 Video Signal Formats for Color

All video signal formats are derived from the primary signals E_R, E_G, and E_B or their gamma corrected derivatives E'_R, E'_G, and E'_B, which are generated by the camera chain. The camera signals are in analog form and have amplitudes proportional to the spectral energy for the primary colors in the scene. At the receiver, the inputs to the receiver kinescope or other display device are modified replicas of these signals. Between the camera chain and the kinescope input, however, the signal is usually transformed to another format—analog or digital—that is better suited to the functions of signal processing, recording, transmission, and broadcasting.

For all formats, a luminance signal E'_Y is generated by matrix circuits in accordance with Eq. (5.1):

$$E'_Y = 0.299\,E'_R + 0.587\,E'_G + 0.114\,E'_B \tag{5.1}$$

The coefficients of the components E'_R, E'_G, and E'_B are based on the relative sensitivity of the human eye to the primary colors. The magnitude of E'_Y is then proportional to the perceived brightness of the scene.

Composite color formats can be either *compatible* or *noncompatible* with the monochrome format. Compatible color transmissions can be received on monochrome receivers (displayed in monochrome, of course) and monochrome transmissions can be received on color receivers. All formats currently used for broadcasting are compatible.

In part this is the result of the U.S. industry's unfortunate experience with the incompatible CBS *field sequential* system. This was approved by the FCC in 1950, but it was a commercial failure and was replaced by the compatible NTSC system in 1953.

5.3.1 Composite formats

The FCC's experience with sequential color demonstrated that a successful format for color broadcasting would have to be compatible with the existing monochrome format. To be compatible for broadcasting and related services, a format had to meet three requirements:

1. The luminance and chrominance information had to be contained in a single channel (hence it is composite).

2. The luminance component had to be sufficiently similar to the monochrome signal so that color broadcasts could be received on monochrome receivers and vice versa.

3. The bandwidth had to be no greater than required for monochrome.

The television industry devoted enormous efforts to the development of a compatible system, and the result was proposed to the FCC by the NTSC in July 1953. It was approved by the FCC for color broadcasting in the United States in December 1953.

The NTSC system has been remarkably durable—in a fast-changing industry it is still in use 40 years later. It is also the approved system for color broadcasting in Japan and in most western hemisphere countries.

Two additional analog color systems, PAL and SECAM, are used in other countries—PAL in Europe (except in France and Russia) and in the countries of the British empire (except Canada), and SECAM in France and Russia. These systems are similar in principle to NTSC in that the color information is transmitted by one or two high-frequency subcarrier(s) (3.58 MHz for NTSC) superimposed on the analog luminance signal.

5.3.2 Component formats

A major disadvantage of subcarrier composite systems is the impossibility of completely eliminating cross talk between the luminance and chrominance components as the result of nonlinearities in the system and overlaps of the sideband spectra of the subcarriers. The nonlinearities cause *differential gain* and *differential phase* distortion of the chroma component—variation in the amplitude and phase of the subcarrier with brightness—and visibility of the subcarrier in the luminance component. The spectrum overlaps cause *cross color* and *cross luminance intermodulation*.

Differential gain and phase distortions are described in Section 5.10. Cross color results when the luminance signal contains frequency components near the subcarrier frequency, and spurious colors are generated in the image. Cross luminance occurs when the signal representing a given point changes from field to field or on the edges of objects so that the subcarrier cancellation is not complete. A dot pattern then occurs in the luminance channel.

Another disadvantage of subcarrier composite systems results from the fact that the noise in transmission systems that use frequency modulation, e.g., microwave and satellites, increases linearly with the modulating frequency. This puts the subcarrier at the high-noise end of the spectrum and can lead to a poor signal-to-noise ratio for the chrominance signal.

Signals employed for editing, point-to-point transmission, and recording are subject to none of the constraints of broadcasting, and it is possible to choose a component format that eliminates the use of a subcarrier with its cross talk and noise problems. Component formats have no cross-luminance or cross-chroma distortion. The most common component format employs three components—luminance or E_Y, and the color difference components, $E_R - E_Y$ and $E_B - E_Y$. They may be transmitted on separate channels or time-multiplexed on a single channel as with MAC (see Section 5.13).

5.4 Analog Color Television System Configurations

Figure 5.1 is a simplified functional diagram that shows the configurations of composite and component color television systems. Both begin with a camera chain that includes a transducer for each primary color, which may be a tube or solid-state device, and signal processing circuitry that generates the primary signals, E'_R, E'_G, and E'_B. The processing functions include white and black level control, aperture correction and image enhancement, color correction by matrixing, gamma correction, and sync and blanking insertion.

In composite systems, the camera chain is followed by an encoder that produces the composite signal by multiplexing the primary signals. This composite signal is transmitted to the receiver, which decodes the signal to recover the original E'_R, E'_G, and E'_B primary signals that are the inputs to the picture tube.

In component systems, the luminance and color difference signals are generated by matrixing the primary signals, i.e., by mixing them in controlled ratios. After matrixing, the component signals may go to a video recorder, to a microwave, fiber-optic or satellite transmission system, or to an A–D converter for conversion to a digital format.

Figure 5.1 Functional diagrams for color television systems

5.5 Broadcast Signal Components

5.5.1 The luminance signal component

The amplitude of the luminance signal component, E'_Y, is given by Eq. (5.1). Its bandwidth varies from country to country (see Table 5.1). Most European countries use UHF for color broadcasting; this accounts for the greater bandwidths available for their systems.

5.5.2 NTSC chrominance signal components

It is possible to conserve spectrum space by deriving two chrominance components, E'_I and E'_Q, called the I and Q signals, from the primary components:

$$E'_I = 0.60\,E'_R - 0.28\,E'_G - 0.32\,E'_B \tag{5.2}$$

$$E'_Q = 0.21\,E'_R - 0.51\,E'_G + 0.30\,E'_B \tag{5.3}$$

The chrominance components, E'_I and E'_Q, have the following characteristics:

1. E'_I is at a maximum for orange or cyan hues while E'_Q is at a maximum for green or magenta.

2. The ability of the eye to discern fine detail in colors is less than for monochrome, and it is less for magenta than for orange. As noted above, this property is utilized in the NTSC for reducing the bandwidths of the chrominance signal components (see Table 5.2).

3. With white light $E'_R = E'_G = E'_B$, and all chrominance components, E'_I, E'_Q, $(E'_R - E'_Y)$, and $(E'_B - E'_Y)$, disappear.

Table 5.1 Color television system bandwidths

Country	System	Bandwidth (MHz)
United States, Japan	NTSC	4.2
Canada, Mexico	NTSC	4.2
Great Britain	PAL	5.5
Germany, Austria, Italy	PAL	5.0
France	SECAM	6.0
Former USSR	SECAM	6.0

4. Their amplitudes relative to the luminance signal are a trade-off between a large value that would lead to excessive cross talk with the luminance component and a small value that would lead to an inadequate chroma signal-to-noise ratio.

5.5.3 PAL chrominance signal components

With video bandwidths of 5 to 5.5 MHz available, the PAL system does not need to be as frugal in its usage of spectrum space as the NTSC system. Accordingly, there is no need to add to the complexity of PAL systems by creating the I and Q signals, and the color difference primaries are used directly for producing the color subcarrier.

The two chrominance components in the PAL system are E'_U and E'_V. Their equations are:

$$E'_U = 0.493\,(E'_B - E'_Y) \tag{5.4}$$

$$E'_V = 0.877\,(E'_R - E'_Y)$$

The bandwidth of the E'_U and E'_V components is 1.3 MHz (to the –3 dB points).

Table 5.2 Component bandwidths

Component	Bandwidth (MHz)
E'_Y (luminance)	4.2
E'_I (orange-cyan)	1.5
E'_Q (green-magenta)	0.5

5.6 Subcarriers

The essence of the difference between the NTSC and PAL color television systems is their choice of subcarrier frequencies, phases, and formats. The subcarrier frequencies in both systems are chosen to be as high as possible while keeping their sidebands within the video passband. Their relationship to the frequencies and phases of the scanning lines and fields must also minimize the cross talk between the luminance signal and the subcarrier and the visibility of the subcarrier in the composite signal. The solutions to these problems differ for each system.

5.7 The NTSC System Subcarrier

5.7.1 The NTSC subcarrier format

The NTSC subcarrier format was the first to be developed (in the United States in the early 1950s), and it is the simplest. It combines amplitude and phase modulation, its amplitude determining the saturation of the image and its phase the hue. The use of phase to transmit hue requires the transmission of a reference phase that is supplied by the burst, a train eight cycles of the subcarrier frequency with a fixed phase on the back porch of the blanking pulse. It is shown in Figure 5.2.

The NTSC subcarrier, e_{scn}, can be derived either from the I and Q signals or from the color difference signals, $(E'_R - E'_Y)$ and $(E'_B - E'_Y)$.

If e_{scn} is derived from I and Q, its equation is:

$$e_{scn} = [E'Q \sin(\omega t + 33°) + E'I \cos(\omega t + 33°)] \tag{5.6}$$

where: $\omega = 2\pi f_{sc}$ and f_{sc} is the subcarrier frequency.

If e_{scn} is derived from the color difference signals, the equation is:

$$e_{scn} = 0.877[0.562(E'B - E'Y)\sin(\omega t) + (E'R - E'Y)\cos(\omega t)] \tag{5.7}$$

Note that the subcarrier vanishes for white light. Also note that its average value is zero so that in a linear system, the luminance signal is unaffected by the chroma. This is known as the constant luminance principle. Stated otherwise, there is no cross talk from chrominance to luminance unless there are nonlinearities in the system.

The chrominance subcarrier e_{scn} and its terms can be represented by vectors with the sine and cosine functions in quadrature. Figure 5.3 shows the phase relationship of these functions to the burst and the significance of the 33° angle in the equation.

The NTSC subcarrier format is subject to cross talk from luminance to chrominance resulting from nonlinearities in amplifiers, recorders, and

Figure 5.2 Color subcarrier

transmission systems that cause variations in gain or phase shift with the brightness level of the luminance signal. These effects are known as *differential gain* and *differential phase*. Differential gain causes distortions of the image saturation and differential phase its hue.

Differential phase shift is a particularly serious problem because the eye is very sensitive to hue distortions. After the introduction of NTSC, the Europeans, who were considering other systems, jokingly said that NTSC meant "never the same color." With the passage of time, advances in technology have greatly reduced the differential gain and phase of video amplifiers, transmission systems, and recorders, and their nonlinearities no longer need be significant problems. They still must be considered, however, in system design and operation, and the vulnerability of the NTSC format to system nonlinearities continues to be a potential problem.

Other forms of cross talk are cross color and cross luminance (see Section 5.3.2).

5.7.2 NTSC subcarrier frequency

The NTSC subcarrier frequency is derived from the scanning line frequency, which in turn is an integral submultiple of the 4.5 MHz audio-video carrier spacing used in broadcast systems. This relationship minimizes the beat between the audio and color subcarrier. It is also desirable to have the line frequency as close as possible to the monochrome line frequency of 15,750.

Figure 5.3 NTSC subcarrier vector diagram

As a result of these considerations, the line frequency is the 286th submultiple of 4.5 MHz:

$$f_{line} = \frac{4.5 \times 10^6}{286} = 15{,}734.26 \text{ Hz} \tag{5.8}$$

The field frequency is related to the line frequency by the number of scanning lines:

$$f_{field} = \frac{f_{line}}{525/2} \tag{5.9}$$

It would be desirable for the field frequency to equal that of the primary power source, 60 Hz (as for monochrome), but it is mathematically impossible to match this frequency precisely while other requirements are being met. The small difference between 59.94 Hz and 60 Hz causes hum bars to move slowly through the image, but this is not a serious problem with modern receivers because of their excellent filtering.

To reduce the visibility of the subcarrier in the luminance signal, its frequency is an odd multiple of one-half the line frequency. With this relationship, the dot pattern resulting from the subcarrier is stationary, and the phase of the subcarrier (for a given hue) at the beginning of each line reverses on successive scans. With the phase reversal, the dots on successive scans have opposite polarity and tend to cancel.

On the basis of these relationships, the equation for the subcarrier frequency, f_{sc}, is:

$$f_{sc} = 455\frac{f_{line}}{2} = 3.579545 \text{ MHz} \tag{5.10}$$

The relationship between subcarrier frequency and line rate must be maintained very precisely to avoid dot crawl. Usually the line rate is established by counting down from the subcarrier frequency by a series of frequency dividers:

$$f_{line} = \frac{2 f_{sc}}{(5)\,(7)\,(13)} = 15{,}734 \text{ Hz}$$

Although the line and field rates are slightly different than those for monochrome television, the differences are so small that they cause no problems with monochrome receivers.

5.7.3 NTSC subcarrier reference phase

If 0° and 180° designate the subcarrier reference phases at the beginning of successive lines, their phases are:

Line	Field 1	Field 2	Field 3	Field 4
1	0°		180°	
2		180°		0°
3	180°		0°	
4		0°		180°
5	0°		180°	

Note that four lines are required before the pattern repeats.

Cross talk of the chroma signal into the luminance channel, which produces a dot pattern in the image, is minimized by the choice of subcarrier frequency and phase. By relating the subcarrier and line frequencies, the dot patterns are stationary and thus less visible. And the reversal of the phase of the subcarrier on alternate frames causes at least a partial cancellation of the visual effect of the dots.

Figure 5.4 NTSC subcarrier frequency spectrum

5.7.4 NTSC signal spectral content

The video bandwidth requirements and the relative amplitude of the luminance and I and Q chrominance components over the video passband for a typical signal are shown in Figure 5.4.

The luminance and chrominance components of the signal can occupy the same portion of the spectrum with a minimum of cross talk because of their fine-grained spectral content. The content of the luminance component was shown in Figure 1.9. It consists of groups of frequency components, each with a center frequency that is a multiple of the line rate and surrounded with a cluster of sideband-like components separated by the field rate. Unless there is an unusual amount of motion in the picture, there are wide spectral gaps between these groups in which additional signal information can be inserted.

The spectrum of the subcarrier consists of similar groups of sideband components. Because of the integral relationship of the subcarrier frequency and the line rate in the NTSC system, the groups of subcarrier sidebands fit in the gaps in the luminance spectrum.

The interleaving of the luminance and chrominance components in the frequency domain is an indication that the chrominance signal will cause a minimum of cross talk between the two signals. In the time domain this is evidenced by the fact that the subcarrier pattern is stationary and that it has opposite phases on successive scans.

5.8 The PAL System Subcarrier

The PAL system was developed in Europe after NTSC had been developed in the United States. Thus, it benefitted from the early experience with NTSC and some improvements were considered.

5.8.1 The PAL subcarrier format

The format of the PAL subcarrier was designed to minimize the effects of differential phase, one of the results of cross talk from the luminance to the chrominance channel. It is similar to the NTSC format except that the phase of the subcarrier component derived from the color difference component, $E'_R - E'_Y$, is shifted ± 90° at the end of each line with respect to the $E'_B - E'_Y$ component—hence the system's acronym is PAL for Phase Alternating Line-rate.

Because of the wider frequency band used for most PAL systems, it is not necessary to employ the I and Q signals with their greater complexity, and the subcarrier is derived directly from the color difference signals, $E'_R - E'_Y$ and $E'_B - E'_Y$. The equation for the PAL subcarrier is:

$$
\begin{aligned}
e_{scp} &= 0.493(E'_B - E'_Y)\sin(\omega t) \pm 0.877(E'_R - E'_Y)\cos(\omega t) \\
&= E'_U \sin(\omega t) \pm E'_V \cos(\omega t)
\end{aligned}
\tag{5.11}
$$

The ± indicates the phase reversal on successive frames.

The phase relationship between the burst and the $E'_R - E'_Y$ and $E'_B - E'_Y$ subcarrier components is shown in vector form in Figure 5.5. Unlike NTSC, the PAL burst does not maintain a fixed relationship with the subcarrier components but shifts 90° at the end of each line. This is sometimes called a *swinging burst* and it provides some cancellation of errors as explained below.

The dotted vectors show how the phase error due to differential phase is reversed on successive scans of the same line. For example, if differential phase causes a shift of the subcarrier phase and the corresponding image hue

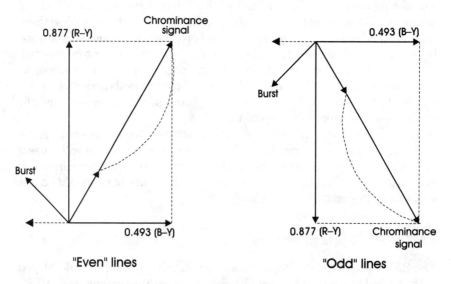

Figure 5.5 Vector diagram of PAL subcarrier

toward the $E'_R - E'_Y$ phase on the first scan, it will cause a shift toward the $E'_B - E'_Y$ phase in the next. Visual averaging of the two hues would then minimize the perceived hue distortion. This results in a number of annoying spurious effects, but these can be avoided by electrical delay-line averaging, in which the chrominance signal of each line is added to an electrically delayed signal of the same line from a previous scan. This technique, however, reduces the vertical chrominance resolution by a factor of two.

As with NTSC, the PAL subcarrier vanishes for white light.

5.8.2 The PAL subcarrier frequency

In the PAL system the field, frame, and line rates are the integral numbers used for European monochrome systems: The field and frame rates, f_{fld} and f_{fr}, are 50 and 25 Hz respectively, and the scanning line rate, f_{line}, is 15,625 Hz, 625 times the frame rate.

Since two phase changes are involved in the PAL subcarrier—the phase of the reference carrier at the beginning of the line and the phase of the $E'_R - E'_Y$ component with respect to $E'_B - E'_Y$—the relationship between line rate and subcarrier frequency used for NTSC would not be satisfactory for PAL. Instead, the subcarrier frequency is an odd multiple of one-quarter the line rate plus one-half the frame rate:

$$f_{scp} = 1135 \frac{f_{line}}{4} + \frac{f_{fr}}{2} = 4.43361875 \text{ MHz} \tag{5.12}$$

The term, $f_{fr}/2$, causes the subcarrier phase on successive scans of the same line to shift by 90°. A rigid relationship between the subcarrier frequency and line rate is also required for PAL systems.

5.8.3 PAL subcarrier reference phase

The relationship between the subcarrier frequency and the line and frame rates results in a 90° phase shift in the subcarrier reference phase at the beginning of each line (as contrasted with 180° for NTSC).

If 0°, 90°, 180°, and 270° indicate the carrier reference phase at the line beginning, and A and B represent the two phase relationships between the color difference subcarrier components, the phase relationships of the subcarrier and its components are shown in Table 5.3.

To achieve the proper phase of the burst for each line, it is necessary to shift its insertion time by one line at the field rate. This is done with a circuit known as a *meander gate*.

Table 5.3 PAL subcarrier phase relationships

Line	\multicolumn{8}{c}{Field}							
	1	2	3	4	5	6	7	8
1	0A		90B		180A		270B	
2		270B		0A		90B		180A
3	90B		180A		270B		0A	
4		0A		90B		180A		270B
5	180A		270B		0A		90B	
6		90B		180A		270B		0A
7	270B		0A		90B		270B	
8		180A		280B		0A		90B

5.8.4 PAL signal spectral content

As a result of the relationship between the PAL subcarrier frequency and the line rate and the phase shift of the $(E'_R - E'_Y)$ components at the end of each line, the PAL chrominance harmonics do not fall neatly between the luminance harmonics as with NTSC. In the time domain, the subcarrier phase shifts only 90° on alternate frames, and complete cancellation owing to phase reversal does not occur. Thus the subcarrier is inherently more visible in PAL than in NTSC, but the effect is mitigated by the higher PAL subcarrier frequency, which makes an uncancelled subcarrier less visible. The relationship between the luminance and chrominance spectrum components is shown in Figure 5.6.

5.9 Comparison of NTSC and PAL

PAL systems are usually used in countries where wider video bandwidth is available, often by the use of the UHF band for transmission. This provides a design flexibility that is not available to narrower band NTSC systems. In addition, the wider bandwidth provides an advantage in signal quality that is independent of the choice of the color system.

PAL systems are sometimes described as being more *rugged*, meaning that they are less vulnerable to hue distortions resulting from phase errors in the transmission system.

On the other hand, NTSC systems are simpler, they do not require quite as much bandwidth, and their subcarriers are less visible. Also, improvements in transmission and recording systems make the greater vulnerability of NTSC to phase errors less significant.

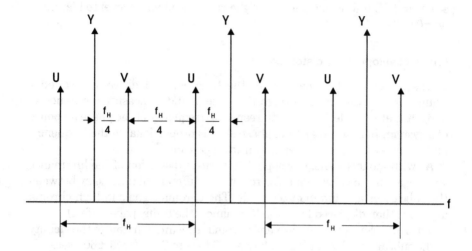

Figure 5.6 PAL frequency components

5.10 Color Signal Distortions and Artifacts

Multiplexing the luminance and chrominance components of a color television signal with a subcarrier is not without problems, and distortions and artifacts (spurious signals produced artificially) result. This section describes some of the most significant that occur in NTSC and PAL systems. More detailed descriptions of these distortions together with suggested tolerances and methods of measurement have been published by the Electronic Industries Association, the FCC, and the CCIR.[1]

5.10.1 Chrominance-to-luminance gain inequality

The amplitude-frequency response of the luminance channel in a color transmission has the same significance as in monochrome with respect to the definition of the image (see Chapter 4). But since the saturation of the image depends on the subcarrier/luminance signal ratio, it is exceedingly important that this ratio be carefully controlled with special precision in the region around the subcarrier, 4.0 or 5.0 MHz. The end-to-end EIA tolerance for a transmission system in this frequency region is ±7 IRE units (on a voltage

---

1 EIA/TIA Standard. "Electrical Performance for Television Systems," EIA/TIA-250-C, Washington, D.C., 1990. FCC Rules, Part 75. CCIR, *Recommendations and Reports, CCIR, 15th Plenary Assy.*, Vol XI, Broadcasting Service (Television), ITU, Geneva, 1982.

scale, an IRE unit is one percent of the range from white level to blanking) or ±0.6 dB.

5.10.2 Envelope delay distortion

Envelope delay is the measure of the time delay of different frequency components as they pass through the system, and is measured in nanoseconds. It is also the derivative with respect to frequency of the phase response of a system. In an ideal system, phase shift is proportional to the frequency, and the delay is constant over the entire spectrum.

As with monochrome, envelope delay causes distortion of the luminance waveform, but in color it can also result in a differential time delay between the subcarrier and luminance signals. The chroma signals in high chroma areas are then displaced from the luminance (the funny paper effect).

The EIA tolerance for end-to-end chroma transmission time in the vicinity of the subcarrier (3 to 4 MHz) relative to 2 kHz to 600 kHz is ±60 nsec.

5.10.3 Differential gain

The transmission system for a color television system must maintain its gain as a function of luminance within close tolerances in the vicinity of the subcarrier frequency. Gain variation at the subcarrier frequency as a function of the luminance signal level is known as differential gain, a form of cross-luminance distortion that distorts the image color saturation.

Tests[2] have shown that differential gain in the range from 10 percent to 90 percent of average picture level becomes noticeable to the median observer at ±2 dB and objectionable at ±5 dB. The EIA voltage or current standard for long-haul circuits is ±10 percent or ±0.8 db.

5.10.4 Differential phase

The eye is even more sensitive to differential phase, another cross-luminance distortion, than to differential gain since it affects the hue of the image. In the early days of color television, the tolerance for differential phase distortion was generally considered to be ±10°. This tolerance was established more on the basis of the state-of-the-art capability of the then-current transmission systems than of the visibility of the distortion. Reducing the visibility of the effects of differential phase distortion was perhaps the primary motivation for the development of the PAL system.

---

2   D.G. Fink, *Television Engineering Handbook*, McGraw-Hill, New York, 1957.

With the passage of time and improvements in technology, much tighter tolerances have become practical, and the most recent EIA standard for an end-to-end transmission circuit is ±3°.

5.11 Subcarrrier Cross Talk Reduction

The presence of the subcarrier on the luminance channel has the potential of creating cross talk in the form of annoying dot patterns on the image. This is called cross luminance distortion. A number of procedures have been developed to minimize the visibility of these patterns.

5.11.1 Relating subcarrier and line frequencies

An important technique that is employed in all subcarrier systems is to relate the subcarrier and line frequencies as described earlier (see Sections 5.7.2 and 5.8.2). This relationship causes the polarity of the dot patterns to reverse or partially reverse on successive scans so that there is a cancellation of their visual effect. It can also cause the dot pattern to be stationary.

When a monochrome receiver is used with a color signal, the color subcarrier may be visible as dots in colored areas of the picture because of imperfect subcarrier cancellation. This effect is often mitigated by poor aperture response in inexpensive monochrome receivers.

5.11.2 Low-pass filters

A brute force reduction of dot visibility used in early color receivers was a low-pass filter in the luminance channel with a cut-off below the subcarrier frequency—3.58 MHz in the case of NTSC. This limited the horizontal resolution and definition, but the high-frequency output of the image orthicon camera tubes used in this period was so low that very little information was removed by the filter.

5.11.3 Comb filters

The signals from photoconductive camera tubes and CCDs (see Chapter 8) have a sufficiently low noise level to permit aperture correction, and the loss of frequency components above 3.58 MHz significantly degrades the picture. To solve this problem, *comb filters* were introduced in the receiver luminance channels (see Chapter 12). As their name suggests, comb filters remove alternate bands of frequency components. In their color television application they remove the chrominance channel frequency components, which are interleaved with the luminance channel components at harmonics of the line frequency (see Section 5.7.4). Comb filters also have the desirable effect of improving the signal-to-noise ratio by nearly 3 dB since they remove the noise from these regions of the frequency spectrum.

Although comb filters are highly effective, they do not completely eliminate the problem of subcarrier visibility. The clusters of chrominance sidebands spread when reproducing edges with sharp color differences, so that they are not completely removed by comb filters. This creates artifacts in the form of a dot pattern along the edges.

A similar spreading of the chrominance sidebands occurs if there is rapid motion in the picture or if there are major signal changes from line to line. The luminance channel is also affected because the sidebands surrounding the line frequency harmonics spread into the areas attenuated by the filters.

If the dot pattern on a tri-color kinescope has approximately the same spacing as the subcarrier dot pattern, a heterodyning effect can occur that causes artifacts in the form of a coarse spurious pattern in the picture (called *moiré*).

5.12 Component Signal Formats

Component signals are used in applications where it is especially important to avoid the artifacts and cross-color and cross-luminance distortions that are characteristic of composite systems. These applications include but are not limited to HDTV and complex program production systems where the signals are subjected to repeated recording and rerecording.

Component systems frequently employ a digital format, which further reduces the distortions that occur with analog formats.

Conventional component systems require three channels (see Figure 5.1), the Y or luminance channel, and two color difference channels. The three channels usually are transmitted simultaneously, but they can also be transmitted sequentially as in the MAC family of formats.

5.13 Multiple Analog Component (MAC) Systems

MAC (multiple analog components) is a term describing a family of time-division multiplex signal formats in which the luminance and chrominance components are time compressed and transmitted sequentially. In common with other component systems, they avoid the problems of cross talk and sensitivity to nonlinearity described in Sections 5.9 and 5.10 for subcarrier systems. They also have a better signal-to-noise ratio because the absence of strong frequency components in the vicinity of the color subcarrier. Synchronizing information can be sent in digital form, thus eliminating the large amplitude conventional and digital audio signals inserted for a portion of the sync intervals.

At the same time, MAC shares the feature of composite systems of including the entire signal in a single channel.

Its disadvantages are its lack of compatibility with broadcast systems and somewhat greater bandwidth requirements. Its principal applications, therefore, are in situations where compatibility is not required and where adequate bandwidth is available. It is particularly useful for satellite transmissions.

5.13.1 The principle of MAC transmission

The principle of MAC transmission is illustrated in Figure 5.7, which shows time relationships in one line of B-MAC, a principal member of the MAC family (see Section 5.11.2).

The 65.5 microsecond line scan time (it is virtually identical for NTSC and PAL) is divided into three segments—34.9 microseconds for the luminance signal, E'_Y, 17.4 microseconds for the color difference signals, $E'_R - E'_Y$ or $E'_B - E'_Y$, and the remainder for blanking, burst, audio, and data (optional). (Alternatively, I and Q signals can be used for transmission of the chroma information.) This requires reduction in transmission time for all video signal components—by 2:3 for the luminance signal and 1:3 for the color differences. The $E'_R - E'_Y$ and $E'_B - E'_Y$ components are transmitted on alternate lines. As a result, four fields are required to transmit a complete picture, and the vertical chrominance resolution is reduced by one-half.

To achieve the time-reduced format, the luminance component is sampled at a rate of 15.5 MHz, high enough to avoid aliasing, and the color-difference components are sampled at one-half this rate or 6.75 MHz. The samples are stored in memories and the MAC signal is generated by reading out the samples at a rate of 20.25 MHz in the sequence Y', $R'-Y'$, Y', $B'-Y'$. The luminance samples for one line then occupy 34.9 microseconds for a time reduction of 3:2 and the color difference samples 17.5 microseconds for a time reduction of 3:1. The sequence of the samples for a single line is shown in Figure 5.7. Note that space is reserved for transmission of the burst, audio, and data service.

The process is reversed at the receiver. The samples in the segments of each line corresponding to Y', $R'-Y'$, and $B'-Y'$, are stored in a memory bank and read out in real time to generate the luminance and color difference signals simultaneously. They are then matrixed and processed to recover the original E'_R, E'_G, and E'_B.

The resolution of the chrominance signal is reduced both horizontally and vertically by this process—horizontally because the time reduction is twice as great and vertically because only alternate lines are transmitted.

To achieve resolution of the luminance component that is equivalent to that of the original signal, the bandwidth must be increased by an amount that is inversely proportional to the time compression. In the absence of digital compression (see Section 3.12), a bandwidth of 6.1 MHz (4.2×1.5) is required to achieve resolution equivalent to that of the original signal. Equal resolution could be achieved in the same bandwidth, but this would require transforming the signal to a digital format and "MAC" would be a misnomer.

Figure 5.7 MAC transmission

5.13.2 MAC formats

A number of MAC formats have been developed to meet the needs of various applications. The complete family includes ACLE, A-MAC, B-MAC, C-MAC, D-MAC, HD-MAC, HD-MAC60, MAC-60, and MUSE. The formats for the video portions of the signals are identical in many cases, and the differences are in the audio and data channels.

B-MAC, described above, is the most common. In its analog form, it requires a bandwidth slightly in excess of 6 MHz (as compared with 4.2 MHz for an NTSC signal).

C-MAC is an occasional variation. It requires a bandwidth of 20 MHz, the additional bandwidth being used for data.

HD-MAC has been proposed for high-definition television in Europe. It operates at 50 fields per second, with 1250 scan lines interlaced per frame. The aspect ratio is 16:9.

Digital Video Systems— HDTV

6.1 Introduction

The analog television standards for NTSC, PAL, and SECAM are 40 or more years old. They have been extremely successful and are still serving well, but they are based on the technology of the 1950s. For at least the last 10–15 years, the technical possibility for designing improved systems has existed and research labs around the world have given this much attention. Because the most obvious improvement is increased resolution, this work has been named *High Definition Television* (HDTV). Although many systems have been proposed, no country other than Japan has adopted a new system.

Early HDTV system proposals were based on analog technology and generally required some means to provide more bandwidth for each channel. This posed serious problems of spectrum allocation for broadcasting and reduction of channel capacity for cable distribution. However, more recent developments using digital technology have shown that HDTV signals can be broadcast or distributed in the same 6 MHz bandwidth used for current television. In addition, the flexibility offered by digital systems allows many new features that greatly increase the potential of HDTV to expand the uses of television.

In the United States, numerous laboratories have worked for many years on different approaches to HDTV. In 1993, as the potential of a digital HDTV system became clear, seven organizations formed an organization called the Digital HDTV Grand Alliance[1,2] (GA) to develop a single system to propose to the FCC for adoption as the U.S. standard. At this writing (1996), the GA proposal is before the FCC. This chapter discusses the issues of HDTV and describes the GA system in detail.

6.1.1 Leading up to the Grand Alliance

Research in HDTV had been going on in the United States for some time, and in 1987 a group of television broadcasters petitioned the FCC to consider setting broadcast standards for an "advanced television" (ATV) broadcast service. The impetus for this was a demonstration in the U.S. of the Japanese MUSE (multiple sub-Nyquist encoding) system. The FCC response was to set up the Advisory Committee on Advanced Television Service (ACATS) to consider system proposals. By the end of 1988, ACATS had received 23 different proposals! At the same time, the Electronic Industries Association (EIA) established the Advanced Television Test Center (ATTC) in Alexandria, Virginia, to test HDTV system proposals. Another testing laboratory, the Advanced Television Evaluation Laboratory (ATEL), was established in Ottawa by the Canadian government to also participate. However, the resolution of 23 different proposals required more than simply testing them (no small task itself), and in March 1990, the FCC announced some basic ground rules:

1. HDTV should be a separate service broadcasting simultaneously with the NTSC service rather than an add-on to NTSC broadcasting that requires an additional channel. Thus, the HDTV service would not have to be compatible with the NTSC service, and there would be a transition period of some years where both services would be broadcast. Further, this means that HDTV must fit into a single 6 MHz channel.

2. Some of the system proposals were for *extended-definition TV* (EDTV), which offered increased definition while maintaining some NTSC compatibility. The FCC said it preferred to first consider HDTV proposals that had increased definition not constrained by any need for NTSC compatibility.

As a result of these guidelines, the system proposers went back to their drawing boards. At this time, General Instrument Corp. made the first all-digital HDTV proposal. This was a surprise, but it caused other proposers to consider digital systems, and by 1991 there were four all-digital system proposals. These and two analog/digital proposals were tested by the ATTC and ATEL and in 1993, the ACATS tried to pick a winner.

The only thing the ACATS could agree on was that the system should be all-digital. However, it endorsed an industry effort already underway to merge the four digital proposals into one—the Grand Alliance. This was formally announced on May 24, 1993, and it comprised seven members: AT&T, General Instrument Corp., the Massachusetts Institute of Technology (MIT), Philips Electronics North America, the David Sarnoff Research Center, Thomson Consumer Electronics, Inc., and Zenith Electronics Corp. Over the succeeding months, the GA system details were crystallized and in 1994, system construction and testing began.

6.2 HDTV System Objectives

It is not sufficient to develop a new television system based on advanced technology and call it HDTV. The new system must provide meaningful improvements in performance and features that broadcasters, program providers, and consumers will perceive as worth the cost and difficulty of a transition to the new system. The full scope of HDTV objectives has evolved over time; the following list gives the objectives that are met by the Grand Alliance system:

1. High definition was the earliest objective for an HDTV system, as indicated by its name. Analog television provides sufficient definition for viewing ratios down to about 4:1, but this limits the size of displays for use in normal-size rooms to about 35" diagonal. Although larger TV displays are being sold and viewed at lower viewing ratios, the limitations of existing analog TV such as scanning lines, interlace flicker, fuzziness, and color edge effects become clearly visible to anyone. The first step of improvement is to use more scanning lines and scan progressively instead of interlacing. Along with this, the horizontal resolution must be proportionately increased to overcome fuzziness.

2. Most viewers appreciate the advantages of a wider aspect ratio as a result of their experience with it in motion pictures and that has become a second inprovement in all HDTV proposals. A wider picture relative to height actually does not require further increase in horizontal resolution because resolution is defined relative to the picture height, but it does imply an increase in bandwidth and data rate.

3. Sound quality has been shown to add significantly to the TV viewing experience. Multichannel sound and high fidelity are important HDTV objectives.

4. Everyone has had the experience of watching noisy TV pictures, maybe because of the use of a rabbit-ear antenna or because of distance from the station. As digital transmission has become part of HDTV proposals, the possibility of robust reception of noise- and distortion-free pictures right out to the fringe area has become a reality.

5. Digital transmission also allows the incorporation of data channels into the HDTV bit stream to support user interactive features to view information in addition to and other than the picture and sound.

6. Because HDTV will be broadcast simultaneously with NTSC for some years, the control of interference between the two types of broadcasting is an important objective.

7. Digital video and audio are finding uses in many fields other than television and the *interoperability* of the HDTV signals with these other uses (such as computers, game machines, or video players) must be supported in the standards.

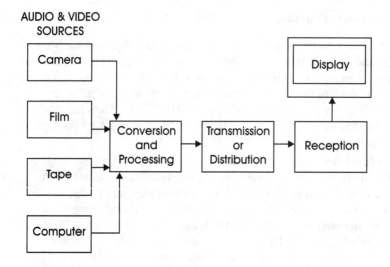

Figure 6.1 Block diagram of a video system showing elements to be considered for HDTV

Taking all the objectives together, HDTV is an impressively performing system with important new features that are recognizable by any consumer. Meeting the above objectives in a 6 MHz broadcast channel or equivalent distribution bandwidth was a tremendous challenge to the GA system designers. However, with digital technology, they were able to do it.

6.3 The Grand Alliance HDTV System

Figure 6.1 is a diagram of a complete video system showing the elements that must be considered in an HDTV system design. The Grand Alliance standard defines only the Transmission or Distribution block, which makes it possible for anyone to design receivers. The other blocks are covered by other standards, such as the SMPTE 274M production standard. However, the use of digital technology makes it possible to interface other standards into the system using appropriate conversion hardware or software.

The flexibility of the GA system is further enhanced by defining the transmission standard in four *layers*. The layers simply separate the standard into four consecutive blocks that have defined interfaces. This not only makes it easier to understand the standard, but it provides means for interoperability with other standards that may have equivalent layers. Figure 6.2 shows the four layers of the GA standard.

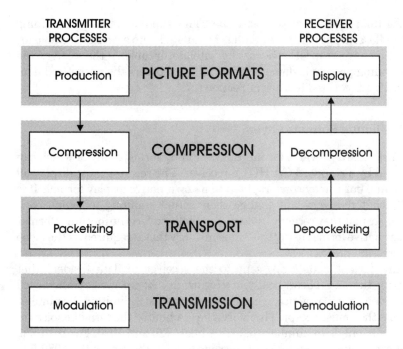

Figure 6.2 Layers of the GA HDTV transmission standard

6.3.1 The GA layered architecture

The layers represent the processing steps of the GA transmission standard. In broadcasting, the layers are traversed top-down, and in a receiver they are processed bottom-up. Thus, the top layer is "Production" for broadcasting and "Display" for receiving. This layer only defines scanning formats that are supported by the transmission standards (there are several formats; see Section 6.4). It does not mean that either the production or display scanning necessarily has to use these formats (in a digital system the conversion between scanning standards is relatively easy), but it does mean that the transmitted signal is defined in terms of one of these formats.

The second layer, Compression, performs audio and video compression in transmission and decompression at the receiver. This layer of the standard defines the video and audio compression techniques that are used; they are based on MPEG-2 for video and Dolby AC-3 for audio. Details are given in Section 6.5.

The Transport layer during transmission assembles multiple data streams into a single composite stream by means of a packetizing technique (see Section 3.7.1) based on MPEG-2. The transport data stream is demultiplexed in the receiver to recover the individual data streams for video, audio, and auxiliary data. Section 6.6 gives the details of the Transport layer.

In the final layer, Transmission, the composite data stream containing video, audio, and possibly other data is modulated onto a carrier suitable for distribution by terrestrial broadcasting, satellite, or cable. At the receiver, the corresponding process is demodulation, which recovers the packetized data stream. This is covered further in Section 6.7.

6.4 The Picture Formats Layer

The GA standard provides for multiple scanning formats, any of which can be decoded by a standard GA HDTV receiver. The receiver must display all the formats, but it may convert them to its own native display format. It is unlikely that receivers will change their actual scanning to match each incoming format they receive. Instead, they will convert the incoming format into their native format using the frame memory that is required by the video decompression process.

Picture format details are sent to the receiver in data headers that accompany the data. Theoretically, any format numbers could be transmitted by the system, but in practice, receivers are simplified by limiting the number of formats they must accept. Therefore, the GA limited the format choices to variations of only two numbers of active lines: 720 and 1080. For both line numbers, a choice of frame rates is provided and in the case of the 1080-line formats, either interlaced or progressive scanning can be used. The 720-line formats are always progressively scanned. Table 6.1 shows the format choices.

6.4.1 Active lines and pixels

The 720-line format always has 1280 active pixels per line, which provides square pixels (equal horizontal and vertical resolution) in the 16:9 aspect ratio. Since transmission of blanking intervals is not required in a digital system (all supporting and synchronizing information is contained in headers), numbers for total number of scanning lines or total pixels per line are not relevant for transmission. Of course, they are important for cameras or

Table 6.1 Scanning formats of the GA transmission standard

Active Lines	Active Pixels/Line	Rate (Hz)	Interlaced
720	1280	59.94/60.0	no
	1280	23.98/24.0	no
	1280	29.97/30.0	no
1080	1920	59.94/60.0	yes
	1920	23.98/24.0	no
	1920	29.97/30.0	no

displays, but that is not part of the GA standard. However, typical numbers for total lines and total pixels, such as were used in the GA prototype system, are 750 lines and 1650 pixels.

The 1080-line format provides for nominally 1920 active pixels per line. These were delivered in the GA prototype system by scanning with 1125 total scan lines and 2200 total pixels per line.

6.4.2 Frame rates and interlacing

Each line format provides for three basic frame rates: 60, 24, or 30 Hz. In the 720-line format, progressive scanning is always used. In the 1080-line format, progressive scanning is used for the 24- and 30-frames/second frame rates, but interlaced scanning is used for the 60 Hz vertical-scan option, giving a frame rate of 30 for that choice. This is done because progressive scanning at 60 Hz vertical would make the data rate too high for the transmission channel, or it would be necessary to make unacceptable picture-quality tradeoffs.

The frame rates of 24 and 30 Hz are used when the system input originates on motion picture film at either of those rates. The 60 Hz rate is used when the input is from an HDTV video camera. These numbers are the frame rates implicit in the transmission. For the film rates of 24 or 30, the receiver would scan its display at an appropriately higher rate to avoid flicker.

Because interoperability of HDTV with NTSC signal sources will be important, at least during the transition period, alternative frame rates are also available based on the 59.94 Hz NTSC field rate. This will simplify the display of NTSC-originated material on the HDTV system.

6.4.3 Video sampling rates

Video sampling rates depend on the actual total lines and total pixels used in the originating production equipment. Using the numbers given above for the 720-line format in the GA prototype system, the video sampling rate for the 60 fps format is:

$$750 \times 1600 \times 60 = 74.25 \text{ MHz}$$

Because the 1080-line format uses interlaced scanning (receivers may convert this to progressive scanning for their own displays), the numbers given above for the 1080-line prototype result in the same sampling frequency as the 720-line format:

$$\frac{1125}{2} \times 2200 \times 60 = 74.25 \text{ MHz}$$

Sampling at 74.25 MHz and collecting only the active pixels and lines with 24 bpp gives a raw data rate of:

$$74.25 \times \frac{1280}{1600} \times \frac{720}{750} \times 3 = 171 \text{ MB / sec.} = 1{,}370 \text{ Mb / sec.}$$

The Compression layer has to compress this to a data rate that can be transmitted in a 6 MHz channel!

6.5 The Compression Layer

As will be shown below in Section 6.7, state-of-the-art digital transmission systems can handle data rates of approximately 3 bits/Hz of bandwidth with acceptable error rates for video or audio. Thus, a 6 MHz television channel could transmit 18 Mb/second. The objective of the compression layer is to reduce the data rate given above to approximately 18 Mb/s, while also leaving room for audio, headers, and auxiliary data. That calls for a compression factor of about 70.

As explained in Sections 3.10 to 3.12, massive compression of sampled video information is possible because of the redundancy that is inherent in most moving pictures. A combination of techniques is normally used for this and the GA video standard specifies such a combination process that falls within the choices available in the MPEG-2 standard. Compatibility with MPEG-2 draws on the extensive research put into that standard as well as enhancing the possibility of proposing the GA system as a worldwide HDTV standard.

6.5.1 Compression processing

Using the terminology introduced in Chapter 3, Figure 6.3 shows the video compression processing of the GA system.

6.5.1.1 Source-adaptive processing Since GA video encoding can accept various video source formats, the first step must be to conform them all to the same format for further processing. All formats are converted to RGB components at this stage.

6.5.1.2 Color-space processing The RGB components are digitized (if they are not already digital) and then converted to a luminance and two chrominance component format using a linear transformation matrix. The purpose of this is to put the signals into a format where bandwidth can be allocated according to the acuity of the human visual system. Therefore, the resolution

Figure 6.3 Video compression processing

of the chrominance components is reduced 2:1 in each direction by filtering and sub-sampling. This accomplishes 1.5:1 compression.

6.5.1.3 Motion compensation In most motion video sequences there is considerable redundancy between frames because some parts of the image do not move at all and other parts of the image may be simple translational moves from the previous frame. This is *temporal* redundancy and motion compensation is the technique for dealing with it (see Section 3.12.2.1). In the GA system, motion compensation is done by the technique known as *block matching*. The current frame is divided into blocks and each block is compared to all possible locations in an adjacent frame. When a near match is found, its location in the previous frame (called the *translation vector*) is stored so that a predicted frame can be built using all the matched blocks moved to their new locations. The predicted frame is then compared to the actual current frame and the difference, which is called the motion-compensation *residual*, is passed on for further compression.

In most motion sequences, the prediction obtained by block matching is very good and the residual is small and easily compressed. However, this is not always true, especially with high-motion sequences and of course, always when scene transitions occur. In these cases, the GA compression system will give up on motion compensation and the frame(s) will be compressed without dependence on other frames. This is called *intra-frame* coding. Of course, intra-frame coding must be used at the start of a sequence, where there is no previous frame, and it is usually used periodically throughout a sequence to facilitate editing or other actions that require starting decompression in the middle of a sequence.

6.5.1.4 Discrete cosine transform coding Intra-coded frames or residual frames are compressed using the DCT process followed by adaptive quantization and statistical coding (see Section 3.11 for explanation of these terms). 8 × 8 pixel blocks are used for the DCT processing. Because the same amount of data comes out of the DCT processing as went into it, no real compression is achieved by the DCT process itself, but the data have been transformed into a structure where it is now easy to separate significant from insignificant information as far as the final viewer is concerned. The remaining steps of quantization and statistical coding exploit the new data organization—this is where the real compression of DCT occurs.

Many of the DCT coefficients are typically zero and zigzag ordering as in JPEG (see Section 3.12.1) is used to group most of them together. Run-length encoding is used to send the long runs of zero coefficients that result. The run lengths and the nonzero coefficients are then coded by statistical coding (Huffman—see Section 3.11.7).

6.6 The Transport Layer

The video compression yields a bit stream that contains only the video data along with the header information and table data required to decode it. However, for transmission that stream must be merged with one or more audio streams and possibly an auxiliary data stream. In some applications, there may be more than one stream of each type (such as with multiple-language audio or with multiple programs in one channel). It is the purpose of the Transport layer to merge these streams and add other information needed for efficient, error-free transmission. A packetizing approach is used (see Section 3.7.1).

6.6.1 The GA transport packet format

The basic packet format is a 188-byte fixed-length block as diagrammed in Figure 6.4. This size was chosen because of several considerations:

1. The packet should be large enough that the overhead due to the transport headers does not represent a significant portion of the channel capacity.

2. The packet size should be consistent with typical block error correction approaches so that packets can be synchronized with error correction blocks. This limits the maximum packet size.

3. The exact number should be chosen to provide several ways to be interoperable with the *asynchronous transport mode* (ATM) format that is expected to be widely used in other data communication systems.

The packet contains a 4-byte "link" header and a 184-byte data field that may include a variable-length "adaptation" header.

Figure 6.4 The transport packet structure

6.6.1.1 The link header   The link header accomplishes the following tasks:

1. Packet synchronization—The first byte in every GA packet is a sync byte that has the hexadecimal value 47H (0100 1111) (see Section 3.2.2).

2. Packet identification—13 bits of the header are assigned as a packet identification (PID) field, which is used for multiplexing and demultiplexing bit streams. This is done by first extracting a packet (PID=0) containing a map of the multiplexing structure that identifies a PID number for each stream in the multiplex structure (see Section 6.6.1.3).

3. Error handling—4 bits are assigned to a *continuity counter* field. It has a value that cycles uniformly from 0 to 15 at the transmitter for all packets with the same PID value. (Some packets may not have data in them. The continuity counter is not advanced for these packets.) If a discontinuity in the continuity counter value is experienced at the receiver, it normally means that data has been lost and the receiver circuits should take appropriate action.

4. Conditional access—The transport format allows for scrambling of data in the packets to control access at the receiver. This can be done independently for each bit stream in the system. Although the GA stand-

Figure 6.5 The adaptation header of the transport packet structure

ard specifies the descrambling approach to be used, it does not specify the descrambling key or how it is obtained at the decoder.

5. Flags—the rest of the bits in the link header are used as flags for specific purposes as shown in Figure 6.4. For example, a 2-bit field indicates whether the packet contains an adaptation header and/or a data payload.

6.6.1.2 The adaptation header When indicated by the *adaptation field control* bits in the link header, an adaptation header follows immediately. If there is an adaptation header, the first two bytes are predefined. The first byte gives the total length (bytes) of the adaptation header excluding the first byte. The second byte contains flags for specific purposes as shown in Figure 6.5. Some predetermined uses for the adaptation header are:

1. Synchronization and timing—In the GA compressed data delivery system, the amount of data delivered for each picture is variable and timing cannot be derived directly from the start of picture data as it could be in a system such as NTSC where the data rates and sizes are precisely known. The digital system has no equivalent to the NTSC sync pulses. Instead, timing information is transmitted in the adaptation headers of selected packets.

   The *program clock reference* (PCR) flag indicates that this specific adaptation header contains a program clock data field. This is a 6-byte field that contains the value of a time counter running from a 27 MHz clock at the transmitter. The receiver should compare this with its own clock to determine whether the receiver is in sync with the program. Ordinarily, it is used only to adjust the receiver clock rate, but in the case of a program switch, another adaptation header flag (*discontinuity indicator*) indicates that the receiver counter should be updated to the new value.

2. Random access—A flag in the adaptation header (*random access indicator*) indicates that this packet is the start of a random access entry point into the bit stream of that packet. Thus, when a receiver is channel switching or searching for a resynchronization point, it can read packet headers and discard whole packets until it finds the random access indicator flag set. To facilitate rapid channel switching, random access entry points should be included as often as possible.

3. Local program insertion—When it is desired to insert another program into a GA transport bit stream, such as at a broadcast head end, there are some special considerations. Of course, it must be done at a random access entry point, but that is not enough. The system must also confirm that the insertion will not cause buffer underflow or overflow at the receiver. The encoding system keeps track of the receiver buffer status expected when the bit stream is decoded and identifies appropriate splice-in points. These are identified by setting the *splicing point* flag in the adaptation header. That flag indicates that this header contains a one-byte field that gives the number of following packets with the same PID that should be counted until the exact splice-in point occurs. At that point a new program can be switched in. The first packet of the new program must contain an adaption header with PCR flags and fields that will tell the receiver to update its program clock.

4. Other features—additional flags in the adaptation header provide for transmitting "private" (not part of the program data) data to the decoder and for future extensions to the header.

6.6.1.3 Packet multiplexing As described above, each separate bit stream (audio, video, or auxiliary data) has its own unique PID on all of its packets. The value PID=0 is reserved for packets that tell the receiver how the other PID values are assigned. PID=0 packets contain the *program association table*, which identifies one or more complete programs and what is the PID of packets that contain *program map tables*, which tell the PIDs for each data stream in the program. This is illustrated in Figure 6.6. Special headers are defined for these table packets.

The program tables contain information that will facilitate a receiver selecting a specific program from a multiprogram channel and synchronizing to it in a short time.

There are numerous other features of the Transport layer that are not covered here. The system is highly configurable and expandable for the number of channels or different data types. Characteristics can be changed dynamically, even within one program when necessary. There are also important features for interoperability with other transport systems or other compression systems.

Figure 6.6 Finding a program with the map tables

6.7 The Transmission Layer

The packetized bit stream from the transport multiplexer is processed for broadcasting or cable distribution by the Transmission layer. This layer sets up the transport packets in a *data frame* structure, adds error detection and correction codes, and does vestigial sideband (VSB) modulation onto a carrier for transmission or distribution. There are two modes of VSB modulation, 8-VSB with trellis coding for terrestrial broadcast and 16-VSB for cable distribution. All are explained in this section.

6.7.1 The data frame

The transmission layer prepares the data for transmission by converting each packet into a *segment* by taking the packet contents less the sync byte and processing it for Reed-Solomon (R-S) error protection (see Section 3.8.4). A new sync byte is added to each segment at a later stage in the processing, so the effective segment size as transmitted becomes 208 bytes including 20 R-S parity bytes. This is illustrated in Figure 6.7.

The data is also divided into data *fields*, which consist of 313 segments. The first segment of each data field contains a *Data Field Sync* pattern, which has five purposes:

1. indicate the beginning of the data field

2. serve as a training reference signal for the automatic equalizer that corrects for multipath transmission errors in an HDTV receiver

3. allow the receiver to decide whether the interference rejection filter should be used

4. used for system diagnostic measurements

5. allow the phase tracker in the receiver to reset its circuitry and determine its loop parameters

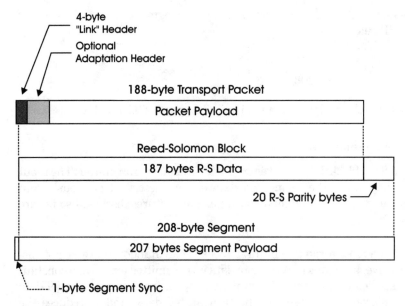

Figure 6.7 From a packet to a segment

6.7.2 Symbols

The VSB modulation is multilevel—for example, 8 amplitude levels for 8-VSB. Thus, each piece of the modulated signal carries more than one bit—8-VSB would transmit 3 bits per piece. The multilevel pieces representing multiple bits are called *symbols*. The GA 8-VSB system actually transmits only 2 bits of data in each symbol. That is because of the trellis-coding that inserts a third bit into the transmission and removes it at the receiver. This is another error-correction strategy. Trellis-coding is explained in Section 3.7.2.

So an 8-VSB symbol actually carries 2 bits of data, meaning that there are 4 symbols per data byte. Therefore, the 208-byte segment described above is modulated in 832 symbols, 4 of which are the segment sync pattern.

The modulation system runs at a fixed symbol rate that is chosen relative to the channel bandwidth. In the GA modulation systems, the symbol rate is

$$fs = \frac{4.5}{286} \times 684 = 10.7622 \text{ MHz}$$

The first term (4.5/286) is the NTSC horizontal scan rate in MHz.

The symbol rate must be maintained precisely (± 10 ppm) to insure that the spectrum of HDTV transmission relates to NTSC broadcast spectra in a way that minimizes interference between the two services. Considering this symbol rate, the 2 bits per symbol, the 21-byte sync and error protection overhead, and the 1 segment per 313 field sync segment, the net data rate of the 8-VSB channel is:

$$10.7622 \times 2 \times \frac{187}{208} \times \frac{312}{313} = 19.29 \text{ Mb} / \text{sec}.$$

For a 6 MHz channel, that is a transmission speed of 3.2 bits/Hz.

6.7.3 The transmission system

Figure 6.8 is a block diagram of a terrestrial broadcast transmitter. The input to this diagram is the segmented data stream described previously that contains audio, video, and auxiliary data. The features discussed so far are shown.

6.7.3.1 Data randomizing   The first block, Data Randomize, was not discussed above. It insures that random data is transmitted even when constant data is applied to the input. This might happen, for example, when a data input is disconnected. Keeping the transmitted data random reduces the possibility that this signal could interfere with other services and it also evens out the load on the transmitter. The "randomizing" algorithm is defined and is used at the receiver to derandomize the data. The same pseudo-random bit sequence is simply XORed with the data at both places. (XORing twice restores the original data; see Section 3.5.3.)

The next block in Figure 6.8 (Reed-Solomon Encoding) was covered above.

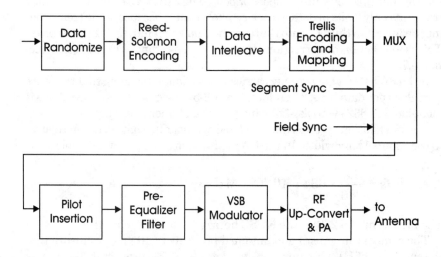

Figure 6.8 An HDTV terrestrial broadcast transmitter

6.7.3.2 Data interleaving Further error protection is obtained by interleaving the data after the R-S encoder. This has the effect of spreading the data bytes from the same R-S block over time so that a long burst of noise is necessary to overrun the R-S error protection. The system employs a convolutional byte interleaver that spreads the data over a 52-segment region. Interleaving is performed only on data bytes, not field and segment sync bytes (these are added after the interleaver). Interleaving is synchronized to the first data byte of each field. The result is that the system will tolerate an error burst of up to 193 μseconds.

6.7.3.3 Trellis encoding and mapping The Trellis Encoding block was covered briefly above, but it should be noted that the trellis coding is also interleaved by having 12 trellis coders that separately process 12 successive symbols. This improves system performance for burst noise. Following the interleaved trellis coders, each 2-bit symbol is represented by 3 bits, which pass to the mapper for conversion to the 8-level analog modulating signal. The operation of the mapper is shown in Figure 6.9.

The insertion of the segment and field sync after encoding is accomplished by multiplexing the sync levels into the 8-level signal from the trellis coder. The sync is transmitted with 1 bit per symbol to improve sync extraction reliability. Modulation levels –5 and +5 are used for this.

6.7.3.4 Pilot insertion The next operation is to provide for inserting a small-amplitude pilot carrier that will assist the receiver in acquiring the signal and maintaining lock in the presence of noise and interference. Since the modulator is balanced to produce a suppressed carrier (zero input produces zero carrier), pilot insertion is accomplished by adding a small DC level to every

Figure 6.9 Operation of the mapper

VSB Carrier

←0.31 MHz

6 MHz Channel

Figure 6.10 GA HDTV transmitter frequency response

symbol before the modulator. This level is 1.25, added to the input symbol levels (±1, ±3, ±5, ±7). The result is a small carrier output—it is 11.3 dB below the data signal power and represents an increase in transmitted power of only 0.3 dB.

6.7.3.5 Pre-equalizer filter To correct for any in-band ripple or band edge rolloff in the high power transmitter, an automatic equalizer is used. It responds to a signal sample that is tapped off the transmitter output signal before it goes to the antenna and is sampled by a reference (ideal) demodulator. The reference demodulator contains its own automatic equalizer whose tap weights can be transferred into the transmitter pre-equalizer to provide precorrection of transmitter frequency response distortion.

6.7.3.6 VSB modulation VSB modulation is amplitude modulation where the lower sideband of the modulation has been suppressed to reduce the required bandwidth by a factor of almost 2. In order to avoid difficult filtering requirements, lower sidebands are not totally removed, but they are rolled off according to a precise (and easily realizable) frequency response curve.

In a typical HDTV transmitter, the 10.76 Msymbols/sec 8-level trellis-coded composite data signal is amplitude modulated on two quadrature IF carriers. These carriers are combined to create the vestigial sideband IF signal by sideband cancellation, a technique that is widely used in other forms of single-sideband transmission. For minimal intersymbol interference, the data signal must be properly filtered before transmission in a 6 MHz channel. The proper system (transmitter plus receiver) response is a linear phase raised-cosine Nyquist function, which gives the frequency response shown in Figure 6.10. This system response is achieved in the system by identical filtering at both the transmitter and receiver, so each is actually a *root* raised-cosine Nyquist filter.

6.7.3.7 Up-conversion The VSB modulation is on an IF carrier that is the same frequency regardless of the actual transmission channel frequency. In the GA prototype, the IF frequency corresponding to the VSB carrier is 46.69 MHz. The modulated IF signal is up-converted to the final channel frequency and amplified to the power level required by the antenna. The amplifiers, of course, must be linear to preserve the signal properties. Further considerations of HDTV transmitters are given in Section 15.16.

6.7.4 Transmission summary

The parameters of the GA transmission system are summarized in Table 6.2. The table shows the features for both the 8-VSB with trellis coding format that was described above and the 16-VSB format for cable transmission. The latter format doubles the data rate per channel, which is possible because cable systems can achieve better S/N ratios and better control of signal levels than broadcast transmission. This means that two HDTV programs could be transmitted simultaneously on one cable channel. The two programs would be multiplexed at the transport level.

6.8 GA HDTV Audio

Sound is just as important as the picture in conveying the message of a video segment. This has not been overlooked in the GA system; a state-of-the-art digital audio system is part of the specification. Five-channel surround sound

Table 6.2 Grand Alliance transmission system parameters

Parameter	Terrestrial Mode	High Data Rate Cable Mode
Channel Bandwidth	6 MHz	6 MHz
Excess Bandwidth	11.5%	11.5%
Symbol Rate	10.76 MS/Sec.	10.76 MS/sec.
Bits/Symbol	3	4
Trellis FEC	2/3	None
Reed-Solomon FEC	T=10 (207,187)	T=10 (207,187)
Segment Length	832 Symbols	832 Symbols
Segment Sync	4 Symbols/Segment	4 Symbols/Segment
Frame Sync	1 per 313 Segments	1 per 313 Segments
Payload Data Rate	19.29 Mb/sec.	38.58 Mb/sec.
NTSC Cochannel rejection	NTSC rejection filter in receiver	N/A
Pilot Power Added	0.3 dB	0.3 dB
C/N Threshold	14.9 dB	28.3 dB

is provided using the Dolby AC-3 digital compression system. This system is similar to the one used in the latest motion picture theaters and produces a spectacular sound effect.

Numerous options are available and the system is designed to give the best performance possible with whatever audio hardware is available at the receiver. The audio bit stream, which can use up to 384 Kb/second of the digital channel capacity, is multiplexed with the video bit stream in the Transport layer using packet interleaving. Additional features allow the system to match the sound level between different programs and to allow end user control of various parameters with the implementation of receiver options.

6.9 Summary

This chapter presented the Grand Alliance HDTV system at a level that exposes most of the system's features but it did not go to the full details of hardware and software. Even so, the system appears to be extremely complex compared to its NTSC or PAL predecessors. However, one must remember that the actual implementation of the system will use custom integrated circuits and most of the complexity will disappear into the ICs. The real details will not be an issue to anyone except the IC designers. Program providers, broadcasters, and cable operators will operate at a level more like this chapter. Consumers will only need to know a small subset of this information to purchase and operate receivers.

The GA HDTV system delivers high definition pictures with high-quality multichannel sound with full digital robustness and does it in the same channel bandwidth that is used for NTSC broadcasting. It is a superb example of the type of advances that will occur in many other fields as the transition to digital technology expands.

References

1. *Grand Alliance HDTV System Specification, Version 2.0*, Grand Alliance, December 1994.

2. "HDTV and the New Digital Television," *IEEE Spectrum*, April 1995, pp 35–80.

Digital Video Systems—
Computers

7.1 Introduction

Capture, processing, and display of video have become important applications of personal computers. PCs are ideal low-cost platforms on which to implement the features of digital video. This is a recent development, brought about by the combination of several trends:

1. The processing power of personal computers has grown approximately 2:1 every two years for the past 25 years.

2. Computer video display technology has advanced in resolution and color capability.

3. Mass storage technology continues to improve performance, increase in capacity, and reduce in cost.

4. Digital video compression technology is maturing.

Each of these factors is a story in itself; all are covered in this chapter.

7.2 The Personal Computer

Personal computers are a development of the late 1970s, when it first became feasible to assemble a microprocessor, digital storage devices, a keyboard, and a video display at a size that would fit on a desktop and at an affordable price. Suddenly, anyone could have computing power that was capable of doing useful tasks right on his or her desktop. Although these early PCs were limited in their processing speed and storage capabilities, they had enough power to attract millions of customers who struggled with all kinds of difficulties to explore the opportunity of having their own computers. Software applications for word processing, spreadsheets, games, and other uses were quickly developed.

However, the remarkable advances of solid-state integrated circuit (IC) technology have fueled constant growth of the powers of PCs, which still goes on today. As PC power advanced, so did the capabilities of the software, which led to new and more important applications. PC users have become accustomed to the idea that the PC you buy today will be available at a substantially lower price in six months and it will be obsolete in two years! That is a manifestation of progress in a highly competitive and volatile market.

It is difficult to appreciate the magnitude of progress in ICs that has occurred at such rates and over such a long period of time. Figure 7.1 shows data for two measures of that progress: microprocessor speed (see Section 7.2.1.2) and memory chip capacity—four orders of magnitude over 20 years! Nothing else in the history of industry has shown such growth. The result has been continued advances in circuit power *combined with* reduction of cost. A $1000 PC today has more power than a multimillion dollar computer of 1970. And the curves are expected to continue going up for many more years.

The same IC features that are utilized in PCs also are driving the conversion of most video equipment to digital technology. Steady cost reduction has brought digital hardware to lower cost than its analog equivalent in most cases.

Figure 7.1 Growth of microprocessor speed and RAM chip capacity over 20 years

7.2.1 PC architecture

The purpose of a television receiver is to display the best possible picture from the signal received. The purpose of a personal computer is different—it is to provide the best possible platform for running the software it receives. The PC's display functionality is just one of the ways a PC communicates with its user. For this reason, early PCs were successful even though their displays were limited to simple monochrome text and they had no ability to output sound beyond beeping. But the architecture of the PC was designed to allow upgrading of video and audio capability when it was necessary and when it made economic sense.

Figure 7.2 is a block diagram of a basic PC. It shows that a PC consists of a main microprocessor (the *central processing unit* or CPU), several types of data storage or memory, and a variety of *input/output* (I/O) devices such as a keyboard, mouse, video display, and sound unit. All are interconnected so that digital data can move freely between them. The CPU is the brain of the system; it performs system control and handles most of the processing functions of the computer.

7.2.1.1 The PC bus   The backbone of a PC is its *bus*, which is a multiwire circuit that makes a parallel connection of all the components of the PC. Having everything in parallel would appear to be awkward until one realizes that each device on the bus is designed to respond only to signals that are designated for it. There are several ways of accomplishing that, but the basic

Figure 7.2 Block diagram of a typical personal computer

idea of them all is to provide separate sets of wires in the bus for data and address information.

For example, a 32-bit bus has 32 parallel wires for data and another 32 parallel wires for an *address*. Each device is assigned its own address, and the device watches the address lines and ignores all data unless it first sees its own address. Then it interrogates the data lines to receive data. This type of interface is called *memory-mapped*, because it is the same access technique used by the computer's main memory.

Important parameters of a PC are its *bus width* and *bus speed*. The width refers to how many bits of parallel data are provided for, and the speed parameter specifies how many cycles of bus activity occur per second. Early PCs had 8-bit data and 16-bit addressing on their buses, and ran at speeds of 1 or 2 MHz. These numbers were consistent with the capabilities of the early microprocessors and other components, but everything in solid-state electronics has progressed since then. Today, 32-bit buses are common, and bus speeds run from 8 MHz up.

Since (at least) one bus cycle is required for each piece of data equal to the bus width, the bus parameters determine the maximum rate at which data can move through the computer. As the need to move more data has grown, especially to support video display activities, various strategies have been developed to get around the main bus limitation. One of these is the *local bus* approach, where a separate short bus is used to interconnect only those devices that need the fastest data access. Local buses are wider and faster than the main bus and they are often used to interconnect the CPU, the display adaptor, main memory, or mass storage.

7.2.1.2 The CPU   A CPU is an integrated circuit chip that performs digital operations such as logic, arithmetic, data movement, and I/O. It is controlled by a *program* that consists of a series of *instructions*. The instructions are digital codes that tell the CPU what kind of operation to do at each step of the program. They are just like other digital data except that they have special meaning to a microprocessor. The program is stored in memory and is read over the bus by the CPU for execution. As a program runs, the CPU may access the bus for other data from memory, mass storage, or I/O.

There are several families of CPUs on the market; they differ in architectural details and in the syntax of their instruction words. Thus, different CPU families require different software. A manufacturer of a CPU family tries to keep new designs in the family *backward compatible* with earlier designs so that existing software will still run on new units. This is important because user's investment in software is often greater than their investment in hardware and they do not want software obsoleted when they upgrade to new hardware. Some current CPU families are the Intel x86 family (used in IBM-compatible PCs), the Motorola 680x0 family (used in Apple Macintosh computers), and several new *reduced instruction-set computer* (RISC) processors such as the PowerPC, Alpha, R4000, etc.

RISC processors are designed with a simplified instruction set that executes faster per instruction. However, RISC instructions individually are less powerful so it may take more instructions to do a particular task than on a CISC (*complex instruction-set computer*) processor. The end result depends on many factors, but RISC machines are at the high end of today's PC performance curve and CISC processors are even incorporating some RISC features in their new designs (without abandoning compatibility, of course).

The previous discussion indicates that speed is probably the most important parameter of a CPU, but that it may be difficult to compare the speed between different CPU families. However, within a family there are two numbers that are significant: the *clock frequency* and the speed in *millions of instructions per second* (MIPS).

Clock frequency defines the internal cycle rate of the microprocessor. Depending on the CPU design, instructions may execute in one or more clock cycles. Some CPUs even execute several instructions at once. The resulting instruction execution rate (MIPS) is therefore a complex function of clock frequency, instruction set, and architecture. Today, clock frequencies are in the 100 MHz range and MIPS ratings are pushing beyond 100 MIPS.

7.2.1.3 Memory   Programs being executed and the data they use are stored in the system's main memory that can be addressed by the CPU. Main memory is usually solid-state *random-access memory*, called RAM. RAM is *volatile* storage—it loses its data when power is removed from the system. Present-day personal computers have 4 MB or more of RAM—the more the better—because when RAM gets filled up, the system must swap data or programs into mass storage, which is much slower. In the fastest systems, the speed of main memory is usually slower than the CPU, so systems may include modest amounts of faster memory, called *cache memory*, to hold the most frequently used data. The latest microprocessor designs also have some on-chip cache memory, which is even faster than off-chip cache.

Most computers have a little built-in software called the *Basic Input/Output System* (BIOS) that provides the first level of interface to the installed hardware. The BIOS is stored in *read-only memory* (ROM), which is nonvolatile solid-state memory that has the data permanently burned into it. A variation of this is called the *flash BIOS*, which is a rewriteable nonvolatile form of memory. By exercising special procedures, a flash BIOS can be updated with new software.

7.2.1.4 Mass storage   Hard disks, CD-ROM, and other devices are *mass storage* units that can hold great amounts of data and programs permanently (nonvolatile). Mass storage may be as much as 1000 times slower than RAM or cache, but its capacity and cost per megabyte is much lower. It is essential as a medium of permanent storage. Mass storage cannot be addressed directly by the CPU; data or programs must first be *loaded* from mass storage to RAM for use by the CPU. Normally, loading occurs before program execution begins, but in the case of motion video, the data is so large that it cannot all fit into RAM. In that case, data must constantly flow from mass storage to RAM as the video is played.

PCs usually have both *nonremovable* and *removable* mass storage. The most common nonremovable storage device is the magnetic hard disk (see Section 9.17), which has a storage capacity from 100 MB to several thousand MB; 1000 MB is a *gigabyte* (GB). Removable storage is used for backup or movement of data between systems. The most common removable mass storage is the floppy disk, which has a capacity in the range of 1 or 2 MB. For larger removable storage, magnetic tape is often used. This has capacities in the same range as hard disks, but it is much slower for access.

The CD-ROM optical disc (see Section 9.18), which is an outgrowth of the audio Compact Disc, is a very important medium for audio and video storage. A single CD-ROM disc has a capacity of 650 MB or more and it can be replicated in quantity for a cost of around $1 per unit. However, the process of recording (called *mastering*) is expensive and the replication of copies takes special facilities. But this is the premier medium for distribution of programs. Thousands of different CD-ROMs are on the market and more are being produced every day.

The optical technology behind the audio CD and the CD-ROM is suitable primarily for program distribution because it is *read-only*. Once a CD has been replicated, its data cannot be changed. This is actually an advantage for distribution use because the distributer is assured that the data will not be modified by anyone.

Drives are available for recording of CDs directly on a desktop PC. Although these are more expensive than hard drives or floppy-disk drives, the large data capacity still makes them a good value for backup and small-scale distribution. Just as with the CD-ROM itself, prices for recordable CD drives can be expected to come down as their usage increases.

Although 650 MB sounds like a lot of data, it represents less than one hour of low-quality video in digital compressed form. With MPEG-2 high quality video or digital HDTV, there is demand for much higher CD data capacity. New optical formats are being developed to increase both data capacity and speed of access (see Section 9.18.1).

7.2.1.5 Input/output For a computer to do useful work, it must communicate with its user, with printers, and often with other computers. This is done via hardware interfaces that are connected to the bus and addressed the same way as other bus devices. Video and audio output devices are in this class;

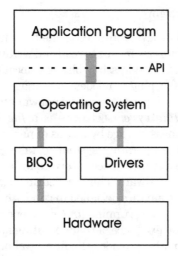

Figure 7.3 Levels of software

these functions have their own architecture that is discussed in detail in Chapter 12.

7.3 Software

A PC is useless without software. Software consists of a list of microprocessor instructions that tell the CPU what to do. Although all PCs have some built-in software (the BIOS), that doesn't provide any useful functionality by itself. Operating system software must be installed and executed to provide even the simplest user interface and application software must be added before the computer will actually do any work. These levels of software are shown in Figure 7.3 and are discussed in this section.

7.3.1 The BIOS

The permanent hardware units in a PC (keyboard, mouse, video display, mass storage, printer, etc.) are physically connected to the bus, but that doesn't tell the CPU how to access them. That is the purpose of the BIOS software, which knows the details of the hardware and provides a generic software interface to it. Different PCs may have different designs for their permanent hardware but each PC contains its own customized BIOS to make the different hardware look the same to the next level of software, the operating system.

7.3.2 The operating system

The BIOS provides a low-level interface to the built-in hardware of the PC. For example, it provides a raw means to read and write data from and to mass storage devices based on addressing physical locations on the medium. However, much housekeeping is needed to manage mass storage through physical locations and most programs would not want to deal with such detail. One feature of *operating system* software is a *file system*, which allows applications to access mass storage by means of named files and directories. It is the purpose of the operating system to take each of the hardware capabilities and create a standardized interface for application programs to use when accessing the hardware. This is called an *application programming interface* (API). All programs are designed to run with a particular operating system (this is often stated as running *under* the operating system) and the program depends on the existence of that operating system's API.

An operating system also provides the first level of *user interface*, which defines the way a user interacts with the computer. One kind of user interface is text-based (called a *command-line interface*)—the user types commands at a keyboard to control the computer. The video display in a command-line interface shows only text. Another interface is graphical (called a *graphical user interface*—GUI). With a GUI, the user sees pictures of objects (called icons) on the screen and he/she interacts with these by use of a pointing device such as a mouse, touch screen, or trackball. Of course, when text input is required in a GUI, the user still operates the keyboard. GUIs are discussed further in Chapter 14.

For example, the *DOS* operating system for IBM-compatible PCs provides a command-line interface. By installing the *Windows* software package on DOS, a GUI is added. (The *Windows 95* or *Windows NT* versions do not require DOS.) The IBM *OS/2* operating system and the Apple *Macintosh OS* both provide GUIs as their basic interface. In the case of OS/2, a command-line interface is available as an option, but the Macintosh OS does not have this option.

7.3.3 Drivers

The first-level interface to hardware provided by the BIOS only applies to the hardware capabilities installed in the basic computer. Many other types of hardware devices could be added to add other capabilities, or standard units may be replaced by different devices. In either case, the BIOS cannot know the details of the new or additional hardware. This is taken care of by an OS feature called *installable drivers*. Drivers are BIOS-level software specific to a device. The OS provides means to add such software so that the API can include the new or different devices. As shown in Figure 7.3, a driver can (if necessary) bypass the BIOS and go directly to its own hardware.

7.3.4 Application software

Specific computer tasks such as word processing, calculating, or drawing are accomplished with application programs. The operating system includes a *launcher* that will load and start applications. The user only needs to select the program and tell the OS to launch it. Each application usually has its own user interface, built in the style of the OS.

Application programs are created by programmers who write the program in languages such as C, Pascal, or Basic, which are designed specifically for application programming. The language provides a syntax that describes the application's structure and processes in a format readable by someone who knows the language. However, these languages are not understandable by microprocessors, which require the much lower-level and detailed syntax defined by their instruction set. The translation can be accomplished in two different ways.

Interpretation is a translation from programming language to machine instructions that occurs while the program is running. An interpreter program reads the programming language and generates instructions that are immediately executed by the microprocessor. This is a very convenient mode of operation, but it is inefficient because both the interpreter and the application must share the microprocessor's cycles. The result is that interpreted programs can be slow. For simple applications and fast processors, this is not a problem. However, complex programs, such as image processing or compression/decompression, will be noticeably slowed.

The other type of program creation is *compilation*. In this case, a compiler program is used during the programming process to convert the programming language code to a separate executable module that can then run by itself. This produces the fastest programs and most major applications are created this way.

7.3.4.1 Multitasking

There are situations where it is necessary to run more than one program at a time. Some operating systems support *multitasking*, which allows multiple programs to run "concurrently." Of course, if there is only one microprocessor available, only one thing can actually be happening at any instant, but computers are so fast that they can switch between two or more programs (tasks) so rapidly that the user thinks everything is actually concurrent.

A good example of the use of multitasking occurs in an application that requires concurrent video, audio, and other computer functions. A single program might be designed to handle these operations concurrently, but it is much better to have the operating system handle it for the program. If separate programs are written for the purposes, the OS can run all of them and manage the prioritization between them to optimally utilize the computer's resources to keep everything running smoothly. OS/2 and Windows NT are operating systems that support this. They also support another mode, called *multithreading,* which allows a single application to set up its own internal multiple tasks (threads) that the OS will manage concurrently.

7.4 Computer Video Formats

Video in a computer can be packaged as individual still images or as motion sequences.

7.4.1 Data files

A computer handles its data objects in the form of *files,* which are blocks of sequential data that serve a specific purpose and are stored together. For example, one file might contain a single still image. Such a file includes a header block that identifies the file contents and its format. The header is followed by the image data that lists all the pixel values for the image. If the image file is compressed, the image data does not represent the pixels directly, but rather it is the input to a decompression processor that will generate the actual pixel values.

Files can be stored in a computer's mass storage or they can be transmitted between computers over digital networks or digital communication links. Because most computers include mass storage (some computers on networks don't have their own storage; they rely on the network for it), transmission of data can occur at any time before the file is to be displayed. However, this assumes that the receiving computer has enough storage to accept all the data. With motion video, especially, the data may be too large to store locally. In that case, the system has to display the data as it is received.

7.4.2 Still image formats

The data for uncompressed still images consists of the values for all the pixels in the image. That sounds simple, but there are actually many ways to deliver the information. It becomes even more complicated when compression is included. As a result, there are more than 100 still image formats in use in PCs.

7.4.2.1 Color lookup tables The bits per pixel parameter (see Section 1.7.2) specifies how many bits are used to carry each pixel value. That has a direct relationship to how many different colors are available to display the image as shown in Table 7.1. When the number of colors is large, as in 15-, 16-, or 24-bpp, the pixel's bits can directly define the color, but for fewer colors it is advantageous to use a *color lookup table* (CLUT). In this approach, the pixel value is an index into a data table that holds the actual color value. The color values in the table can have more bits than the index does, so colors can be selected from a larger palette. For example, 8-bpp images can only use 256 colors at a time, but if the CLUT has 18 bits in each color value position, those 256 colors can each be selected from a palette of 262,144 colors.

When an 8-bpp image is captured, its 256 colors are chosen for the best match to the actual colors in the source image. There are a number of algorithms for that process and the results can be suprisingly good. The most difficult images to reproduce this way are ones that have a lot of smoothly shaded areas such as faces or desert landscapes. These may show *contouring*, which is the artifact that results from having too few colors to reproduce the shading.

7.4.2.2 Color dithering A further improvement in CLUT images is available with the use of *color dithering*. This is a process that depends on the viewer not being able to see individual pixels. Adjacent pixels are deliberately selected to have different values that the eye will combine to create the desired color. In the case of smooth shading, dithering will break up the color boundary lines so that contouring is not so easily seen. Dithering is most effective when the viewing conditions are such that the viewer cannot see individual pixels. When individual pixels are visible, the viewer will see the dithering and that spoils the effect.

7.4.2.3 Still image capture Still images can be captured from motion video or from hard copy. Motion video capture requires a video digitizing board. Hard copy capture can be done with a video camera and a digitizing board, but a better method is to use an *image scanner*, which is an electromechanical device that optically scans images directly into digital formats.

A block diagram for a video digitizing board that connects to a PC's system bus is shown in Figure 7.4. Such boards often provide both composite and component inputs as shown in the figure. Composite inputs are decoded to RGB and each component is separately digitized. In order to deal with capturing a single frame from live video, the board has its own buffer memory to temporarily hold the captured image. This is necessary because single-frame capture cannot be interrupted once it is started and the PC bus cannot be guaranteed to be free of interruptions.

Table 7.1 Bits per pixel vs. number of colors

Bits/Pixel	Colors
4	16
8	256
15	32,768
16	65,536
24	16,777,216

Figure 7.4 Block diagram of a PC video digitizing board

The user will usually want to view the captured image before it is saved to hard disk, which means that the image in the digitizer's frame memory must be displayed. This is easily accomplished when the digitizing functionality is built on the same board as a display adaptor. In that case, the temporary memory can be the display memory. If digitizing is on a separate board, the temporary image must be moved over the bus to the display adaptor for viewing.

When a captured image is to be saved on the computer's hard disk, it is sent over the PC bus to mass storage along with the appropriate header information for the chosen format.

The sampling clock for the A/D converters must be synchronized with the analog video input. Ordinarily, the sampling frequency is chosen so that the proper number of horizontal pixels are generated during the analog video active scanning interval as shown by Eq. (7.1):

$$f_S = \frac{f_H \, R_H}{C_H} \tag{7.1}$$

where: R_H = desired horizontal resolution (pixels)

f_H = *analog horizontal scanning frequency*

C_H = active horizontal scan fraction

For example, for 640 × 480 resolution and NTSC video, the sampling frequency is:

$$f_S = \frac{15{,}734 \times 640}{0.84} = 11.988 \text{ MHz}$$

Vertical pixel choices are limited by the number of active lines in the incoming video. For interlaced scanning, either one or two fields can be captured. If two fields are captured to get a full frame, there can be artifacts when the source image is moving. Notice also that a moving source image will be blurred because of the electrooptical integration that occurs in the camera imager.

7.4.2.4 Pixel formats CLUT pixels are pure binary numbers that index into the lookup table, but direct-color pixels must be formatted to define the encoding of the RGB values. This is usually done by allocating groups of bits for each of the three colors. For example, 24-bpp pixels use 8 bits for each color and each one is a pure binary number that specifies the intensity of that color. 16-bpp pixels cannot use the same number of bits for each color (16 does not divide evenly by 3), so green is usually given an extra bit because green is the visually brightest color. This gives a bit format of 6-5-5 for G, R, and B, respectively.

Some systems use 15-bit pixels (5-5-5) and the 16th bit is defined as a flag that goes along with each pixel. (Since all computer hardware is byte-oriented, it would be too awkward to handle actual 15-bit pixels.) The extra bit in each pixel can define (for example) how that pixel would be combined with the corresponding pixel of another image when two images are mixed. This idea is extended in some systems that actually use 32-bit pixels, which allows 8 bits to be assigned to the control of pixel-mixing operations. These extra 8 bits are called an *alpha-channel*.

7.4.2.5 Compressed still images The JPEG still image compression standard was described in Section 3.12.1. It is an international standard and is in wide use. However, it is quite compute-intensive and it takes several seconds to decompress an image in any except the fastest systems. This can be overcome with dedicated hardware, and some display adaptors have special hardware or a DSP chip to assist JPEG decompression.

Note that the processing time required for still image compression is usually not an issue because compression happens only once and at a time when a user is not waiting to view the image. Decompression, on the other hand, occurs during an application and the user has to wait until it is finished to see the image.

There are still image compression techniques available as options in some of the other image formats. Most of these are simpler than JPEG, they operate more quickly, but they do not offer as much compression at equivalent quality levels. The details of these techniques are beyond the scope of this book.

7.4.3 Motion video formats

Compression *must* be used for full-screen motion video on personal computers because uncompressed video data rates (see Table 1.2) are too high. However, unlike television, motion video does not have to be full-screen, in fact, most of the time it is not. Computers that have GUI operating systems (now most of them) define rectangular areas of the screen as *windows* and can display different information in each one. Windows can be made any size and, of course, the smaller they are the fewer pixels they contain. One way to get uncompressed motion video to work is to reduce its window in size until the data rate becomes manageable. Unfortunately, a window defined that way becomes pretty small and is often referred to as "postage-stamp video."

Another strategy for digital motion video is to reduce the frame rate. A frame rate of 30 per second is considered normal and can satisfactorily reproduce most motion effects. However, if there is not much motion in the scene (such as in a talking-head shot), the frame rate can usually be reduced to 15 frames/second. That is effectively 2:1 compression. Below 15 frames/second, any motion becomes jerky and is not satisfactory.

7.4.3.1 Motion video decompression

When video compression is used, the limit on playback window size is not the data rate but the decompression processing time. Decompression can be done in software alone or with special added hardware. With dedicated hardware, motion video can be full-screen, but with software-only decompression, video is usually limited to a partial-screen window. However, most of the industry focus is on software-only decompression and new algorithms are constantly being introduced and existing ones are being improved. The latest algorithms work well on the latest and fastest machines and, as the curves of future CPU speeds show, software motion video playback will only get better.

One advantage of software-only decompression is that the algorithms can be easily changed since that only requires a change of software, which is loaded for each video playback anyway. When an algorithm is in hardware, it is "cast in silicon" and cannot be improved by updating. The compression field is so dynamic that this is a serious problem. It can be partially solved by building into video display adaptors new functions that support generic motion video decompression tasks at a lower level than algorithms. That concept is being aggressively pursued by several of the display adaptor manufacturers.

7.4.3.2 Capturing motion video

The video digitizing adaptor shown in Figure 7.4 is not satisfactory for motion video capture. That is because it captures into a RAM buffer, which would grow to impractical size to handle any reasonable amount of motion video. Capture of motion video really has to be to hard disk to support video segments longer than a few seconds.

Hard disk capture changes the problem to a data rate one because hard disks are not fast enough for raw video rates. Table 7.2 gives some numbers for different window sizes and frame rates of uncompressed raw video. Considering that current hard drives cannot achieve sustained data rates higher than about 2 MB/second, it can be seen that only 160 × 120 (postage-stamp) video could be captured.

The solution to this problem is to perform limited compression on the fly during capture. (Full compression takes far too much processing time unless dedicated hardware is used on the capture board.) However, simpler algorithms can be used, but this is a disadvantage because the simpler algorithms may introduce too much quality degradation. It becomes a tradeoff and different boards have made different choices.

The best choice is to capture to hard disk at the highest quality possible and then perform full compression as a separate step. The captured frames can be read from the hard disk one at a time and the computer can take as much time as needed to compress them before writing them back to disk as the compressed bit stream. The only disadvantage of this is that the compression step may take a long time, such as an hour of processing to compress a minute of video. In most cases, this is not a problem, but if a lot of video is being handled, special equipment is probably indicated to speed up the compression step.

7.5 Computer Video Standards

Motion video algorithms for computers are still being developed at a great rate. The only official standard is MPEG, described in Section 3.12. MPEG is a high-performance standard but it is highly compute-intensive and definitely requires dedicated hardware for compression and preferably requires hardware for decompression. However, because of its worldwide standard status and its performance, it is the best choice for applications where the necessary hardware can be made available.

Other algorithms are proprietary to their owners, but this is not a large problem because they are mostly software-based and the software is widely available. It is not necessary to know the contents of a software package in order to use it. Current algorithms such as *Indeo*, *QuickTime*, *CinePak*, *Truemotion S*, etc., are successful for that reason.

Because computers are programmable, they can support many different video formats through software. A fully-documented standard for a video format is not required if the appropriate software is widely available. How-

Table 7.2 Motion video data rates for different window sizes (24 bpp)

Window Size (pixels)	Frames/second	Data Rate (MB/second)
640 × 480	30	27.6
320 × 240	30	6.9
320 × 240	15	3.5
160 × 120	15	0.9

ever, this leaves the industry at the mercy of video format distributors who may not always support the software they distribute. In that sense, true standards allow anyone to develop and distribute software (or hardware) that meets the standard. That is a step toward a more mature industry.

7.6 Digital HDTV on Computers

With the development and expected broad proliferation of digital video in the form of HDTV, there is sure to be an impact on the computer market. The pressure this is putting on the CD-ROM field has already been mentioned in Section 7.2.1.4. Digital HDTV will also up the ante for motion video quality on PCs, which will lead to demand for higher bus data rates, larger mass storage capacities, and higher CPU speeds. These are ongoing trends in the PC business anyway—digital HDTV is just one of the ways such higher powers will be used.

It seems reasonable that the computer industry should embrace the digital HDTV standards for higher-resolution computer screens. Hardware will be available in the mass market and it would cost far more to develop different, noncompatible standards. This is a scenario that will probably play out in the next few years.

Video Cameras

8.1 Introduction

The camera is the most basic element of a video system, and its most fundamental component is the *imager*, a device that converts the optical image of a scene into an electrical signal. Cameras are designed around the characteristics of their imagers, and a description of cameras must begin with imager technology. It has changed enormously during the past 45 years, and its development has included three breakthroughs, one of which—the shift from storage tubes to solid-state devices or CCDs—is just being completed.

8.1.1 A brief history of imagers

The first imagers were *instantaneous scanners*. They formed a raster on a live scene or film frame by scanning it with a bright spot of light. A phototube picked up the reflected or transmitted light from the raster and generated the electrical signal. The Nipkow disk, the earliest imager, was a mechanical scanner. It was succeeded by the Farnsworth image dissector, an electronic scanner that showed a brief promise. *Flying spot scanners* produce a raster on a kinescope and focus it on a frame of film. All instantaneous scanners suffer from inherently low sensitivity that make them impractical for live pickup.

The first breakthrough in imager technology came with the introduction of *photoemissive storage tubes* in which incident light from an image of the scene causes electrons to be emitted from a photosensitive surface in a pattern that corresponds to the brightness of the image. Because they store the light energy in an electric-charge image of the scene for an entire frame before it is removed by the scanning process, they have sufficient sensitivity for live scenes. Sensitivity is improved a thousand-fold or more compared to instantaneous scanners.

The *iconoscope*, developed by V. K. Zworykin of RCA prior to World War II, was the first practical *photoemissive* storage tube, and it proved the feasibility of all-electronic television. But it produced a noisy picture, it was difficult to operate (the term "shading" resulted from one of its operational problems), and its sensitivity was marginal for live pickup.

The iconoscope was succeeded for live pickup by the *image orthicon*, a product of World War II research. It had greater sensitivity, and it superseded iconoscopes except for film pickup. The image orthicon had good sensitivity, but it was also difficult to operate (although easier than the iconoscope) and its signal-to-noise ratio was marginal.

The defects of the iconoscope and image orthicon led to the second breakthrough, the replacement of photoemissive tubes with *photoconductive* types. Photoconductive tubes make use of changes in the apparent electrical resistance of a *photoconductor* when exposed to light. They are smaller, cheaper, easier to operate, and more noise-free than the iconoscope or image orthicon. The first commercial photoconductive tube was the *vidicon*, which was introduced in the United States for film pickup in the early 1950s. Its sensitivity was marginal for live pickup, but its use for film continued for more than thirty years.

Ten years later, a new photoconductive tube, the *Plumbicon* was introduced. It had sufficient sensitivity for live pickup while retaining the other desirable features of photoemissive tubes, and by the early 1970s it had replaced the image orthicon. Another photoconductive tube, the *Saticon*, was introduced in 1974.

As of this writing (1996), the third breakthrough is reaching fruition as photoconductive tubes have nearly been replaced by solid-state imagers, specifically *charge-coupled devices* (CCDs). This is another step in the replacement of vacuum tubes by solid-state devices in the electronics industry. Storage tube imagers are still in use, however, and this chapter describes them briefly because of their historical interest and to provide a frame of reference for the performance of CCDs.

Table 8.1 summarizes the introduction dates and technologies of the principal types of storage tubes and CCDs. The introduction dates are not precise but indicate the approximate time period when the imagers became standard commercial items.

8.1.2 Camera categories

There are two major television camera categories, live and film.

Live cameras generate video signals from the optical images of indoor and outdoor scenes. They are classified as studio, outside broadcast (OB), and portable. Studio cameras are designed to provide optimum performance in a controlled indoor environment, usually at the expense of size, weight, and

Table 8.1 Principal television imagers

Name	Introduced	Type	Application
Iconoscope	1939	Photoemissive	Film/live
Image orthicon	1946	Photoemissive	Live
Vidicon	1952	Photoconductive	Film/live
Plumbicon	1963	Photoconductive	Film/live
Saticon	1974	Photoconductive	Film/live
CCD	1980	Solid state	Film/live

portability. The performance of OB cameras often approaches or equals that of studio cameras, but the packaging of their components is designed so that they are *transportable* if not *portable*. Cameras designed for electronic news-gathering (ENG), electronic field production (EFP), or satellite newsgathering (SNG), provide portability, even to the extent of being hand held. In most cases, a portable camera is combined with a video recorder—the combination is called a *camcorder* (see Chapters 9 and 10).

Film cameras (called *telecine* cameras) generate television signals from film or slide images. The utilization of film as a source of video programming is diminishing as an increasing number of programs are produced and recorded on tape, a result of the improved performance of CCD cameras and recorders. Also the development of portable high-performance TV cameras and video recorders has caused a total shift from film to video tape for on-the-spot newsgathering. But both film and TV have advantages as production and storage media, and 35-mm film is still preferred by some for the production of prime-time programs and commercials (see Section 4.17.5). As a result, vital applications for high-performance telecines will continue.

8.1.3 Live camera configurations

The apparatus configuration required to produce a video signal from an optical image is known as a *camera chain*. For studio cameras, it consists of the camera head together with the rack or console-mounted unit or units that perform the control, signal processing, composite signal multiplexing, and monitoring functions. For OB cameras the control units may be transportable. ENG cameras may be self-contained or have a portable control unit that is used for set-up and optionally for operation.

Figure 8.1 is a functional diagram of a studio color camera chain. The basic functions of ENG and studio cameras are similar, but the ENG camera must have the ability to control operating parameters such as gain and sensitivity automatically because the operator cannot be required to make any technical adjustments. The camera is set up initially with the use of external monitoring equipment, and its stability and automatic features permit it to operate for considerable periods without manual change of the set-up adjustments.

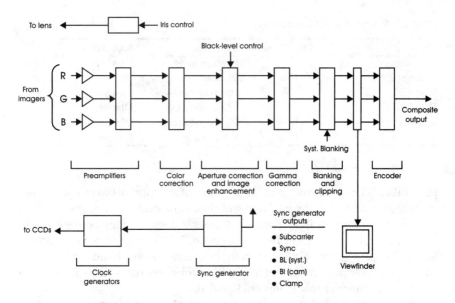

Figure 8.1 Analog color camera chain—functional block diagram

8.1.4 Film camera configurations

Film camera systems must cope with the incompatibility of the standard 24 frame per second film rate and 25 frames per second for most PAL systems or approximately 30 per second for NTSC. It was once believed that the 4 percent difference would be barely noticeable for either video or audio (except to trained musicians with absolute pitch). More recent European experience has indicated the difference is noticeable to listeners to dialogue because of the change in voice quality. To avoid this, digital algorithms have been developed to restore the audio to its original frequencies.

Since frame rate conversion was not mandatory in PAL systems, flying-spot scanners and continuous projectors were the preferred film pickup devices. When CCD technology was developed, the flying-spot was replaced by a single-line CCD device. These approaches provide a high-performance telecine system that is easy to set up and maintain.

For NTSC systems, the projector/camera combination must have the facility of converting from 24 to 30 frames, and a number of complex electronic and mechanical systems were developed to accomplish the conversion. They are now being replaced by digital frame store devices similar to those used for standards conversion (see Chapter 4). The signal is generated at 24 frames per second, converted to digital form, stored in a digital memory, and read out at 25 or 30 frames per second, thus making even the small correction necessary for PAL systems.

8.2 Photoconductive Storage Tubes

Although photoconductive storage tubes have been replaced by CCDs for most applications, they are described briefly here for historical purposes.

The *vidicon* was the first of a family of commercial photoconductive tubes that eventually included *Plumbicons, Saticons, Newvicons, Chalnicons,* and *silicon diode* tubes. The term "vidicon" is sometimes used generically to indicate any photoconductive tube.

The vidicon utilizes the photoconductive properties of antimony trisulfide, and it is an excellent tube for film pickup. However, when its operating conditions are adjusted for high sensitivity it suffers from severe lag or smear, and its use for live pickup is primarily limited to industrial applications.

The Plumbicon is a *heterojunction* photoconductive tube that uses a lead oxide semiconductor as the photosensitive material. It greatly reduced the vidicon's lag problem and was dominant for live pickups until it began to be replaced by CCDs.

The Saticon is also a photoconductive tube that uses a selenium alloy semiconductor as the photoconductive material. It was introduced later than the Plumbicon and enjoyed certain advantages although it did not replace the Plumbicon. Whereas the response of the Plumbicon to red light is higher after special doping of the semiconductor, the response of Saticons is higher in the blue end of the spectrum.

In addition to these three principal types of photoconductive tubes, there were others that were used for special purposes. They include the *Newvicon*, which uses zinc selenide as the photoconductor, the *Chalnicon*, which uses cadmium selenide, and *silicon diode* tubes. These tubes have greater sensitivity, particularly in the red and infra-red portions of the spectrum, and they are used mainly for surveillance cameras.

8.2.1 Photoconductive tube construction and operation

The construction of photoconductive storage tubes is shown in Figure 8.2. The faceplate, mounted on the front of the tube, is an optically flat glass disk with its inner surface coated with a thin layer or layers of photoconductive materials. A ring of conductive material is sealed to its circumference forming a structure known as the *target*. The target is sealed to the front end of the tube and a connection to the conductive ring is the *signal electrode*.

An *electron gun* at the rear of the tube generates a finely focused beam of low-velocity electrons that scans the target in a raster. An optical image is formed on the photosensitive surface of the target and produces a charge pattern having density variations corresponding to the brightness variations of the image. The scanning beam then converts this pattern to an electrical signal.

Faceplate with
Photoconductive
Layer

Electron Gun

Incident
Light

Signal Electrode

Figure 8.2 Photoconductive tube construction

8.2.2 Photoconductive Tubes for HDTV

HDTV puts severe demands on the performance of imagers, and the industry responded by developing storage tubes with significantly improved performance. One example was a photoconductive tube employing a HARP (high-gain, avalanche rushing amorphous photoconductor) target.[1] This tube offered a major improvement in sensitivity and aperture response as compared with Plumbicons and Saticons. The improvement was not great enough, however, to match the growth of CCD technology, however, and the HARP tube has not found wide application.

8.3 Charge-Coupled-Device (CCD) Sensors

The development of integrated circuit technology allowed the interconnection of thousands of tiny semi-conductors in a rectangular array to create imagers, known as *charge-coupled devices* or *CCDs*.[2] These have important advantages over storage tubes including wider dynamic range, absence of lag, absence of need for operator registration controls, longer life, and superior basic performance characteristics.

The use of solid-state imagers in television began on an experimental basis in the 1960s and was accelerated by the invention of the CCD in 1969. Cameras employing CCD imagers are now almost universally in broadcast and related television applications and also in nontelevision applications such as still imaging systems, electronic photography, and ultra-sensitive detectors for astronomy and military surveillance.

A CCD sensor is a pixel-sized solid-state photosensitive element that generates and stores an electric charge when it is illuminated. It is a building block for a CCD imager, which is a rectangular array of sensors upon which an image of the scene is focused. In most configurations, the sensor includes the circuitry that stores and transfers its charge to a shift register, which converts the spatial array of charges in the CCD imager into a time-varying video signal.

8.3.1 Sensor construction

CCDs have achieved a high stage of technical development as the result of intensive research and development efforts that cover the gamut of technologies from semiconductor physics to solid-state manufacturing techniques. A complete description of the wide variety of designs that these efforts have produced is beyond the scope of this volume. The description of a representative design follows, and the reader can consult the references for further information.

Four functions—sampling, photosensing, charge storage, and charge transfer—are performed by CCDs (see Figure 8.3). The construction of the sensor depends on the location of the elements that perform these functions as determined by the architecture of the imager in which it is used (see Section 8.4.3).

Figure 8.4 is a functional diagram of a CCD element of a type used in an imager with interline transfer CCD architecture. It includes the photosensor itself and the semiconductor circuit elements that control the storage and read-out of the electric charges that it generates. The functions of these elements are described below.

Figure 8.5 is a cross-sectional view of a CCD sensor showing the details of its construction and its complexity. This design is representative only, and the construction of different models varies widely.

The structures of CCD sensors are complex, and they present challenging manufacturing problems as a result of their small dimensions—each pixel is typically less than 0.001 inch on a side. The manufacturing difficulties are increased further by the fact that imagers include 500,000 or more sensors,

Figure 8.3 CCD operation

Figure 8.4 Functional diagram of a CCD sensor element

nearly all of which must work perfectly to avoid an unacceptable level of defects in the picture.

8.3.2 Sensor operation

The operation of this sensor and its associated circuitry can be understood by reference to a "map," Figure 8.6, of the electrical potentials of its components. Electrons freed by the photosensor during a television field are stored in a potential "well," that is collocated with the sensor and positively charged.

Figure 8.5 Typical CCD element cross section

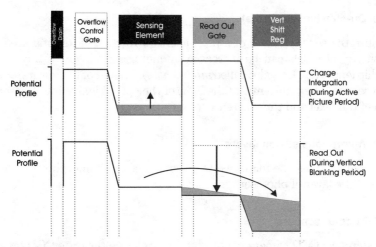

Figure 8.6 CCD potential map

If this portion of the picture is very bright, the well will be filled to overflowing, and the excess electrons will flow into a drain through the overflow gate as shown in Figure 8.7. At the end of the field, the potential of the read-out gate, which is slightly more negative than the overflow gate when closed, is lowered and the electrons pass into the shift register.

The operation of the sensor including the photosensor, its gates, its shift register, and its wells is controlled by its design, and commercial CCDs offer a wide variety of choices of performance specifications.

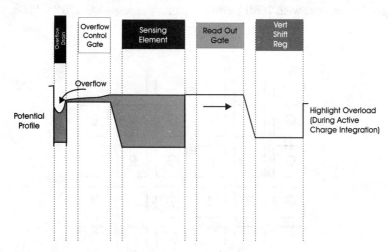

Figure 8.7 Operation of overflow gate

8.3.3 On-chip lenses (OCLs)

Because the sensitive area of a pixel is only a fraction of its total area, the sensitivity of CCDs can be increased by mounting a layer of tiny lenses, on-chip lenses (OCLs—also called *microlenses*), in front of its sensors, one for each sensor. They concentrate the light of the optical image into the sensor area of the pixel and thus increase its sensitivity.

8.3.4 Amorphous silicon layers

Sensitivity can also be increased by placing a layer of amorphous silicon in front of the layer of photosensors.

8.3.5 Tricolor sensors

The standard CCD camera configuration employs a separate CCD for each primary color (see Figure 8.1). For applications that do not require the highest picture quality, however, it is possible to achieve satisfactory results with single-chip cameras that employ single tricolor sensors.

The construction of a tricolor sensor is shown in Figure 8.8. Each pixel has three side-by-side photosensors, one for each primary color and each a part of a column of similar photosensors. The signals from these R, G, and B columns are combined to form the luminance and color difference signals as described in Chapter 2.

Since the columns are one-third the pixel width, the aperture response is excellent, although not as good as with a three-chip camera with spatial offset. For example, the specified limiting horizontal resolution of a commercial single chip camera with 768(h) x 494(v) picture elements is 470 TV lines. The

Figure 8.8 Tricolor CCD

specified limiting resolution of a three-chip camera, each with the same number of picture elements is 700 lines.

Similarly, the sensitivity of the single-chip camera is lower because of the smaller total sensor area and because the filters absorb more than 2/3 of the incident light. Again for example, the specified sensitivity of the single-chip camera described above is 2000 lux (lumens) at f/5.6 while the sensitivity of the three-chip camera is twice as great, 2000 lux at f/8.0. This is because the three-chip camera uses prism optical separation of the color components that does not absorb much light.

8.4 CCD Imagers

A CCD imager consists of a matrix of semiconductor sensors, one sensor per image pixel, mounted in a rectangular array on a silicon substrate. They develop electric charges that are proportional to their illumination by an image of the scene focused on the array. The charges are stored temporarily in potential "wells" (see above) and are then transferred to *shift registers* during the field blanking interval, which transmit them to the imager output in the proper time sequence to generate the video output signal. The purpose of the transfer function, the conversion of the electric charge pattern on a spatial array of photosensors into a video signal, is similar to that of a scanning beam in a photoconductive tube.

8.4.1 Comparison with photostorage tube imagers

The reasons for the replacement of the storage tube by CCD imagers can be seen by inspection of Tables 8.2 to 8.4, which compare the operating features and performance of typical imagers of each type. Most of the comparisons are favorable to the CCD.

8.4.2 Photosensor arrays

The number and dimensions of the photosensors in the imager array are basic factors in establishing its performance. The number of rows of sensors must at least equal the number of active scanning lines, and it is preferable that these numbers be equal.

The number of vertical columns of sensors establishes the imager's Nyquist limit (see Section 1.5), the television line number where horizontal aliasing begins. Aliasing will occur for horizontal frequency components in the image that exceed one-half the number of vertical columns of photosensors per picture height.

The pixel count, the number of rows (v) and columns (h), is a basic specification of a CCD imager. A variety of counts are available commercially. The dimensions of the sensing area and the number of active sensors in each

Table 8.2 Comparison of storage tubes and CCD imagers

Parameter	Comparison
Resolution	Tubes have somewhat higher resolution than CCDs designed for broadcast use. Resolution of HDTV CCDs equals or exceeds that of tubes.
Definition (MTF)	Major advantage for CCDs, especially those designed for HDTV.
Lag and burn-in	Major advantage for CCDs.
Smear	Major problem for CCDs but solutions exist.
Dynamic resolution	Advantage for CCDs because of absence of lag and the ability to include electronic shuttering.
Signal-to-noise ratio	S/N for tubes depends on pre-amplifier while noise in CCDs is generated internally (see Section 7.5.3); moderate advantage for CCDs.
Artifacts	Advantage for tubes.
Sensitivity	Advantage for CCDs.
Overload blooming	Advantage for CCDs.
Colorimetry	Improved red response of CCDs gives advantage to CCDs.
Sensitivity to vibration	Advantage for CCDs.
Registration	Advantage for CCDs.
Ease of operation	Advantage for CCDs.
Life	Advantage for CCDs.
Power consumption	Advantage for CCDs.
Size	Advantage for CCDs.

Table 8.3 Representative S/N ratios

Imager	Bandwidth	S/N (dB) Unweighted	Weighted
CCD—NTSC	8 MHz	62	68
Phototube—NTSC	8 MHz	56	64
CCD—HDTV	30 MHz	54	Not defined

Table 8.4 Representative imager sensitivities

| Imager | Normal Illumination (3200°K, 90% reflectance) | |
	Sensitivity	Exposure Index (see Sec. 8.5.4)
CCD	f/8, 2000 lux	537
Photoconductive tube	f/4.5, 2000 lux	170
Color film (medium speed)		100
Color Film (high speed)		500
	Minimum Illumination	
CCD	f/1.4, 7.5 lux, 18 dB gain	
Photoconductive tube	f/1.2, 10 lux, 24 dB gain	

dimension for imagers that are representative of designs for NTSC and SMPTE 274M HDTV service and which were the basis of the analysis in Chapter 6 are:

| System | CCD Dimensions (mm) | | Number of Sensors | |
	H	V	H	V
NTSC	8.8	6.6	818	513
HDTV 274M	14.0	7.8	1920	1080

The dimensions of the sensors are a trade-off between sensitivity and aperture response. Maximum sensitivity is achieved when the dimensions of the sensor approach those of a pixel. On the other hand, the aperture response can be increased significantly by reducing the sensor dimensions (see Chapter 6). Spatial offset (see Section 8.4.3), a commonly used technique for improving the aperture response while reducing or eliminating aliasing, requires the use of sensors having a horizontal dimension less than one-half the width of the pixel.

8.4.3 CCD transfer and read-out architectures

The essence of CCD imager operation is the conversion of a spatial charge pattern on an array of photosensors into a time-varying video signal as illustrated in Figure 8.3. The three commonly used architectures for this purpose are frame-transfer (FT), interline-transfer (IT), and frame-interline-transfer (FIT); they are shown in Figures 8.9 to 8.11.

8.4.3.1 Frame-transfer architecture Frame-transfer architecture is illustrated in Figure 8.9. The CCD structure is divided into three sections, a sensor array, a field storage array, and an output register. Charges are allowed to

Figure 8.9 CCD frame-transfer architecture

accumulate at each photosensor for a complete frame. During the vertical banking interval, a command from a clock causes the charges in each column of pixels in the sensing area to be shifted to a corresponding column in the storage area. This frees the sensors to accumulate charges from the illumination of the next frame. From the storage area, and again at the command of a clock, the charges are shifted one line at a time to the output register that generates the video signal.

The entire sensor area is available for the sensors, and frame-transfer imagers do not suffer the sensitivity loss of interline-transfer imagers due to masking a portion of the sensor area for the storage areas.

On the other hand, the frame-transfer architecture has the problem of smear—vertical streaks above and below bright spots in the image. It is caused when the charges in the image array move across the optical image during the transfer of charges at the end of each frame. This happens very quickly, but the sensor array elements pick up small spurious charges during the transfer that cause streaks in the picture that are most visible in picture highlights.

This type of smear can be prevented by a rotating mechanical shutter that blocks the light from the image during the transfer period. This is a very effective solution, but it has the problem of introducing a mechanical component in an otherwise all-electronic system.

Smear can also be prevented or reduced by the use of interline-transfer architecture (see below).

8.4.3.2 Interline-transfer architecture Interline-transfer architecture, which is commonly used in consumer cameras, is illustrated in Figure 8.10. The sensors in vertical columns of sensors are connected to storage elements (potential wells) in alternate columns. After exposure of a single frame and when commanded by a clock, the photo-generated charges in the pixels in each column are transferred along their rows to the neighboring vertical transfer register. This frees the sensors to accumulate the charges for the next frame as determined by their illumination.

The effective exposure time for each photosensor can be reduced by the use of a clock-controlled electronic shutter that transfers the electrons that are generated during only a portion of each frame. As with photographic cameras, this reduces the blurring of objects in motion but at the expense of sensitivity.

Figure 8.10 CCD interline-transfer architecture

Another clock command causes the charges to be shifted from the vertical to the horizontal register. From this register they are read out as the video signal.

In the basic form of this architecture, the vertical register columns must be covered with opaque masks and only 30 to 50 percent of the imager surface can be photosensitive. This is an inherent drawback of interline-transfer architecture.

8.4.3.3 Frame-interline architecture Frame-interline architecture, Figure 8.11, was developed to provide the sensitivity advantage of frame transfer, but with resistance to smearing from strong highlights without the necessity of a mechanical shutter. It is the preferred architecture in most professional cameras. The unique feature of frame-interline-transfer architecture is a row of selection gates between the image and storage areas. The gates are biased so that charges in excess of a predetermined level are drained from the system before being transferred to the storage area.

The sensors for CCD imagers with frame-interline architecture are more complex and hence such imagers are more costly.

Figure 8.11 CCD frame-interline architecture

Another problem with frame-interline architecture is that the signal from the output of every photosensor must be switched in sequence during each frame, and a very high switching clock rate is required—for the 1036 × 1920 CCD it is 74.25 MHz. The input impedances of the CCD circuits are capacitive, and driving them at this frequency takes considerable current. This problem can be alleviated by dividing the photosensor columns into two sections, each section including alternate columns. The sections are then clocked alternately at a frequency of 37.125 MHz, and their outputs are combined to produce a signal that is identical to that produced if all the photosensors were in a single group.

8.4.4 Spatial offset

Spatial offset employs a camera construction that establishes a special relationship between the positions of the columns of R, G, and B sensors in the three imagers as illustrated in Figure 8.12. The columns of green sensors are alternated with columns that contain the red and blue sensors. As noted above, this requires that the width of the sensors be less than one-half the width of a pixel.

The columns of green, red, and blue sensors are mounted on separate imagers, and they are registered by bonding them with extreme precision to the prism block in the camera (see Section 8.7.2). Such precision in manufacture is a challenging problem, but it has been solved, and no registration adjustments are provided. This greatly simplifies the set-up and operation of the camera—an important additional advantage of this technology.

The operation of spatial offset can be explained by noting that the signals from the R, G, and B imagers are added in the luminance channel, and because

Figure 8.12 Spatial offset CCD

of the spatial separation of the R and B from the G sensors the phase of their sampling sideband signals are reversed. If the sum of the amplitudes of the R and B sidebands equaled the G, there would be complete cancellation, and aliasing (which is caused by the overlap of the sidebands) would disappear. In practice, and depending on the color content of the scene, the cancellation is not complete, and some residual aliasing occurs at image frequency components between the Nyquist limit and twice this number, i.e., the sampling frequency.

The use of narrower sensors for sampling also increases the aperture response to the higher frequency components in the image and hence its definition.

8.5 CCD Imager Performance

8.5.1 Performance criteria and standards of comparison

Chapter 6 described the performance characteristics of television *systems*; this chapter describes the performance of CCD imagers, a principal system component.

Performance criteria described in Chapter 6 include definition or sharpness (as measured by the aperture response or MTF), resolution, signal-to-noise ratio, sensitivity, dynamic range and gamma, spectral response, aliasing, smear, lag, artifacts, and other defects. Since the performance of CCDs of different design and under different operating conditions varies considerably, the comparisons are representative rather than precise.

8.5.2 Image definition and resolution

Criteria for evaluating picture definition or sharpness described in Chapter 6 include the modulation transfer function or MTF (aperture response), limiting resolution, and the equivalent line number or N_e. The modulation transfer function is the most complete specification, but it is presented in graphical form that is not as convenient or as easily interpreted as a single number. Although N_e is a single number and is an excellent basis for analysis, it is not widely used. Limiting resolution is an incomplete specification, and it has the additional disadvantage of ambiguity in systems that involve sampling. These include all television systems in which the scanning process samples the image in the vertical dimension, and CCD imagers in which discrete photoconductors sample the image in both dimensions. For these reasons, the MTF is used in this chapter as the basic criterion for the definition of CCD imagers.

8.5.2.1 CCD MTF   Equation (6.5) for the MTF of a sampling process becomes Eq. (8.1) for a CCD imager.

$$\text{MTF}_{(\text{H, V})} = \frac{\sin x}{x} \tag{8.1}$$

where: $\text{MTF}_{(\text{H,V})}$ is the modulation transfer function in the horizontal or vertical dimension

$$x = \pi \, \frac{W_{(\text{H, V})} \, f_{(\text{H, V})}}{F_{(\text{H, V})}}$$

$F_{(\text{H,V})}$ = spatial sampling frequency along the horizontal or vertical dimension = the number of sensors per picture height

$W_{(\text{H,V})}$ = (sensor width or height)/(sensor spacing)

$f_{(\text{H,V})}$ = spatial frequency component of the image

Since the number of rows of sensors equals the number of scanning lines, $\text{MTF}_{(\text{V})}$ is equal to MTF for the scanning pattern.

The $\text{MTF}_{(\text{H})}$ for CCDs that are representative of NTSC and HDTV systems is shown in Figures 4.11 and 4.13.

8.5.2.2 CCD resolution   CCDs utilize a sampling process in both dimensions, and the ambiguities, which are inherent in specifying and measuring the *vertical* resolution of all television systems because of scanning line sampling, are present in CCDs for both vertical and horizontal resolution. In fact, there is a question as to the significance of limiting resolution as a criterion of CCD performance.

Nevertheless, the industry is so accustomed to using limiting resolution as a primary specification of imager performance that it is done for CCDs as well.

The *horizontal* limiting resolution of a CCD imager, which usually is specified as the resolution of the camera, can be approximated by multiplying the number of sensors per picture height by the Kell factor. For the 493 by 768 sensor described in Section 8.4.4, and with a Kell factor of 0.7, the vertical resolution would be 359 lines (513 × 0.7) and the horizontal, 429 lines (818 × 0.7 × 0.75).

The horizontal resolution can be nearly doubled by the use of spatial offset. For example, the specified limiting horizontal resolution of a CCD camera with spatial offset and 818 sensors in each line (Sony, Model DXC-M7) is 700 lines as compared with 404 without spatial offset.

8.5.3 Signal-to-noise ratio

Very little noise is generated within photoconductive storage tubes, and their signal-to-noise ratio is determined almost completely by wideband noise from the preamplifier. High signal-to-noise ratios (as compared with photoemissive tubes) are therefore characteristic of photoconductive tubes.

Unlike photoconductive tube imagers, most of the noise from solid-state imagers is generated within the device rather than in the external circuitry. This is indicated by the relative signal currents, which in CCDs are measured in microamperes as contrasted with nanoamperes in photoconductive tubes.

There are two basic noise sources in CCD imagers. The first is coherent, fixed-pattern noise caused by imperfections or irregularities in design or manufacture, e.g., nonuniformity in the dark current (see below) of individual sensors or cross talk from the high-speed clocking signals associated with the imaging process. This type of noise, while troublesome, is not inherent and can be greatly reduced or eliminated by suitable design and manufacturing practices.

The second noise source—thermally generated and random variations in the signal current—is inherent in all electronic communication systems; in CCD imagers it is generated as an inevitable by-product of the collection, storage, and transfer of electrons from and within the CCD. It is produced at three levels—the input circuitry, the transfer circuitry, and the output

Figure 8.13 Fixed-pattern and random noise (*SMPTE Journal*, July 1990, p. 534)

circuitry. This type of noise predominates except in the darkest areas of the scene as shown in Figure 8.13.

The magnitude of the thermal noise power in a CCD camera can be determined by measuring the *dark current* of the imager, its output when the lens is capped. The S/N ratio is calculated by comparing this power to the power level of the peak-to-peak signal, i.e., black level to reference white.

Noise power must be measured over the entire bandwidth of the camera, and since thermal noise is "white" (see Section 6.16) its magnitude is proportional to the video bandwidth. All else being equal, the noise power in an HDTV system with a 30 MHz bandwidth will be 3.75 times or nearly 6 dB greater than the noise in an NTSC system with an 8 MHz bandwidth.

Although thermal noise is always present in CCD imagers, its magnitude can be reduced by design advances. The S/N of CCD imagers has steadily improved in recent years, and it typically exceeds that of storage tube imagers. Representative S/N ratios of current imagers were shown in Table 8.3. These ratios exceed the standards specified in Table 4.3.

8.5.4 Sensitivity

There are a number of definitions of the sensitivity of television imagers.

The definition of greatest interest to scientists is the *quantum efficiency*, the percent of incident light quanta that create a useful output of electrons from the photosensor.

Two other definitions are more useful for engineering purposes:

1. The speed of the optical system required to produce an output signal of normal amplitude and signal-to-noise ratio with a specified scene brightness.

2. The minimum illumination required to produce satisfactory (but not optimum) signal quality with the camera lens set at maximum speed and full gain in the video channel.

An alternative criterion of sensitivity that allows comparison between electronic and photographic imaging is the *exposure index*, E_I, borrowed from photography.

E_I is given by Eq. (8.2), which combines the lens speed, exposure time, and illumination in a single number.

$$E_I = K \frac{f^2}{I \, X_T} \tag{8.2}$$

where: K is a constant = 280 when I is measured in lux

f = the lens stop

I = the illumination to produce a reference white signal

X_T = the exposure time

A high exposure index indicates high sensitivity. Increasing the photosensor size increases the sensitivity of CCD imagers because larger photosensors collect more light from the optical image. The increased sensitivity is achieved at the expense of a lower MTF at higher optical frequencies [see Eq. (8.1)] and the design of CCDs requires a tradeoff between sensitivity and image definition. Fortunately, recent improvements in CCD technology such as on-chip lenses (see Section 8.3.3), an amorphous silicon layer over the sensor array (see Section 8.3.4), and hyper-HAD (hole accumulated layer) sensors,[3] have increased sensitivity significantly, and this tradeoff no longer requires major compromises in definition.

The sensitivities of representative CCD and photoconductive imagers and of a medium-speed color film were shown in Table 8.4.

8.5.5 Spectral response

The spectral response of a CCD imager of recent design as compared with photoconductive tubes is shown in Figure 8.14. The CCD response exceeds that of phototubes except in the extreme blue end of the spectrum. Note that Figure 8.14 shows the response of the *imager*. The colorimetry of the *camera* is adjusted by placing red and infrared filters in its optical system.

The major difference between the spectral response of CCDs and phototubes underscores the importance of including the color temperature of the illumination as part of the sensitivity specification of a camera.

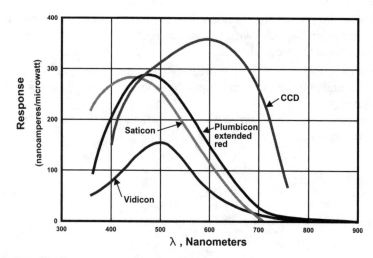

Figure 8.14 Spectral response of CCD imagers

8.5.6 Aliasing

Because CCD imagers employ sampling, aliasing is one of their inherent properties (see Section 6.8.4). A number of effective methods have been developed to eliminate aliasing with CCD imagers or to reduce it to an acceptable level. These include spatial offset (see Section 8.4.4) and optical filters that remove the high frequency components from the optical image (see Section 8.7.3). An even better approach is to use more pixels in the CCD (both horizontally and vertically, if possible) than that required by the scanning standards and digitally converting to the standards at the output. Of course, this will significantly increase the cost of the CCDs, which may not always be acceptable.

8.5.7 Transfer function and knee control

The transfer function (see Section 4.14.1) of CCD sensors within their dynamic range is essentially linear. Because of their excellent signal-to-noise ratio, which permits operation down to low light levels, the dynamic range of CCD imagers is excellent, approaching 1,000 to 1—more than is needed in a practical system. Further, because of their excellent signal-to-noise ratio, deliberate alterations of their transfer functions for gamma correction and range compression can be made without the introduction of excessive noise.

The SMPTE HDTV production standard 274M specifies a transfer characteristic that is linear at low light levels and has a gamma of 0.45 at higher levels, specifically:

$$V_C = 4\ L_C \text{ for } L_C \langle 0.0228 \tag{8.3}$$

$$V_C = 1.1115\ L_C^{0.45} - 0.1115 \text{ for } L_C \rangle 0.0228$$

where: V_C = the video signal output as a fraction of the reference white signal

L_C = the light input at reference white

Linear operation is specified at low light levels to limit noise amplification. Although it is not specified by this equation, it is normal practice to introduce signal compression at light levels above reference white to avoid system overload (see Section 8.8.4).

8.5.8 Lag

Lag usually originates in the camera imager and is caused by the failure of the scanning process to eliminate the stored charge pattern completely during a single scan. As a result, motion in the image, whether caused by movement in the scene or of the camera, leaves a trail.

Lag was a relatively unimportant problem with image orthicons but was a serious problem with vidicons. Unfortunately the vidicon operating mode that gives it maximum sensitivity (high target voltage) also produces maximum lag. As a result, vidicons had limited use for live pick-up, mainly in closed-circuit applications where image quality was not critical. For the pick-up of film programs, however, sufficient light is available to reduce the lag of vidicons to an acceptable level, and they were widely used for that purpose.

The Plumbicon and other later tube imagers had a much lower lag than vidicons at low light levels, and they were widely used for studio and field cameras. Even the Plumbicon, however, had a residual signal of about three percent on the third field following the removal of the image from the photosensitive surface.

One of the major advantages of solid-state CCD imagers is that they are relatively free of lag. This property, together with short exposure time, received a dramatic demonstration in the 1984 World Series when the stitches on a rapidly spinning pitched baseball were made visible.

8.5.9 Smear and other defects

Vertical smear can be caused by the process of transferring the charges from a CCD storage area to the signal output circuit (see Section 8.4.3.1). *Horizontal smear* is caused by inadequate frequency or phase response of the video channel. It is an extreme example of low definition, the edges are widened, and because of poor phase response the trailing edges of pulses may differ from their leading edges.

Blooming, *streaking*, and *comet tails* are primarily defects of storage tubes. They are related to lag in that they are spurious signals that result from the transducer's response to excessive light levels. They are caused by overloading of the photosensitive surface or storage mechanism with extremely bright highlights.

8.6 Camera Chain Configurations

8.6.1 CCD camera types

Four broad types of CCD cameras are available that vary widely in performance and cost.

The simplest and most inexpensive employs a single CCD chip for all three primary colors (see Section 8.3.5). It is used in the consumer market for camcorders and in lower performance professional applications.

The second type is designed for use in broadcast and other NTSC and PAL systems. It employs three CCD chips, one for each primary color and is available in both studio and field versions.

The third type, the camcorder, combines the functions of a camera and a recorder. Camcorders are available in a range of models, from professional to consumer, with great variations in features, performance, and cost. The consumer versions have enjoyed wide popularity.

The fourth type, which is now (1996) in the process of development, is designed for use in HDTV systems. Its performance requirements, and accordingly its price, are considerably higher.

8.6.2 Studio camera functional configuration

The functional diagram of a CCD studio camera is similar to that of a camera employing tube imagers as shown in Figure 8.1, except that the sync generator drives clocks rather than deflection yokes.

The lens, almost invariably a zoom lens, focuses an image of the scene on the R, G, and B CCDs through a beam-splitting prism that separates the light into its three primary components. After preamplification, which may include such functions as camera blanking, the outputs of the three imagers are subjected to additional processing functions including aperture correction and image enhancement (see Section 2.10), correction or *masking* by matrixing (see Section 3.8.2), and gamma correction. In addition to gamma correction, many cameras have an adjustable knee in their transfer characteristic that limits the output to a fixed maximum level to avoid overloading the system (see Section 8.5.7).

8.6.3 ENG camera configuration

The functional diagram of an ENG camera does not differ widely from that of a studio camera, but there are important differences in the locations of the functions.

An ENG camera must be self-contained and be able to operate without connection to the outside world except for its video output. This requires an internal sync generator and color encoder. (Sync generators of the 1940s required two racks of equipment!) It should have a *genlock* output so that other cameras can be synchronized with it. CCD cameras, with their lower power requirements, can operate from batteries. Of equal importance, they must be equipped with automatic features so that they can operate over a reasonable range of lighting conditions without adjustment for a reasonable period of time. An ENG camera operator should not be distracted from shooting to make adjustments. Manual control from a *camera control unit*, CCU, should be optional.

8.6.4 Camcorders

The development of camcorders, which combine the functions of cameras and recorders, has been made possible by major reduction in size and improved performance of these system components. Camcorders are described more fully in Chapter 10.

8.7 Optical Components

8.7.1 Lenses

The technology of television camera lenses has advanced almost as rapidly as electronic technology during the past 30 years. Thirty years ago, the quality of continuously variable focal length *zoom lenses* was not adequate for color cameras and early color cameras used fixed-focal length lenses mounted in a turret. The focal length could be varied only in fixed increments and by interrupting the signal. The widespread use of zoom lenses began in the early 1960s, and they have greatly enhanced the versatility of cameras for program production. Their quality has improved steadily, and their performance now approaches that of high quality fixed-focal-length lenses.

Zoom lenses are available for both studio and ENG cameras, with speeds ranging from f/2.2 to f/1.2 and focal length ratios ranging from 12:1 to 30:1. Lenses for ENG cameras are much smaller, in part because of lower optical speeds and in part by the use of smaller imagers.

The zoom function is controlled electrically, and the focal length can be adjusted either by the cameraman or remotely.

8.7.2 Optical beam splitters

Early color cameras employed image orthicon imagers that required a complex array of lenses and mirrors to split the light into its three primary components. The advent of the smaller Plumbicon tube made it possible to employ prisms rather than mirrors for light splitting, and their use was continued with CCDs.

Figure 8.15 shows a camera optical system. Prism light splitters simplify the optics, they occupy less space, and they are about one f/stop faster than mirror systems. The advantages of prism systems are so great that they are almost universally used.

The key technology of a prism beam splitter is the *dichroic* surface, which reflects light in one region of the spectrum and transmits the rest. The dichroic surface is coated with a monomolecular film that is ¼-wave thick at the center of the reflected spectral band. The phase of reflected light in this portion of the spectrum is shifted 90° in each of its two traversals of the surface and 180° upon reflection, thus emerging in phase.

In systems that employ spatial offset to reduce or eliminate CCD aliasing (see Section 8.4.4) the imagers are cemented to the output ports of the prism beam splitter. They must be located with extreme precision because the registration of the primary color images is determined by their relative

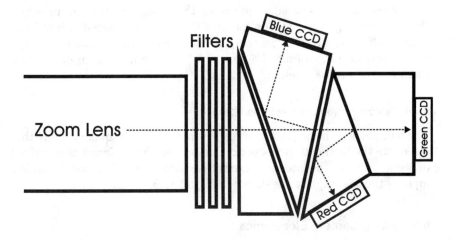

Figure 8.15 Prism light splitter

location. The location tolerance is measured in microns, and a sophisticated manufacturing technology has been developed to achieve it.

8.7.3 Color and neutral density filters

A variety of optical filters is used in modern color cameras; they include spectral trim filters, filters for daylight and sunlight, neutral density filters, and optical low-pass filters.

The spectral pass bands of the outputs of the dichroic prism in the beam splitter are broad, and it is necessary to employ trim filters to sharpen their edges. With trim filters, fairly close conformance to the standard primaries can be achieved by purely optical means, although additional electrical correction (masking) by matrix circuits is usually employed.

Filters are also used to adapt the colorimetry of the camera to the color temperature of the illumination. Typically, the color temperature of incandescent light is 3200°K, while daylight is 5600°K. The human eye with its marvelous power of adaptation automatically compensates for this over a wide range of color temperatures, and the difference between natural and artificial illumination is not noticed while viewing a scene directly. This power is lost when viewing a television image, and pictures of a scene illuminated by incandescent light from a camera adjusted for daylight will appear reddish. Color filters are used to compensate for differences in the color temperature of the scene illumination.

A typical complement of color filters adapt the camera to scene illuminations of 3200° and 5600°, incandescent and daylight. This complement permits the camera to be switched from daylight to artificial illumination without major readjustment of the controls.

Lens assemblies are usually equipped with neutral density filters to reduce the need for large variations in f/stop of the lens and hence depth of focus with widely varying scene brightnesses. They are rated by the amount of light that they transmit, typically 1/4 and 1/16. They are most often used under conditions of intense sunlight.

8.7.4 Optical frequency low-pass filters

The reduction of aliasing by the use of low-pass optical frequency filters that attenuate the high optical-frequency components in the image is described in Section 4.9.1.2. The filter is placed between the camera lens and the optical input of the beam splitter as shown in Figure 8.15.

8.8 Analog Camera Electronics

8.8.1 Preamplifier

The requirements for storage tube and CCD preamplifiers are quite different because of the different levels of signal currents at their outputs. The low output from storage tubes requires special considerations in the design of the preamplifier. By contrast, most of the noise in CCD imagers originates in the imaging device, and the output level from the device is high enough so that no special precautions are required in the preamplifier.

Camera blanking, the removal of the signal during the blanking intervals, can also be added at the preamplifier. This is augmented by the addition of *system blanking*, which determines the *set-up* or system black level, later in the chain.

8.8.2 Aperture correction and image enhancement

The gain-frequency characteristics of aperture correction and image enhancement circuits are described in Sections 4.9 and 4.10. Aperture correction increases the gain of the higher frequency signal components sufficiently to make the aperture response of the camera flat to about 250 lines in NTSC systems and 400 lines in HDTV systems. Additional correction, known as image enhancement, can be used in the 200 to 300 line region to increase the sharpness of the picture further, but it must be applied with care to avoid unnatural overshoots.

In one refinement of the aperture correction process, it is applied only in the midrange of the gray scale. This avoids degradation of the S/N in the dark areas of the picture where noise is most visible and also reduces a tendency for bright skin tones to appear rough.

It is important that the aperture correction gain enhancement be achieved without phase shift to avoid enhancing overshoots. The *transversal filter* is one means of achieving this.

Contours-out-of-green is another technique for increasing picture sharpness. Enhancement signals are generated from the green signal at each edge in the image by separating its high frequency components. The green channel is used to generate the enhancement signal because it has the best signal-to-noise level. The "white" contours—changes from dark to light—are added at the aperture correction circuit while the "black" contours are added after gamma correction.

8.8.3 Color correction (linear matrixing)

Electronic color correction is employed for two purposes–to correct residual errors in the colorimetry of the optical system and to introduce deliberate distortions of the colorimetry for aesthetic reasons. Electronic correction is accomplished by matrixing the color signal components, and is called *linear matrixing*.

8.8.4 Gamma correction and adjustable knee

Gamma correction is usually introduced in each of the three color channels to compensate for the high gamma of the receiver picture tube. This stretches the signal in the black area where noise is very visible, and the uncorrected signal must have a good signal-to-noise ratio. The SMPTE Standards 240M and 274M [see Eq. (8.3)] specify the gamma correction for HDTV production systems.

Additional gamma correction can be applied to the luminance (Y) channel of the encoder. This has the advantage that it does not upset the color balance of the signal.

To handle extremely large variations in scene brightness, some cameras provide an adjustable "knee" in the transfer characteristic. The characteristic follows the normal gamma curve up to an adjustable breakpoint, above which increases in illumination produce only a minor increase in signal level. This is a form of range compression (sometimes called *contrast compression),* and it avoids overloading the system in subsequent stages while retaining some detail in the brightest areas. Such a characteristic is illustrated in Figure 8.16; it follows the SMPTE Standard 240M over its normal operating range, but with a knee above which there is a major reduction in its slope. The knee is set slightly above reference white, and the image contrast is greatly compressed but not eliminated above this range.

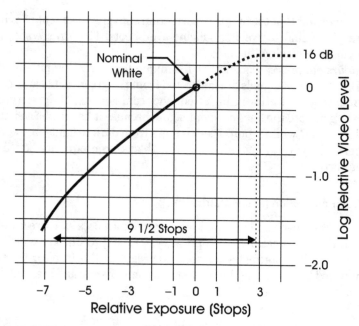

Figure 8.16 SMPTE 240M transfer characteristic

8.8.5 Sync, blanking, and clipping

The sync generator can either be external, as in a studio installation where cameras, recorders, and editing and processing equipments are synchronized by a common source, or internal, as in a self-contained field camera. The *genlock*, a device that derives its timing pulses and subcarrier from an external signal, is a third possibility.

The system blanking and clipping function establishes the final system black-level set-up. This is 8.5 percent for NTSC, the original monochrome standard, but 0 for PAL.

8.8.6 Encoder

Encoding, the conversion of the three primary color signals into composite NTSC, PAL, or SECAM signals, is the final function of the camera chain. The characteristics of the encoded signals were described in Chapter 2.

Figure 8.17 is a block diagram of an NTSC encoder. The gamma corrected primary color signals, E'_R, E'_G, and E'_B, are passed through matrices to form the E'_Y, E'_I, and E'_Q signals.

Sync is added to the E'_Y signal to form the luminance signal.

After filtering and the necessary delays, the E'_I and E'_Q signals modulate subcarriers, which are delayed by 57° for the I component and 147° for the Q, with balanced modulators to generate the color subcarrier. The color subcar-

Figure 8.17 NTSC color encoder

rier is combined with the Y' luminance signal to form the composite color signal.

The burst is also added by keying the subcarrier with the *burst gate* signal. The subcarrier can be generated by an atomic clock, a temperature controlled crystal oscillator, or derived from an incoming signal with a genlock.

8.8.7 Viewfinder

Electronic viewfinders have been a standard feature of studio cameras from their earliest days, and they are functionally superior to the optical viewfinders on film cameras. Viewfinders on modern cameras provide a number of useful features in addition to their basic function:

- Cursor—A remotely controlled cursor provides a means of communication between the director and camera person.
- Safety zone—An area encompassing approximately 90 percent of the image marking the useful area of the picture.
- External video connection—A connection that makes it possible for the camera person to see the output of the camera in relation to other signals. Warning and status signals can also be displayed.
- Color—Viewfinders are now available in color.

8.8.8 Triax cable

The first color camera heads were connected to the control unit with three large and heavy cables. With the smaller size, lower power requirements, and greater stability of solid-state components, and the use of photoconductive and CCD imagers, it has been possible to concentrate more of the circuitry in the camera head and reduce the number of connections between camera head and camera control unit. Thus, all the video, control, and power connections can be transmitted in a single small cable. The control signals are digitally multiplexed, and the video circuits are transmitted by subcarriers as shown below:

Signal	Transmission Mode
R video	AM subcarrier—8.33 MHz
G video	AM subcarrier—24.98 MHz
B video	AM subcarrier—41.64 MHz
Sound channels	FM subcarriers—1.7 MHz, 2.7 MHz
Viewfinder back video	FM subcarrier—59 MHz
Cue channel	FM subcarrier—70 MHz
Control channels	PCM

As the signal electronics of cameras becomes digital, it is possible to perform all triax multiplexing digitally and some newer cameras are doing that. Fiber-optic cable has also been introduced for camera cable applications.

8.9 Digital Camera Electronics

The camera signal processing described in Figure 8.1 and Section 8.9 includes operations that deal with sensor limitations, scene characteristics, and general matching of the signal to the properties of the system that follows. They are fundamental processes that must be done in any camera, whether analog or digital. The digital revolution, which is making radical changes in every aspect of video technology, has now come to camera signal processing.

Digital processing offers a number of significant advantages and it is replacing analog processing in more demanding applications. These advantages include:

- Greater stability and repeatability of adjustments. The drifting of analog circuits is a long-standing problem (although there have been great improvements in circuit stability in recent years). With digital processing, circuit parameters are stored in a memory and stay constant for any desired length of time. If it is necessary to make changes, the memory is updated with the new parameters.
- Precision of camera alignment and matching. Cameras can be aligned and matched precisely by equalizing their parameter values.
- Range and flexibility of camera adjustments. Camera parameters can be adjusted over wider ranges and with greater flexibility for optimum control of image quality.

- Highlight handling.[4] The output of CCD imagers from highlights can be 600 percent or more of the reference white level and it is desirable to reproduce image detail over this entire range. With analog processing, the signal is squeezed into the available gray scale range by the brute force combination of compression and clipping. This results in severe loss of detail in the image highlights. With digital signal processing, detail can be reproduced over the entire range with a combination of analog and digital compression. For example, the CCD signal from a 600 percent highlight can be compressed in the analog preamplifiers ahead of A–D conversion to about 220 percent of reference white. This is sometimes called "pre-knee" compression. This signal is then converted to a 10-bit digital signal which can be further compressed by variable amounts over the contrast range—for example, a high degree of compression can be applied to signal levels between 130 and 220 percent of reference white and a lesser amount of compression can be used at lower signal levels.

8.9.1 A–D conversion

It is desirable to do all camera signal processing digitally, especially because operations that are often handled separately in analog circuits can be done all at once in a digital circuit. Therefore, A–D conversion should be done as early in the system as possible, usually in the R, G, and B channels right after the preamplifiers. Then, all subsequent processing is digital, including the encoder. When the rest of the system is analog, a single D–A converter after encoding provides the proper analog output. Of course, if the rest of the system is digital, there is no need for any conversion at this point.

The quantization levels in camera signal digitization should be consistent with the signal-to-noise ratio of the CCD sensors used so that the digital channels will handle the full capability of the sensors. For modern broadcast camera sensors, this means that 10 to 12 bits/pixel must be used. This is true even though the dynamic range may later be reduced to suit the capability of displays, where 8 bits/pixel is usually enough.

8.9.2 Basic digital processing

Figure 8.18 shows a digital processing system for a fully digital camera.[5] This may be compared to Figure 18.1 for an analog camera.

A–D conversion occurs immediately after the preamplifier in each channel. Then black shading is performed. This is an automatic process that involves capping the camera lens and reading the sensor black signals into a frame memory for each channel. The contents of that memory is then continually read out to correct the black level of each pixel while the camera is in operation.

Figure 8.18 Block diagram of an all-digital camera processor

The next step is to set color balance by adjusting the gain of each channel. This is performed by use of a RAM *lookup table* that contains the desired pixel value for each of the possible quantizing levels (a 10 bits/pixel system would have $2^{10} = 1024$ locations in this memory). In a lookup table, the input pixel values are used as the address into the memory and the memory read values are the output. The values in the RAM are calculated and updated by the system CPU during blanking intervals. Adjustments are thus immediately applied to the system.

Contrast compression and gamma are applied in the same way through a RAM lookup table. In this case, the RAM values represent the desired nonlinear transfer function. Again, the system CPU calculates the curves and manages the updating of the RAM.

Aperture correction is applied by extracting signals before the nonlinear processing, creating the appropriate correction signals, and adding them to the channel after the nonlinear processes. This is to avoid enhancement of noise in the nonlinear processes. Aperture is provided both horizontally and vertically, and various algorithms can be implemented. For example, in an NTSC camera, the contour signal can be dynamically derived from either the red or green channel depending on which has the greatest amount of detail present. This avoids the problem of loss of red detail when contours are derived from green only, while still not enhancing registration or lateral chromatic aberration errors.

The presence of the system CPU and its bus allows all control functionality and algorithms to be implemented in software. This provides the same flexibility for offering options or updating as enjoyed by personal computer users.

8.10 Camera Operating Controls

Early color cameras had an enormous array of operating controls, and many stations employed a video operator for each camera to keep it in adjustment. The advent of photoconductive storage tubes, CCDs, solid-state circuitry, automatic controls, and general advances in technology have greatly simplified the control problem. Automatic exposure controls that adjust the lens aperture make it possible for the camera to operate over reasonable ranges of scene illumination without adjustment. Digital technology has increased the stability of control circuits so that they do not require frequent adjustment. As a result, hand-held ENG cameras frequently operate without any readjustment of the controls for considerable periods of time and multicamera studios can now be handled by a single camera operator. Digital technology has also made it possible to put a number of preset control positions in a memory to be recalled at will.

On the other hand, the ever-increasing demand for the highest possible picture quality has necessitated more sophisticated control of the cameras' adjustments, including precise camera matching. Further, the adjustments may be dictated as much or more by aesthetic judgment as by rigorous technical considerations. The adjustment of the controls by the appearance of the picture rather than by objective test procedures is called *painting*. Needless to say, a carefully adjusted color monitor must be used as the reference.

Table 8.5 shows the operating controls (as opposed to set-up controls that would be adjusted during routine maintenance procedures) for a typical studio camera. Function controls that have the option of presetting or automatic operation are indicated. The "painting" category includes adjustments that may be influenced by subjective judgments as well as objective measurements.

8.11 Camera Specifications

Table 8.6 is a tabulation of published specifications that are representative of current broadcast studio cameras.

Table 8.5 Camera operating controls

	Options	
Item	Automatic	Preset
Basic Picture Quality		
Master gain		X
Master black		X
Exposure		
Iris	X	
Color filter		
Neutral density filter		
Registration		
(not required with CCDs)		
Painting		
White balance	X	X
Black balance	X	X
Flare compensation	X	X
Shading		X
Aperture response		X
Image enhancement		X
Gamma adjustment		X
Knee adjustment	X	X
Color balance		X

8.12 Television Film Camera Systems

8.12.1 A brief history

Before it was possible to record video signals on magnetic tape, film was the only practical medium for the storage of television program material. Television film cameras and their companion projectors were critical components of television station equipment complements, and in many stations most local programming originated on film. The tape recorder, first introduced in 1956, provided an alternative medium with the additional feature that recordings could be made easily and quickly by individual stations. The advantages of the tape format increased steadily as progress was made in recording technology, and a transfer from film to tape is often made before its contents are put on the air. Television film cameras are now used for this purpose as well as to generate signals that are broadcast directly. Film continues to be an important production and storage medium, and film cameras will continue to be an essential part of the television industry's equipment complement for the foreseeable future.

Table 8.6 Representative published camera specifications

Sensitivity (Section 8.5.4)	
Plumbicon	f/4.5, 2000 lux
CCD	f/8, 2000 lux
Minimum Illumination (Section 8.5.4)	
Plumbicon	10 lux, f/1.2, 24 dB gain
CCD	7.5 lux, f/1.4, 18 dB gain
Optical Filters (Section 8.7.3)	
Color	3200°K, 5600°K
Neutral density	¼, 1/16
Sensor Array (NTSC) (Section 8.4)	
Total	818(h) × 513(v)
Active	768(h) × 493(v)
Architecture	Spatial offset
Horizontal Resolution (Section 8.5.2.1)	
Plumbicon	700 TV lines
CCD	700 TV lines
Signal-to-noise Ratio (Unweighted)[1] (Section 8.5.3)	
Phototube	56 dB
CCD	60 dB
Registration (Section 8.4.4)[2]	<0.05%

1 It is customary to specify the unweighted signal-to-noise ratio with gamma correction, aperture correction, and image enhancement removed. This assures that specifications are on a comparable basis.

2 This means that the maximum separation of the images of the three colors is 0.05 % of picture height or about 0.25 television lines (NTSC).

8.12.2 Frame rate conversion

In countries where the standard TV frame rate is 25 per second, film frame rate conversion is often not used. The film is simply speeded up by 4 percent from its normal rate of 24 per second. The difference in the picture rate is barely noticeable, and the increase in sound pitch is noticeable only to the relatively small number of individuals who have absolute pitch. However, with digital technology, frame rate conversion is easy to accomplish so that even this small discrepancy can be corrected.

In 30 frame-per-second countries such as the United States, the difference in frame rates is too great to be ignored, and standards conversion is necessary. The conversion of the 24 frame-per-second film frame rate to the 30 frame-per-second TV rate is a basic function of film camera systems in these countries.

Standards conversion was originally effected by (1) the use of a flying-spot scanner with a projector that generated a continuous image of film frames that lap-dissolve into each other; or (2) by a storage tube camera with an intermittent projector employing a 3:2 pull down.

The most commonly used film systems employed 3:2 projectors and storage tubes, usually vidicons or plumbicons. The film frames were pulled down at alternate intervals of 2/60 and 3/60 seconds for a total of 1/12 second or two film frames for each cycle. The storage capability of the camera made it unnecessary for the exposure and read-out to be simultaneous, and five television fields were generated from the images of two film frames, thus effecting the rate conversion.

8.12.3 The CCD/continuous-projector system

Two major technical developments brought a third system into use that will probably replace the others. It combines a continuous projector, a CCD camera, and a digital standards converter (see Chapter 4).

The continuously moving film is illuminated by a light source, and its image is projected onto a horizontal row of red-, green-, and blue-sensitive solid-state sensors arranged in triads, one triad for each pixel. Horizontal line scan is accomplished by sampling the sensor voltages sequentially. Vertical scanning is accomplished by the motion of the film. The output line-scan signals, which are generated at the 24 fps film rate, are converted to a digital format and stored in a memory. Standards conversion is accomplished by reading them out of the memory at the desired television frame rate.

The functions performed by the electronics for a CCD film camera are similar to those for live cameras but with some modifications necessitated by the line-scan format and standards conversion.

References

1. M. Kurahigi, N. Egami, K. Tanioka, and K. Shidara, "Super-Sensitive HDTV Camera Tube with the Newly Developed HARP Target," *SMPTE J.*, July 1988.

2. Koichi Sadashige, "An Overview of Solid-State Technology," *SMPTE J.*, February 1988.

3. L. Thorpe, "Focus on CCDs," Part I to Part V, *TV Technology*, September–December 1987, February 1988.

4. L. Thorpe, E. Tamura, and T. Iwaski, "New Advances in CCD Imaging," *SMPTE J.*, May 1988.

5. Leacock, Tom, et al. "An HDTV Digital Camera Processor," *SMPTE J.*, pp. 580–585, September 1994.

Professional Video Recorders

9.1 Introduction

Video program production as it exists today would be impossible without video recorders. This chapter covers some of the fundamental technologies used in nearly all video recorders. It then concentrates on high-end recorders used in broadcasting and program production.

9.1.1 A brief history of video recording

For more than ten years following the introduction of commercial television at the end of World War II, the problem of developing a satisfactory video recorder baffled the industry's major research laboratories. Such a product was desperately needed, but the technical difficulties proved too daunting until 1956. In that year the situation changed dramatically when a small group of young engineers at the Ampex Corporation announced a combination of technologies that provided a satisfactory magnetic tape recording solution called the *quadruplex* recorder.

The key component of the quadruplex recorder was the headwheel, a wheel about two inches in diameter with four magnetic recording heads spaced 90° apart around the rim. The direction of tape travel was parallel to the axis of the wheel and the 2-inch-wide tape was partially wrapped around the rim of the wheel. The tape was pulled past the wheel at a speed of 15 inches/second, and most of the head-to-tape motion was imparted by the transverse rotation of the wheel at a rate of 240 revolutions/second. The rotation resulted in a head-to-tape speed of 1,560 inches/second, far higher than would be practical for a longitudinal recorder. The combination of the rapid transverse motion of the head from the wheel's rotation and the much slower longitudinal motion of the tape produced a series of recording tracks, almost normal to the tape edges and occupying a time interval of approximately 16 scanning lines.

Quadruplex recorders were capable of producing excellent pictures when skillfully maintained and operated, but they were subject to a number of problems that led to a rapid deterioration of picture quality if maintenance was less than perfect. The most serious was "banding"—variations in the appearance of the 16-line bands recorded by adjacent heads due to small differences in their frequency or phase response. In addition, the quadruplex format lacked flexibility and could not be used for slow-scan or stop-frame applications. Aided by year-by-year design improvements, the quadruplex recorder had a long life, and it was in use for more than twenty years until the late 1970s.

However, another type of rotary-head recorder remained in the background—the *helical-scan* format. In the 1970s, these recorders were introduced into consumer and semi-professional markets and became successful.

The helical-scan format solved many quadruplex recorder problems, but it introduced new ones. The most serious was its *time-base instability* resulting from the long recording tracks on the elastic tape medium. This instability could be tolerated in consumer-level recorders but not in complex high-performance broadcast systems. The problem was eventually solved by the digital time-base error corrector, which provided an effective and economical means of stabilizing the time base. (Analog time-base correctors were used in quadruplex recorders, but that technology could not achieve the range of correction required for helical-scan recording.)

After digital time-base correction was introduced, the replacement of quadruplex by helical scan was almost inevitable, even for the most demanding broadcast and teleproduction applications. The transition was not sudden, owing to the huge libraries of tapes recorded in the quadruplex format, but it was nearly complete by the mid-1980s. As of this writing (1996), helical-scan recorders are almost universally used.

The helical-scan format is remarkably versatile, and variations of it now exist for the entire gamut of video recorder uses, from inexpensive mass-produced VCRs to the most complex and sophisticated broadcast and teleproduction systems.

9.1.2 Types of video recorders

The video recording market has grown to include a broad range of products with a broad range of applications. One way to classify the products is by their end use:

Home

Education and training

Broadcasting and production

Multiformat production

These categories are listed in order of signal performance capability, which, not surprisingly, is in inverse order of price. However, there are other measures of system capability besides price–performance and these mean that even the lowest-price units require extremely sophisticated design and production. The following paragraphs discuss each of the categories above.

9.1.2.1 Home video recorders The home environment is not friendly to sophisticated electronics. Any home product must be completely reliable without the help of any operator skill and it must do that over a period of many years. If it does not, the consumer faces a difficult and expensive service problem, which causes most people to give up and the equipment ends up in a closet. If this happens too soon in the product life, the consumer will feel taken and the product will quickly get a bad reputation and will not sell. Reliability is the most important design objective after price.

The signal performance of a home recorder can have some compromises compared to other categories of video recorder. The pictures must look good and the audio must sound good, but exact performance numbers are secondary. What this means in terms of numbers is shown in Table 10.1.

9.1.2.2 Recorders for education and training This market is sometimes called "semi-professional" and includes other classes of use that require the same attributes. These users can afford no compromise in reliability or ease of use but they require higher signal performance. That is because recording must be good enough to support the production-postproduction style of program creation and tapes will be duplicated to multiple users. At least three generations of recording will often occur in producing and distributing tapes. Other features such as computer control interfaces also become important to these users. At the same time, the market can only afford modest price increases compared to the home market.

9.1.2.3 Broadcast recorders Signal performance comes first with broadcast video recorders. Standard specifications must be met and elaborate system integration features are required. Skilled technicians will be available, but the need for their time must be minimized. For all this, the broadcaster will pay significantly more to obtain the high-performance reliable equipment that they need. Broadcast equipment prices are often 10 or more times higher than the lesser markets—this is a result of the more exacting requirements and also because the number of products sold will be much lower.

9.1.2.4 Production master recorders Producers of high-quality programs are interested in delivering their programs in as many formats as possible in

Figure 9.1 Magnetic recording

order to achieve the widest distribution. Conversion to other analog video formats may be required, such as going from NTSC to PAL (or back), video disc, motion picture film, or any other video format. Since the cost of recording is small in the total cost of the production, producers will pay to guarantee that the recording medium will not limit distribution choices. Component digital recorders (see Section 9.11) were first developed with this kind of use in mind.

Production recorders have the highest possible performance and are capable of as many generations as required for any type of production. This chapter discusses the technology of broadcast and production recorders; Chapter 10 will cover the home and educational products. However, there is a lot of technology overlap between the classes, so there are many cross references between these two chapters.

9.2 Principles of Magnetic Tape Recording

Magnetic recording is accomplished by providing a means of relative motion between a plastic tape that has a coating of magnetic material and a *head* structure that is capable of generating a magnetic field for recording and sensing a magnetic field for playback. This is shown in Figure 9.1. The signal current passes through a winding around the core of the record head assembly, which provides a magnetic circuit for the lines of force that are induced by the signal current. The lines of force, whose density depends on the magnitude of the signal current, fringe into the tape at the low perme-ability gap, leaving a pattern of alternating N–S poles on the magnetic surface of the tape. The pattern is called the record.

Figure 9.2 Record zone. (*From K. Blair Benson,* Television Engineering Handbook, *McGraw-Hill, New York, 1986, Fig. 15.39.*)

Figure 9.2 shows the detail of the magnetizing force, H, in the vicinity of the pole-tip gap. The solid curve is the locus of points where H equals the coercivity of the magnetic tape coating, H_C. The coercivity is the value of H that is required to magnetize a tape or to return its magnetization to zero. The area within the contour $H = H_C$ is known as the record zone, and no magnetism is produced outside this zone. The depth of magnetization and the diameter of the zone is less for high-coercivity tape, and this results in the ability to record shorter wavelengths and higher frequencies.

The reverse process occurs on playback. Magnetic lines of force from the pattern on the tape pass into the core of the head and induce a voltage in the winding that is proportional to the rate of change, the derivative with respect to time, of the magnetic field strength.

$$e = C \frac{dB(t)}{dt} \tag{9.1}$$

where C is a constant

$B(t)$ is the magnetic field strength across the gap

If $B(t)$ is a sinusoidal function, the voltage is proportional to the frequency and it increases by 6 dB per octave until effects described in Section 9.7.1

reduce the efficiency of the energy transfer and the response declines with increasing frequency.

9.3 Video Head Design

Video recording imposes stringent requirements on the performance of the video record/playback heads, and improvement of their design has been an ongoing effort that has continued during the past 35 years. In combination with improvements in the tape itself, head development has been one of the most important factors contributing to the steady improvement in recorder performance.

Among the requirements that video recording heads must meet are the following:

1. They must be able to operate effectively at very high frequencies—10 MHz and upward for FM modulated analog signals and 108 Mbs and upward for PCM modulated digital signals. Among other constraints, this requires that the gap between pole tips be somewhat less than the wavelength λ of the highest recorded frequency f as calculated by Eq. (9.2):

$$\lambda = \frac{\text{head - to - tape writing speed}}{f} \tag{9.2}$$

    Typical modern recorders operating at an FM carrier frequency f of 8 MHz and a writing speed of 1000 inches/second have a recorded wavelength of 125 μinches (0.000125 inch). The head gap length must be smaller than this value and must be accurately maintained. The mechanical precision required is awesome.

2. The pole tips must be sufficiently narrow to record and play back on very narrow tracks to provide a high recording density. Typical present-day values range from 0.001 to 0.005 inch.

3. The head material must have high *permeability*, the ratio of the magnetization flux density, B, to the magnetizing force, H, so that a satisfactory flux density is induced with a modest magnetizing force.

4. They must have a high ohmic resistance to minimize losses from eddy currents.

5. They must be capable of being machined or otherwise formed with very close dimensional tolerances.

6. They must withstand the mechanical environment of heat-to-tape contact and provide a long service life. This means that they must be resistant to wear from abrasion and to mechanical damage from cracking or chipping.

Figure 9.3 Magnetic recording head. (*From K. Blair Benson,* Television Engineering Handbook, *McGraw-Hill, New York, 1986, Fig. 15-43.*)

The major research and engineering effort that has been devoted to head materials and designs has resulted in an array of different designs, many of which have proprietary features.

Because the requirements for pole-tip and core-structure materials are different, many heads have a composite construction with one material for the pole tips and another for the core. Figure 9.3 shows the essential features of a typical composite head design. It has a supporting core structure to which pole tips are bonded. As with the resistance of electric conductors, the reluctance of magnetic conductors is inversely proportional to their cross section, and this is large except in the area of the pole tips. The pole tip is narrow to generate a narrow recorded track.

It is also possible to produce heads from a single material that combines suitable mechanical and magnetic properties. The material selection is more difficult, but manufacturing becomes easier and this approach is used in most head designs for high-volume production.

In composite head designs, the core material is usually ferrite, a mixture of iron oxide (Fe_2O_3) and manganese, nickel, or zinc, which has high permeability and resistivity. Ferrites come in two varieties, hot-pressed and single crystal. Single crystal ferrites can be machined or otherwise shaped more accurately, but the crystallographic orientation must be carefully controlled.

The pole-tip material has the additional requirement that it must be resistant to wear by abrasion and mechanical damage. This was a problem

Table 9.1 Magnetic properties of head materials

Material	Permeability*	Resistivity ($\mu\Omega$/cm)†
Alfesil‡	100–1,000	90
Ferrite		
Hot-pressed, Mn, Zn	300–500	10^5
Single-crystal	300–500	10^5

* Permeability = $\dfrac{\text{flux density } B}{\text{magnetizing force } H}$

For a complete definition, see Fink and Christiansen, p. 6-65. Permeability is an important specification for head materials because it indicates the flux density and signal voltage generated by the magnetic pattern on playback.

† High resistivity is desirable to minimize eddy current losses.

‡ The resistivity of alfesil is increased and its permeability reduced by fabricating from powders.

SOURCES: K. Blair Benson, *Television Engineering Handbook*, McGraw-Hill, New York, 1986; Donald G. Fink and Donald Christiansen, *Electronic Engineers' Handbook*, McGraw-Hill, New York, 1989.

for ferrites that was reduced by a glass bonding technique in which a layer of glass is bonded to the surfaces in the vicinity of the gap. It also led to the use of iron/silicon/aluminum alloys known by such names as alfesil, alfecon, and sendust. These alloys had the problem of low resistivity, but this was overcome by fabricating them from powders in which the particles were held together by an insulating cement. This reduces the permeability and a balance must be struck. The electrical and magnetic properties of representative head materials are tabulated in Table 9.1.

9.4 The Helical-Scan Principle

The process of presenting the tape to the heads is basically a mechanical one and mechanisms for this purpose are called *tape transports*. They handle the unreeling and rereeling of tape, and provide a *tape path* where the tape passes over video, audio, and auxiliary recording heads.

A *capstan* drive is used for accurate control of tape speed. A capstan is a rotating metal shaft that contacts the tape under pressure from a rubber *pinch roller*. The capstan is driven by a servo motor whose speed is electronically controlled to achieve the exact tape speed and synchronization.

The preferred means for achieving the relative motion between tape and heads at speeds suitable for video recording is the helical-scan method, illustrated in Figure 9.4. The tape is wrapped in a helix around a drum on which record/playback heads are mounted, and the drum rotates in a direction opposite to the longitudinal tape motion. The result is a series of diagonal tracks recorded across the tape as shown in Figure 9.5, which is for the SMPTE Type C format. In addition to the diagonal tracks used for video and

Figure 9.4 The helical-scan principle

sync (video and sync are recorded separately in this format), longitudinal recording is used for the audio tracks and for the control track on which signals are recorded for synchronizing the tape deck.

9.5 Tape Transport Configurations

A variety of helical-scan tape transport configurations have been utilized in commercial recorders to meet the varying requirements of different applications.

The tape can be stored on reels for reel-to-reel recording or in cassettes (see Section 9.5.1). Cassettes are easier to handle and load but their tape capacity is not as great as open reels. Capacity limitation has become less of a problem with improved and thinner tapes, and the trend is toward cassettes. All consumer recorders use cassettes because of their ease of loading. The SMPTE Type C recorder, the most widely used professional format, is reel-to-reel, but cassettes are becoming increasingly common as their performance is improved.

A basic classification of helical-scan systems is the way the tape wraps around the drum. *Omega wrap* (as in Figure 9.4) and *alpha wrap* are named for their resemblance to the Greek letters and the classification is based on whether the tape path loops all the way around the drum (α) or whether the tape approaches and leaves the drum along a radius (Ω). Most recorders employ omega wrap and one or two video heads per channel.

With two heads, the tape only wraps a little more than 180° around the drum but wth single-head recorders the tape must wrap entirely around the drum except for a very small gap. On recording, the rotation of the drum is

Figure 9.5 SMPTE Type C record format. (*From SMPTE/ANSI Standards,* V98.18M-1983, V98.19M-1983, V98.20M-1979 (Src), RP85-1979, RP86-1979.)

synchronized with the video signal so that the head crosses the gap (when recording is interrupted) during the vertical blanking interval.

Another classification of helical-scan systems is based on the fraction of a TV field recorded in one pass of a head, field-per-scan, or segmented scan. Field-per-scan is advantageous because no head switching occurs during the picture and slight differences between heads are more easily tolerated. Segmented-scan formats require two or more heads per channel and very precise matching of heads or the use of digital recording.

9.5.1 Tape cassettes

A cassette is an enclosure that holds two tape reels and a tape that can be reeled from one reel to the other by an external drive that engages the cassette reels. This concept was originally developed for audio recorders, which have stationary heads. In audio recorders, the tape is presented to the heads by having an opening in the side of the cassette that exposes the tape and simply presses that part of the cassette against the heads. The only other step is to provide a means to engage the capstan drive for the tape, which is usually done by dropping the cassette over the capstan shaft and using a rubber pinch roller outside the cassette to press the tape against the capstan. This is illustrated in Figure 9.6(a).

The cassette principle is more difficult to implement for video recorders because the tape has to wrap around the scanning drum as well as press against stationary heads and capstan. The usual approach is to use a two-head-per-channel configuration and provide a mechanism that draws the

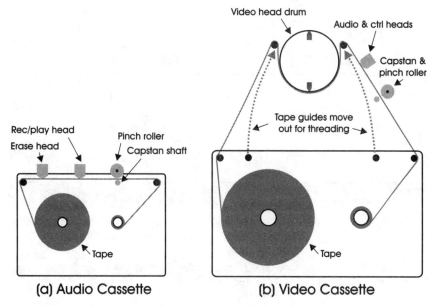

Figure 9.6 Operation of tape cassettes

tape out of the cassette to the scanner. Figure 9.6(b) shows this. In spite of the difficulties, all new video recorders, both professional and home, in the last five years use cassettes. The widely-used acronym, VCR, means video *cassette* recorder.

9.6 Magnetic Recording Tapes

Magnetic recorders originally used a fine iron wire as the recording medium, but this was superseded by tape soon after the tape format was developed by German engineers during World War II. As with magnetic recording heads, magnetic tape has been the subject of an enormous amount of research since World War II, not only for its uses in audio and video recording but also for data processing and storage.

A cross-section of a magnetic recording tape is shown in Figure 9.7. It consists of a plastic tape base film coated on the front with a magnetic material and on the back with a conductive backcoating to prevent the build-up of static charges.

9.6.1 Tape base material

The tape base is a polyester chosen for its strength and dimensional stability with respect to temperature, humidity, and time. The thickness of the base film is chosen to be as thin as possible to minimize the reel diameter but thick

Figure 9.7 Magnetic recording tape. (*From K. Blair Benson,* Television Engineering Handbook, *McGraw-Hill, New York, 1986, Fig. 15-47.*)

enough for mechanical strength and to prevent *print-through*, which is the tendency for recordings to interfere between adjacent layers of the magnetic coating when the tape is rolled up on the reel. The smoothness of the magnetic coating surface must be controlled carefully also—as smooth as possible to provide close contact with the recording head but still rough enough to provide proper friction for tape handling.

9.6.2 Tape magnetic material

The layer of magnetic material is a powder suspended in a binder that bonds it to the base film. The preparation and application of this layer involves processes, often proprietary, that differ from manufacturer to manufacturer.

The desired electromagnetic properties are high permeability and coercivity (see Figure 9.2). These may seem to be in conflict since permeability is a measure of the ease with which a material can be magnetized while coercivity is a measure of the threshold magnetizing force. An ideal material is one in which the threshold is high so that the depth of magnetized layer and the minimum wavelength is low but in which small increases in magnetizing force above the threshold produces large increases in magnetization.

Three magnetic materials, ferric oxide, chromium dioxide or CrO_2, and metal particles have been commonly used for the magnetic layer. Each has advantages that make it suitable for particular applications. A fourth material, iron oxide doped with cobalt, has desirable magnetic properties but suffers from strong temperature dependence. The permeability and coercivity of these materials are tabulated in Table 9.2.

Table 9.2 Magnetic properties of tape materials

Material	Permeability	Coercivity
Ferric oxide	5,000	300–350
Chromium dioxide	6,000	300–700
Metal particles	10,000	1,000

SOURCE: Donald Christiansen, *Electronic Engineers' Handbook*, McGraw-Hill, New York, 1989.

Ferric oxide was originally used for the magnetic tape and is still the most common. It is relatively inexpensive and is satisfactory for consumer applications that do not require the ultimate in performance.

Chromium dioxide has another desirable property, a high saturation magnetization, the maximum possible value of B. With the increased magnetization, a signal-to-noise ratio 5 to 7 dB higher can be achieved than with ferric oxide. A disadvantage is that it is more abrasive.

Metallic particles suspended in lacquer to reduce rusting have high coercivity, high output signal level, high signal-to-noise ratio, and a wider frequency response. It is the common choice for applications with demanding performance requirements. The recording characteristics of these materials are different, and they require different head designs for optimum results.

9.7 Tape and Record/Playback Head Performance

This section describes the combined performance of magnetic tape and record/playback heads with respect to frequency response and linearity for analog signals and bit rate for digital—the signal criteria that are directly determined by the tape and head characteristics.

9.7.1 Frequency response

Achieving an adequate frequency response for the FM-modulated video signal was one of the most difficult problems faced by the pioneers in video tape recording technology. The bandwidth of the original quadruplex recorder RF channels was limited to 6.5 MHz, an enormous technical feat at the time. High-band recorders, introduced in 1964, had a bandwidth of 10.0 MHz, another breakthrough. Present professional analog recorders have bandwidths to 15.0 MHz or higher. (Consumer-type recorders typically have bandwidths of 7 MHz; see Chapter 10.)

The equivalent frequency response (half the bit rate for PCM digital channels) of digital video recorders is even greater. The SMPTE D-2 standard for composite NTSC signals specifies a $4f_{sc}$ sampling rate with 8-bit words (see Table 3.2). This requires a bit rate of 114.6 Mbs, approximately equivalent to an analog bandwidth of 57.3 MHz. The D-2 recorder uses two parallel tracks to achieve this bit rate.

The standard sampling rates for component signals are 13.5 Mbs for the luminance signal and 6.75 Mbs for each of the two color-difference signals. The corresponding bit rates are 108 Mbs and 54 Mbs. Combining these signals serially on single track would require an equivalent bandwidth of 108 MHz, an impractical rate even with the most modern technology. As a consequence, in the D-1 recorder, these signals are recorded with four parallel tracks.

At lower frequencies the frequency response of the tape-head combination increases at the rate of 6 dB per octave as shown by Eq. (9.1). However, as the frequency is increased further and the wavelength [Eq. (9.3)] approaches the gap length in the direction of tape travel, the other effects described below reduce the response so that it falls off with frequency.

The fall-off in frequency response at higher frequencies has four sources:

1. The finite gap width. The response from a gap of length d is:

$$e = k \frac{\sin\left(\dfrac{2\pi d}{\lambda}\right)}{\left(\dfrac{2\pi d}{\lambda}\right)} \tag{9.3}$$

where: k is a normalizing constant

d is the length of the pole-tip gap

This equation has the $\sin(x)/x$ format that describes many aperture processes. Note that there is a null where $d = \lambda$. There are practical limits to how small the gap length can be made. Gap lengths smaller than about 40 μinches (1.0 μm) are difficult to maintain in mass production. Therefore, there is an upper limit to the bandwidth of a single recording channel—around 20 MHz. If a system requires more channel bandwidth than this, some means of multiplexing two or more channels must be used. This is necessary in most digital video recorders (see Section 9.11).

2. The depth of recording. The deeper the recording, the wider the record zone and the poorer the response at high frequencies. The response begins to fall off at the point where its wavelength is about one-third of the record depth. Since the recording depth is less with high coercivity tape, its high-frequency response is better. Record depth can also be controlled by limiting the thickness of the magnetic coating.

3. The spacing between the head and the recording layer. High-frequency response falls off with separation between tape and head according to Eq. (9.4):

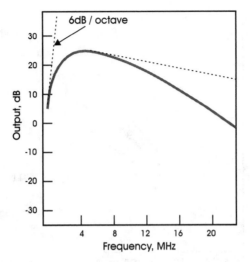

Figure 9.8 Representative frequency response of a magnetic recording head

$$\text{Separation Loss (dB)} = 54.6\frac{d}{\lambda} \tag{9.4}$$

For example, at a wavelength of 100 μinches, there would be approximately 5 dB of separation loss for a head-to-tape average spacing of 10 μinches. Such a small number, which is comparable to the tape surface roughness, is achievable only by forcing the head into actual contact with the tape. All video tape recorders use in-contact recording, which means that some amount of head and tape wear is always present. (Note that computer hard disks operate with a deliberate amount of separation between head and disk—see Section 9.17.)

4. Azimuth loss. This occurs when the gap is not aligned at the same angle with respect to the recording track on playback as when recording. It depends on the wavelength, the track width, and the angle of misalignment. Heads must be precisely aligned to control this effect.

In each of these cases, the high-frequency response is a function of the relationship between the relevant critical dimension and the wavelength. Figure 9.8 shows the frequency response of a typical head-tape combination. The increase in high-frequency response that has been achieved in recent years is the cumulative result of technology improvements that affect all of these parameters. Gap lengths have been reduced by precision fabrication techniques. Tape writing speed has been increased by moving the head past the tape rather than vice versa. Improved tape materials have increased in coercivity and permeability. Improved tape manufacturing techniques have reduced surface roughness and permitted closer head-to-tape spacing.

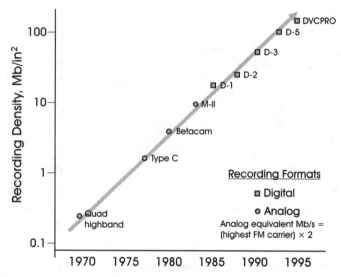

Figure 9.9 Progress in recording density

9.7.2 Linearity

The linearity of the magnetic recording process is determined by the shape of the highly non-linear B/H magnetization curve. The nonlinearity is so severe that direct recording of a baseband analog audio or video signal is not satisfactory. The problem is solved for analog audio signals by adding a high-frequency bias waveform to the audio waveform. The problem is solved for analog video signals by the use of frequency modulation. In digital recorders, the linearity issue is dealt with by special modulation methods (see Section 9.12.3).

9.7.3 Recording density

As the fundamental performance of video recorders has improved over the years, the amount (area) of tape has become less for a given recording time at a given quality level. This is expressed by the parameter of *recording density*. In digital recording, it is expressed in terms of the number of bits recorded in a square inch of tape. The same parameter can be used for analog recorders by applying the approximation that an analog bandwidth of W is equivalent to a bit rate of $2W$ bits/second. Figure 9.9 shows how recording density has changed with time. This graph clearly displays how the recording industry has achieved steady improvement in performance over many years. It is similar to the graph of Figure 7.1, which shows steady improvement in the solid-state integrated circuits industry.

Improvement in recording density obviously reduces the volume of tape needed for a given playing time. Over the years, this has made possible cassette tape handling (early tape reels were so large that a two-reel cassette would have been impractical). It also has reduced tape costs and storage space for tape libraries.

9.8 Analog Recording

9.8.1 Overview

Video signals are analog by nature, and until the late 1970s they were universally recorded in analog form. The attractions of digital recording of analog video signals were numerous, however, and advances in basic recorder technology made digital recording practical as well. Digital recording has important advantages for many applications and it now coexists with analog.

The SMPTE has established three standards for analog helical-scan formats with performance satisfactory for professional use:

Type C: This was the original standard for professional helical-scan recorders, which was approved in 1983 after reconciling the proposals of competing manufacturers. It specifies a reel-to-reel format with 1-inch tape.

M-II: This is a standard that specifies ½-inch metal particle tape contained in a cassette and is used in the same applications as Type C. M-II is an upgrade of the VHS home format (see Section 10.4). Two video tracks are used to record luminance and chrominance separately.

Betacam: This standard also employs ½-inch tape in a cassette. It is an upgrade of the Betamax home VCR format and is widely used for ENG, particularly in camcorders, combined cameras and recorders. Like the M-II, it also has two video tracks and records analog components.

9.8.2 Analog signal modulation

Direct magnetic recording of analog video signals is not satisfactory because of inherent nonlinearity. The linearity problem is solved by the use of a high-frequency carrier frequency modulated by the video signal. FM for video recording was introduced by Ampex in the original quadruplex product, and it has continued to be the predominant modulation format for analog signals.

Frequency modulated video signals differ from FM communication signals in three important respects:

1. The carrier frequency is only a little higher than double the highest frequency component of the signal.

2. The frequency deviation relative to the width of the baseband is much lower.

3. The DC component (see Figure 1.7) in the signal is retained by setting the frequency representing the sync tip at a fixed value regardless of the content of the signal.

The SMPTE Type C recording standards are an example of FM recording. The specified instantaneous frequencies in MHz are:

	NTSC	PAL
Peak white	10.0	9.3
Blanking	7.9	7.8
Sync tip	7.06	7.06

The useful picture information is transmitted by a carrier deviation of only 2.1 MHz (7.9 to 10.0 MHz) for NTSC and 1.5 MHz (7.8 to 9.3) MHz for PAL. The deviation for the complete composite video signal, including sync and blanking, is 2.94 MHz for NTSC and 2.24 MHz for PAL. With a maximum video signal frequency component of 4.2 MHz, the modulation index for the composite NTSC signal is only 0.7. This compares with a typical modulation index of 2.5 for microwave and satellite systems. The small modulation index is required to minimize the lower second order sidebands $(f_c - 2f_m)$ that would overlap the baseband or even extend into the negative frequency region (phase reversed, sometimes described as "folded back").

The frequency spectrum of an FM carrier modulated by a composite color signal in accordance with SMPTE Type-C standards is shown in Figure 9.10. Both sidebands of the FM signal are recorded, but most recorders suppress part of the upper sideband on replay to improve signal-to-noise ratio. This must be done with a precise response curve to avoid differential gain distortion. The removal of part of one FM sideband introduces an amplitude modulated component that is removed with a limiter (see Section 15.4.2).

9.8.3 Signal-to-noise ratio

The characteristics of FM video signals make the calculation of signal-to-noise ratio complex and imprecise. The problem is complicated further by the fact that significant noise components can be generated by three sources, the tape, the head, and the electronics. A mathematical analysis of S/N in FM video recording systems is beyond the scope of this volume, but it is important to consider the factors that affect it:

1. The low modulation index adversely affects the S/N in two respects: S, the signal power, is proportional to the square of the modulation index; and the *FM improvement factor* (see Section 17.8.2), the ratio of the S/N of the demodulated baseband signal to the S/N of the carrier, is much smaller than in communication systems.

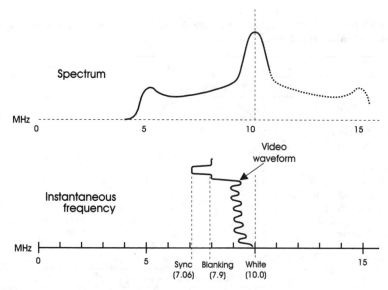

Figure 9.10 Frequency spectrum, SMPTE Type C video record

2. The S/N of the demodulated signal decreases linearly with the carrier amplitude down to a critical threshold at which point the S/N drops very rapidly and the signal becomes unusable.

3. As with FM communication systems, the signal-to-noise ratio can be improved by preemphasis of the high-frequency components of the video signal before recording with complementary deemphasis on playback. The preemphasis standard is CCIR 567 (based on Report 637).

4. Also in common with all FM systems, the amplitude of the frequency components of noise generated by the recording and playback process (not noise in the original baseband signal) rises linearly with frequency. This leads to a greater difference between weighted and unweighted noise than with flat or white noise (see Section 4.15.3) because high-frequency components are not as visible. A typical weighting factor is 12 dB.

5. The quality of the tape has a major effect on the S/N. Metal particle tape with its high coercivity improves the signal-to-noise ratio by several dB as compared with conventional iron oxide tape.

The unweighted signal-to-noise ratios currently being achieved on professional helical-scan recorders is approximately 49 dB (see Table 9.3).

Table 9.3 SMPTE Recording Standards and Typical Performance, Type C, M-II, and BetaCam analog recorders

	Type C	M-II	BetaCam
	Standards		
Tape handling	Reel-to-reel	Cassette	Cassette
Track geometry	Full field	Full field	Full field
Tracks/TV field	1	2	2
Tape width, in.	1	0.5	0.5
Tape type	Oxide	Metal particle	Metal particle
Tape speed, in./s	9.6	2.6	4.6
Tape usage, sq.in./s	9.6	1.3	2.3
Number of record heads	2*	4	4
Video track width, mils	5	2.3	2.3
Azimuth recording	no	yes	yes
Drum diameter, in.	5.4	2.44	2.93
Drum rotation speed, rev/s	59.94	29.97	29.97
Video writing speed, in./s	1000	230	276
	Typical Performance		
Playing time	126 min., 11.75-in. reel	90 min., cassette	90 min., cassette
S/N ratio, unweighted			
Composite, dB	49	47	47
Luminance, dB		49	49
Bandwidth, MHz			
Composite	4.1	4.2	4.1
Components			
Luminance		4.5	4.1
Chrominance		1.5	1.5
Differential gain, %	<4	<3	<3
Differential phase	<4°	<3°	<3°

* Entire active picture recorded on one head, second head records only the vertical blanking interval.

9.9 SMPTE Type C Recorders

9.9.1 Transport geometry

The SMPTE standard layout[1] of the Type C record drum assembly is shown in Figure 9.11(a) (plan view) and 9.11(b) (elevation). The assembly includes a rotating upper drum on which the head pole tips are mounted and a slightly smaller stationary lower drum that maintains contact with the reference (lower) edge of the tape to provide accurate alignment as the tape follows its helical path around the drum.

The upper drum makes a complete revolution each TV field or approximately 3,600 rpm for the NTSC system, and each track records the complete active field. This gives the Type C recorder two important advantages over segmented systems: The problem of matching heads is eliminated and

Figure 9.11 Type C record drum. (*From Donald G. Fink and Donald Christiansen,* Electronic Engineers' Handbook, *3d ed., McGraw-Hill, New York, 1989, Fig. 20-69.*)

full-quality variable speed playback is possible (variable speed operation of a segmented system results in a bar across the picture).

There are three pairs of heads on the upper drum, each pair having one head for video and one for sync. The head pairs have different purposes. The first pair performs the basic record/playback functions. The second pair provides simultaneous playback of the recorded material for monitoring the recording. They are sometimes called the "confidence heads." The third pair provides the facility for selective erasure of portions of the record for editing purposes.

A servo system synchronizes the rotation of the drum with the field rate so that vertical blanking occurs during the short time gap at the end of a video track as the head completes one track and begins the next. During the gap in output from the video head, the output is switched to the signal from the sync head, which is interrupted at a different time because of its 30° offset location on the drum. Some early recorders used a single head and reconstructed the sync and blanking signals during this gap, but recording the vertical sync signals on a separate track provides more positive synchronization.

9.9.2 Video, audio, and control track records

Figure 9.5 shows the record pattern for the Type C recording standard. The following dimensions should be noted:

The width of the video and sync tracks is only 0.005". To maintain this tolerance on playback would be extremely difficult without automatic tracking (see Section 9.9.4.2).

The video tracks are more than 16 inches long, and the timing errors introduced by an elastic medium of this length necessitates the use of time-base error correction. The video track is nearly parallel to the reference edge of the tape, the slant angle is only 2.5°. (This is not shown to scale in Figure 9.5.)

The tape speed (9.6 inches/second) is a little faster than would be indicated by the video track width and the track angle so that there is an unrecorded space between tracks on the tape. This is known as a *guard band* and it is necessary to avoid cross talk between adjacent tape tracks, especially when there is some mistracking. (More recently developed record standards have solved this problem a different way—see Section 9.14.1.)

The sync tracks are about 16 lines long, sufficient time for the transition of the video pole tip from one track to the next during the vertical blanking interval. The sync signal is switched off during the remainder of the field.

9.9.3 Record electronics

Figure 9.12 is a functional diagram of the record electronics for a Type C recorder. In this design, the video input is AC coupled, and DC restoration is provided by the automatic frequency control (AFC) circuit that maintains the instantaneous frequency of the blanking pulse to a reference value at the FM modulator. The gain of the video amplifier is adjusted to set the frequency deviation of the modulated carrier at the values specified in Section 9.8.2. The video signal is preemphasized and signal excursions that exceed the established white level are clipped.

The gain of the record amplifier is adjusted to bring the magnetizing field, H, to the level that produces maximum carrier output from the confidence head. This is called *optimizing* the record current. The record amplifier is carefully designed to have a frequency response consistent with the magnetic characteristics of the head and it must have good amplitude linearity to avoid distorting the FM signal and generating spurious frequency components.

The record electronics also generates the control signals to the drum servo that synchronizes the rotational speed and phase of the drum with the recorded signal and to the capstan servo that controls the tape speed. The synchronizing signals include the sync pulses and the identification of the first field in the sequence of four for NTSC or eight for PAL (see Sections 5.7.3 and 5.8.3).

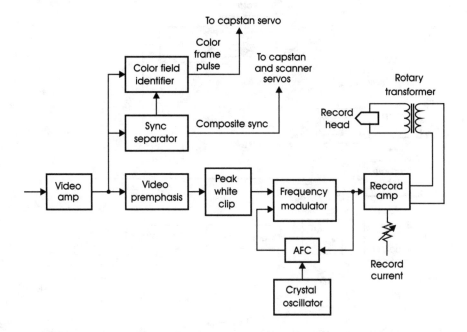

Figure 9.12 Record electronics. (*From K. Blair Benson,* Television Engineering Handbook, *McGraw-Hill, New York, 1986, Fig. 15-60.*)

9.9.4 Playback electronics

The electronics employed by playback systems are more complex than those used for recording since they perform a large number of functions that support and enhance the basic playback function. These include the detection and correction of drop-outs (momentary interruptions of the FM carrier caused by tape defects that interfere with head-to-tape contact), time-base and velocity error correction, and slow-scan and stop-motion modes. Figure 9.13 is a simplified functional diagram of the playback electronics of a helical-scan playback system.

9.9.4.1 Signal channel Because of the rotation of the head drum, the record/playback head is connected to the preamplifier by means of a rotary transformer. The head/transformer/preamplifier combination has a major effect on the signal-to-noise ratio, and it must be designed carefully for compatibility of the components. The input stage of the preamplifier may be an FET or a bipolar differential amplifier. Its input impedance and the turns

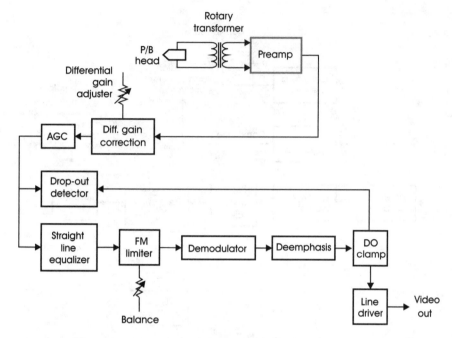

Figure 9.13 Playback electronics. (*From K. Blair Benson,* Television Engineering Handbook, *McGraw-Hill, New York, 1986, Fig. 15-61.*)

ratio (step-up) of the transformer are among the parameters that must be matched.

The preamplifier is followed by an equalizer with rising response at higher frequencies like an aperture corrector. Typically its response is a constant plus a cosine function beginning at 180° (–1.0) at 0 MHz and continuing to 360° (+1.0) at the upper end of the frequency band. This compensates for the losses in the record/playback process shown in Figure 9.7. The equalizer is followed by an automatic gain control (AGC) amplifier and a drop-out (DO) detector.

The AGC amplifier is connected to the limiter and the demodulator in the main signal channel through a *straight-line equalizer circuit*. This circuit converts the FM signal to a vestigial-sideband format and it improves S/N ratio by attenuating the upper sidebands that have more noise coming from the tape. The video signal from the demodulator passes successively through a low-pass filter to remove spurious signal components, a deemphasis circuit to complement the preemphasis in the recorder, and the drop-out corrector.

The sync and burst waveforms are separated from the output signals and are used to synchronize the output of the recorder with the studio sync systems by means of the drum and capstan servos and the time-base corrector.

Tape

Figure 9.14 Tracking correction. (*From K. Blair Benson, Television Engineering Handbook, McGraw-Hill, New York, 1986, Fig. 15-108a.*)

9.9.4.2 Automatic scan tracking A means must be provided to cause the playback head to follow the long and narrow recorded tracks. This is accomplished by a technique known as automatic scan tracking.

The video playback head is mounted on a *bimorph*, two thin flexible strips of a piezoceramic material bonded together (see Figure 9.14). The strips bend when a voltage is applied across them, and they are mounted so that this causes the head to move at right angles to the recorded track. During the fabrication of the bimorph, the strips are polarized so that a bias voltage is required to maintain them in an undeflected position. Deflection in either direction is achieved by adding the deflection voltage to the bias.

Mistracking of the head is detected by a reduction in its output, but this does not indicate the direction of the error. One method of identifying the direction of the error is to apply a dither voltage, typically 500 Hz, to the bimorph strips. The resulting waveforms are shown in Figure 9.15. When the head is on-track, the detected envelope has double the dither frequency. When it is off-track, it has a component at the dither frequency but of opposite phase for locations of the head above and below the track. The phase difference is detected by a synchronous detector, and a correction voltage is generated to return the head to the track. (A similar principle is employed for the tracking of laser disks.)

Although it is not necessary to employ scan tracking for playback at normal speeds, it is an additional assurance.

9.9.4.3 Synchronization It is usually necessary to synchronize the scanning pattern and burst phase of the recorder playback with the studio sync generator so that recorder signals can be combined with other studio signals for mixing and editing. Three events must be timed and synchronized: the

Figure 9.15 Mistracking detection. (*From K. Blair Benson,* Television Engineering Handbook, *McGraw-Hill, New York, 1986, Fig. 15-109.*)

beginning of the four-field subcarrier sequence, the beginning of each line, and the burst subcarrier phase.

The synchronization of fields and lines is performed by the servos that govern the rotation of the drum and capstan (the shaft that imparts the linear motion of the tape). The servos are controlled by comparison of the timing of signals on the control track with the sync generator output.

9.9.4.4 Time-base correction   The precision required for synchronization of burst frequency and phase is measured in nanoseconds and is orders of magnitude greater than can be achieved by electromechanical servos alone. Providing this precision is one of the functions of the time-base corrector. As described earlier, the digital time-base correction was a key technology in the development of helical-scan recorders for professional use. The operation of time-base correctors is described in Section 3.6.1. The time-base correction circuitry may be incorporated in the recorder, or it may be a separate unit.

9.9.4.5 Drop-out compensation   The first step in compensating for drop-outs is detecting them. The amplitude of the envelope of the FM carrier is continuously monitored, and a drop in level of the order of 16 dB or more compared to the average carrier level is considered to be a drop-out. The DO detector produces a pulse lasting for the duration of the dropout. This pulse is then transmitted to the drop-out compensator, DOC, which fills the gap in the signal produced by the drop-out by interpolating the signal from a corresponding position on an adjacent line. Since the immediately adjacent line is in an alternate field, the phase of its color subcarrier is reversed, and this would cause an objectionable color error. This problem can be solved by interpolating a signal from an adjacent line on the same frame, two lines removed, or by processing the subcarrier from the adjacent line to reverse its phase.

9.9.4.6 Slow-scan and still-frame operation

The helical-scan format is particularly adapted to slow-motion and still-frame operation. These effects are achieved by slowing or stopping the tape capstan drive while continuing the drum rotation at its normal speed. Since the angle of the track with respect to the reference edge of the tape is determined by the relationship of the tape and rotational speeds, changing the tape speed automatically causes mistracking.

Figure 9.16 shows the relationship between the record tracks and the playback path under three conditions:

1. Tape motion stopped; still-frame operation
2. Tape motion slowed moderately; slow-motion operation
3. Tape motion slowed or accelerated greatly; "picture-in-shuttle"

When the tape motion is stopped on playback, the head does not follow the record track perfectly because the track angle has no tape-motion component. In the absence of tracking correction, the end of the playback paths will coincide with the ends of the preceding record tracks. This mistracking is corrected by the automatic scan tracking system described in Section 9.9.4.2. The mistracking in the slow-scan mode is even smaller, and this, too, can be corrected by automatic scan tracking systems.

The picture-in-shuttle mode can be used either to greatly accelerate or to reverse the tape motion. The mistracking is so severe that tracking correction is not possible or practical, but the resulting picture contains enough information to be recognizable for cueing or editing, even when the tape is run backward or moved at 20 or 30 times its normal forward speed. Picture-in-shuttle, therefore, is a useful mode for rapid search of picture content for the location of a particular frame as required for these purposes.

9.10 ½-inch Cassette Recorders

Major improvements in tape performance, particularly the introduction of metallized tapes, together with improved heads and circuitry, have made it possible to produce recordings on ½-inch tape with a quality level that approaches that of 1-inch tape systems.

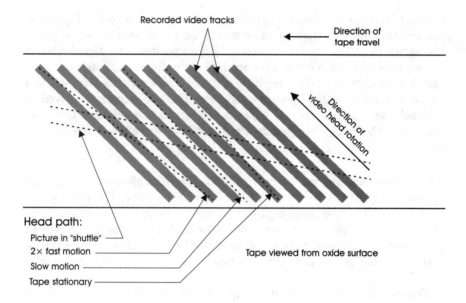

Figure 9.16 Slow-motion, still-frame, and picture-in-shuttle record tracks

When the narrower tape became practical, the logical next step was to adapt the cassette configuration for many applications. ½-inch cassette recorders of professional quality are now widely used, both in portable configurations for electronic news gathering (ENG) and in studio configurations.

Two SMPTE standards, M-II[2] and Betacam, have been established by the SMPTE for professional-quality, ½-in cassette recorders (see Section 9.8.1). Some of the significant recording parameters and typical performance specifications of the Betacam and M-II formats are shown in Table 9.3.

The Betacam and M-II records have parallel dual tracks recorded by closely spaced heads using the azimuth recording method (see Section 9.14.1). The luminance signal, Y, is recorded on one track, and the color-difference chroma signals, R-Y and B-Y, on the other. The components, R-Y and B-Y, are combined for recording by time-division multiplex (see Section 9.10.1).

All signal components employ frequency modulation. However, a carrier frequency lower than that used for composite signals can be tolerated for component recording because the absence of a high-energy color subcarrier greatly reduces the cross talk between the baseband and modulated carrier sidebands. A lower deviation can be tolerated because of technical advances that have improved the signal-to-noise ratio, particularly the use of metal tape.

Figure 9.17 is a drawing of the M-II record format. Note the parallel pairs of Y and C tracks. Also note that, unlike Type C, the head rotation and tape

Figure 9.17 M-II record format. (*From Panasonic Corporation,* Instruction Manuals, 17-11 recorders.)

travel are in the same direction. To facilitate the cassette configuration, which is difficult to implement with single-head recorders, both of these recording systems use two heads per channel. The tape only wraps 180° around the drum, so automatic tape threading is much easier.

9.11 Digital Recording

The nearly transparent characteristics of digital video recording are particularly important in complex editing systems and other situations where many rerecordings or "generations" are necessary. Its relative freedom from signal-to-noise and linearity degradations as compared with the analog format make it almost a necessity for these applications.

Digital recording was hindered initially by the high bit rate requirements of high-quality video, but this problem has been overcome by steady improvements in tape and head technology. Another problem was the establishment of standards. This has been particularly difficult not only because of the many choices that are available for digital recording, but also because of the growing number of broadcasters, manufacturers, and engineering laboratories whose views had to be considered. The problem was addressed seriously in the late 1970s when representatives of all industry segments began standardization meetings under the aegis of the SMPTE in the United States and the EBU in Europe.

9.12 Digital Recording Standardization Policy

Early in the standardization process, it was realized that there were serious conflicts of interest between competing manufacturers with respect to equipment standards. Fortunately it was possible to resolve many of these conflicts by standardizing the tape and its record rather than the apparatus for achieving it—a major policy change. This assured interchangeability of tape records (the principal purpose of standardization) while retaining the benefits of competition.

In the United States, the SMPTE has led the effort to establish standards for digital recording and created the Digital Television Tape Recording group, DTTR, for the purpose of establishing standards for 525-line systems. At the same time, the EBU appointed a group called MAGNUM to establish standards for 625-line systems.

9.12.1 Standard formats

When digital recording became technically practical, the industry reaction was to go all the way in picture quality improvement and employ a component format that would eliminate the artifacts, cross color, cross luminance, and other defects of the composite NTSC, PAL, and SECAM systems.

The first digital recording format to be standardized was for component signals and was designated D-1[3] (see Table 3.3). The sampling rate is 13.5 MHz for the luminance component and half that or 6.75 MHz for each of the two-color difference components. These rates have come to be known as "4-2-2" sampling. That name resulted from the fact that a proposed sampling rate for composite format was four times the subcarrier frequency. This was considered for the luminance channel of the component system, and the nomenclature remained even after a different rate, 13.5 MHz rather than 14.3 MHz, became the standard.

The D-1 format is an excellent choice for production use and has made possible all-digital production facilities, but it is overkill for general broadcast use. The component format also makes D-1 machines difficult to integrate into broadcast facilities that are still mostly analog. Therefore, there was a need for a digital machine that would easily fit into analog facilities and still provide the advantages of digital recording. This need could be met by digital recording of composite video. The D-2[4] composite format was the answer.

The D-3[5] and D-5 formats came along later—they are composite and component formats, respectively, and they use ½-inch tape, which allows smaller, lower-cost equipment with good performance. The D-6 format is a component recorder for 1125/60 and other HDTV signals. Each of these is discussed further below. Technical parameters for all the standards are given in Table 9.4 and tape record diagrams are shown in Figure 9.18.

9.12.2 Scope of standards

Standardizing the tape and the recording formats rather than the apparatus requires detailed specifications to assure interchangeability. For example, the D-1 format is defined by five SMPTE standards:

SMPTE 224M — Tape Record

SMPTE 225M — Principal Properties of Magnetic Tape

SMPTE 226M — Dimensions of Tape Cassettes (D-2 also)

SMPTE 227M — Signal Content of Helical Data Records of Associated Control Records

SMPTE 228M — Signal Content of Cue and Time Code Longitudinal Records

9.12.3 Data rates

The data rates given in Table 9.4 are the raw rates that result from the A-D conversion—sampling frequency is multiplied by bits per sample for each video channel. However, the data rate actually recorded in the tape channel is higher than these numbers because of the addition of audio data and the overhead bits needed for error management.

9.12.4 Channel coding for digital recorders

As with digital transmission, recorders require some form of modulation to optimize the use of the record–replay channel. This is often called *channel coding*. Because of the extreme nonlinearity of magnetic recording, the multilevel amplitude-modulated codes described in Section 3.7.2 for transmission channels are not applicable. Instead, recording codes are based on frequency or phase modulation. Because the record–replay process differentiates the signal, it is convenient to think in terms of transitions between recording states (reversal of record current) that produce pulses on replay.

Other considerations of code selection include minimization of the DC component, self-clocking on replay, and control of the minimum and maximum separation of transitions. This latter consideration is called *run-length limiting* (RLL) and it determines the bounds of the frequency spectrum for the signal. The details of channel coding are beyond the scope of this discussion, but the code used in each recording standard will be discussed

9.12.5 Error management in digital recorders

Error correction and concealment are of critical importance in digital recording since the ability to make multigenerational copies of high quality is one of digital's most important advantages.

The error-detection and -correction techniques described in Section 3.8 for digital transmission channels are applicable to digital recording, including Reed-Solomon coding and interleaving. R-S coding is applied twice, by operating orthogonally on a large block of data such as a video field. For example, the *outer code* may operate on the rows of data and the *inner code* then operates on columns of the data array. In addition, digital recorders use a form of random interleaving, called *shuffling*, which is applied between the outer and inner R–S coding steps.

The purpose of shuffling is error concealment. If a large defect causes oveload of the error management system, the errors will all appear together in the picture or (with interleaving) they will form a pattern determined by the R–S blocking algorithm. Shuffling, which is random interleaving, can spread the residual errors randomly, making them much less visible.

9.13 The D-1 Component Format

The D-1 tape width is 3/4", and it is cassette-mounted. The video format is 4-2-2 component, which means that the sampling rate is 13.5 MHz for the Y or luminance signal. (See Section 4.7.1 for considerations involved in the choice of this rate.) The bandwidth and signal-to-noise ratio exceed the performance of analog formats. The absence of cross color, cross luminance, and color artifacts that is characteristic of the component format is an important feature.

Because of the high total video data rate of 216 Mb/s resulting from the 4-2-2 sampling, four parallel video channels are used on the tape. Each video track consists of two video sectors, each having a length of 77.71 mm, separated by a group of four audio sectors, one for each audio channel and each having a length of 2.55 mm. The total length of each track is 170 mm.

The audio segments are located in the center of the tracks because the center of the tape is less prone to dimensional changes and physical damage. The performance of the audio system is excellent. The modulation is digital with a 48 kHz sampling frequency. Quantization is 16 bits per sample, and the result is frequency response to 20 kHz, less than 0.05 percent distortion, and a 90 dB dynamic range. Noise is virtually unmeasurable.

The D-1 system uses a modified NRZI (nonreturn-to-zero, inverted) channel code. A normal NRZI code has a wide spectrum and poor clock content; both of these are corrected by modifying the code to add transitions to the channel signal.

9.14 The D-2 Composite Format

The sampling rates that have been used for composite signal transmission are $3f_{sc}$ and $4f_{sc}$, where f_{sc} is the color subcarrier frequency. $4f_{sc}$ or 14.4/17.7 MHz (NTSC/PAL) gives superior performance and is used by the D-2 standard. D-2 is also a ¾-inch tape format using the same physical cassette as D-1 but with different tape in it.

From the information in Table 9.4 and Figure 9.18, it can be seen that the D-2 format uses a larger head drum than D-1, but there are fewer heads on it. This is because D-2 is a two-head-per-channel system, but there are only two channels instead of the four used in D-1. The drum rotates slower and less tape is used because composite recording (even at $4f_{sc}$) has a lower data rate than 4-2-2 component recording. The result is a considerably more economical system to purchase and operate. For this reason, the D-2 format is widely used.

D-2 channel coding is *Miller-squared* (Miller^2), which is a modified FM code.

9.14.1 Azimuth recording

The D-2 format (and many other formats developed since 1985 or so) uses a technique that eliminates the guard bands between video tracks (see Section 9.9.2) without causing cross talk between tracks. This technique works when there are two or more heads per track, and it depends on the heads being deliberately aligned in azimuth so that their gaps are not perpendicular to the track. As explained in Section 9.7.1, misalignment of the head gap with the track causes a serious loss of signal when the track is replayed by a properly aligned head. Therefore, if adjacent tracks on the tape are recorded with substantially different (but standardized) head alignments, on replay, one head will experience azimuth loss whenever it plays any of the adjacent track due to mistracking. This technique is called *azimuth recording* and it allows guard bands to be eliminated.

9.15 The D-3 Composite and D-5 Component Formats

As professional-level analog recorders came into use with the ½-inch tapes that were originally developed for the consumer market, it was natural that digital formats would also be developed for ½-inch tape. The D-3 and D-5 formats resulted. As with other ½-inch tape recorders, the equipment is smaller and somewhat less expensive than the larger tape systems.

Both D-3 and D-5 are based on the same tape cassette (D-3) and the same scanner dimensions, which provide for two heads per channel and nominally 180° tape wrap. The D-3 (composite) is a two-channel system, so with record, play, and two-track erase heads it has 10 total heads. The D-5 (component) system has four channels to achieve the necessary greater bit rate of component recording and it has a total of 18 heads on its scanner.

D-3 and D-5 channel coding is a *group code*, called 8,14. The numbers mean that 8-bit data bytes have been remapped by the coding to 14 bits. This would appear to be a large amount of overhead but, in fact, it actually reduces the

Figure 9.18 Record patterns for SMPTE standard digital formats

Table 9.4 SMPTE Digital Recording Standards and Typical Performance

	D-1	D-2	D-3	D-5	D-6
Video format	Component	Composite	Composite	Component	HDTV Component
Tape width	¾-in.	¾-in.	½-in.	½-in.	¾-in.
Cassette	D-1 standard	D-2 standard	D-3 standard	D-3 standard	D-2 standard
Tape speed	11.3 ips	5.2 ips	3.3	6.6 ips	20.7 ips
Tape usage	8.48 sq.in./s	3.9 sq.in./s	1.65 sq.in./s	3.3 sq.in./s	15.5 sq.in./s
Number of channels for video	4	2	2	4	8
Number of record heads/channel	1	2	2	2	2
Total heads on video drum	12	10	10	18	34
Video track width	1.6 mil	1.6 mil	0.8 mil	0.8 mil	0.8 mil
Azimuth recording	No	Yes	Yes	Yes	Yes
Drum diameter	2.95 in.	3.8 in.	3.0 in.	3.0 in.	3.8 in.
Drum rotation speed	150 rps	89.9 rps	89.9 rps	89.9 rps	150 rps
Video writing speed	1,404 ips	1,080 ips	847 ips	847 ips	1,773 ips
Video sampling rate—Y	13.5 MHz				
Video sampling rate—Chroma	6.75 MHz				
Total video data rate	216 Mb/s	115/142 Mb/s	115/142 Mb/s	216 Mb/s	1.2 Gb/s
Video channel coding	R-NRZI	Miller^2	8,14	8,14	8,12
Number of audio channels	4	4	4	4	10
		Typical Performance			
Cassette record time	94 min. (L) cassette				64 min. (L) cassette
Video bandwidth		6 MHz			
Luminance	6.0 MHz	n.a.	n.a.		30 MHz
Chrominance	2.75 MHz	n.a.	n.a.		15 MHz
S/N each channel, unweighted	56 dB	54 dB			
Cross-color distortion	None				
Cross-luminance distortion	None				
Color artifacts	Minor				

data rate, to be recorded. This happens because there are 64 times as many possible codes in a 14-bit word than in an 8-bit word. Therefore, only 256 codes can be selected from the 16,384 available to achieve special features, for example, always have at least two like bits together. That by itself cuts the channel data rate in half because only bit transitions are recorded. The code choice is actually much more complex than that in order to control DC content and maximum and minimum run lengths.

9.16 The D-6 HDTV Component Format

The D-6 format was developed for component recording of 1126/60 or 1250/50 HDTV production signals. This calls for a total bit rate of 1,200 Mb/s (1.2 Gb/s), which was achieved by using the D-2 scanner dimensions in an 8-channel configuration. The scanner is also speeded up by about 70 percent. The same tape cassette used for D-1 and D-2 is used, but special tape must be loaded for this use.

D-6 channel coding uses an 8,12 group code.

9.17 Computer Hard Disk Video Recording

Computer hard disks are used for video postproduction editing (see Section 11.5). This technique is rapidly taking over the editing business, but it is only one of the possible applications for video on hard disks. To understand this potential, the following section describes hard disk technology and compares it to digital tape technology.

9.17.1 Hard disk technology

A hard disk uses the same magnetic recording principles described in Section 9.2, but the magnetic medium is a coating on a rotating disk rather than a tape. The disk substrate can be either flexible (floppy disk) or rigid (hard disk). A single head operates on the disk; it is mounted on a moving arm that allows the head to take specific positions along a radius as shown in Figure 9.19. To create the necessary relative motion between medium and head, the disk rotates at a speed in the range from about 60 to 150 rps.

During record or replay, the head remains stationary when processing a single track (one revolution of the disk) but it jumps rapidly from track to track if more than one track's worth of data is required. All these actions occur under computer control. Because the movement of the head from track to track takes approximately 1 millisecond or less, data from any track on the disk surface can be rapidly accessed.

Because the head can be stationary on a single track for long periods of time, it is not practical for the head to be in contact with the magnetic surface as is usual for tape recording. A deliberate and controlled separation is produced by designing the head to "fly" on the film of moving air created by the disk rotation. The design of the head mounting that will do this is a real art and it has taken many years for the technology to progress to its situation today where head-disk separations of less than 20 μinches are routinely maintained. This requires precise control of the disk surface, the head materials and mounting, and the environment in which the disk runs. Even a single dust particle getting between disk and head will cause a disk "crash" that may destroy both head and disk surface. Modern hard disk drives are sealed in their own atmosphere and can operate continuously for many years without crashing.

A disadvantage of a hard disk is that it is limited to the recording capacity of a single disk surface per head. Even though many hard disk drives have multiple disks (called *platters*) with a head for each surface, the data capacity is still far less than a similar-sized tape cassette. Present capacities are in the range of 1 to 10 GB, which doesn't go very far for video. However, the technology is continuing to improve and the production volumes are increasing rapidly, leading to higher disk capacities and lower costs in the future.

9.17.2 Hard disk applications

Most hard disk video applications use video compression to get reasonable playing time from practical disk capacities. For example, recording the Grand Alliance HDTV compressed video at 19 Mb/s gives a playing time of 7 minutes per GB. That's not too bad, considering that it is a medium that has almost instant random-access to any frame of the video, it is easily editable, and it's going to get better in the future.

Figure 9.19 Computer hard disk drive: (a) drive, (b) detail of flying head

Because several GB of storage is now available in a small, removeable package and video compression can be boiled down into one or two custom ICs, it is practical to consider a camcorder that uses a hard disk for its storage. Editing can be done in the field or the disk can be taken back to the studio for editing and replay. This is just the beginning of the use of general-purpose digital storage devices for video (see Section 9.19.3).

9.18 Optical Video Recording

Analog optical video recording has been on the market for nearly 20 years in the laser video disk. Although it was originally introduced as a consumer product, it never caught on in that market but it has seen wide use as a random-access prerecorded video store in education and industrial training. When teamed with a computer, a laser video disk player was, for many years, the only viable way to have interactive video.

9.18.1 The Compact Disc

The Compact Disc (CD)—a trade name, note the different spelling of "disc"—was introduced to the audio market in the early 1980s as a replacement for records and cassettes of prerecorded music. Because it is digital, it offers nearly perfect sound reproduction and the small disc makes a very convenient package. As the market grew, the price of CD players dropped to less than $100 and it is now the preferred medium for prerecorded audio.

It was a natural extension to apply the CD for distribution of computer data. Because most CDs are permanent recordings, the computer version is called the CD-ROM (CD-Read-Only Memory). The CD audio players were adapted for computer interfacing and are now available built-in on most new personal computers.

Figure 9.20 shows how the CD-ROM technology operates in replay. A recorded disc has a pattern of tiny pits in its surface that represent bits in a FM format. The optical system senses the difference between the pits and the lands (between pits) to produce the playback signal. As indicated by the track-to-track spacing of only 60 μinches, the density is extremely high. Because of that, the mechanical head positioning used by hard disks is not good enough and the optical playback system must have servos to maintain both focus and tracking (not shown in Figure 9.20).

Focus servoing is accomplished by sensing the symmetry of the focused spot and controlling a magnetic actuator that moves the objective lens. A typical tracking servo optically splits the laser beam into three spots using a diffraction grating. The spots are positioned so that the center spot reads the track while the other two spots read along the left and right edges of the track. The system is on-track when the two side spots deliver equal output. This

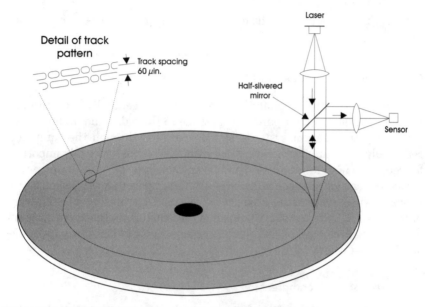

Figure 9.20 CD-ROM technology, showing replay hardware

information controls a moving mirror or other means for adjusting the radial tracking.

With the high area density of optical recording, the data capacity of a single CD-ROM surface is 680 MB. The CD-ROM standard was introduced in 1983—a new standard called the *Digital Video Disc* (DVD) with up to 20 times that capacity is now coming to market. It has been developed specifically for prerecorded video applications but there surely will be many other uses in both the computer and consumer markets.

When originally introduced, CD-ROMs could only be produced by special *mastering* facilities that were far too expensive for most users to own. Mastering facilities replicate discs by a pressing process similar to that used for vinyl records and they can produce thousands of copies at costs less than $1 per disc. Now, however, CDs can be recorded with equipment that costs as little as $1,000 and even this number is coming steadily down. Recordable CDs are made one at a time, so for mass replication of CD-ROM discs it is still appropriate to use a mastering facility.

The CD-ROM originally had a fixed data rate of 1.2 Mb/s, which limited its use in video to fairly low-quality compressed video. At that data rate, the maximum playing time is 74 minutes. A lot of development of compression techniques has been directed to that data rate and the results are sometimes remarkable, but much better video requires higher data rates. Recent CD-ROM drives have been speeded up by 2×, 4×, 6×, or more. However, since the data capacity is fixed, the playing time drops in direct proportion to the speedup. This explains why the industry is developing larger-capacity optical discs.

9.19 Trends in Professional Video Recorders

The objective of professional video recording has generally been to make systems that are as transparent as possible to the video and audio signals. With truly transparent systems, signals can pass through the equipment repeatedly without accumulating distortions, which is especially important for recorders used in production and postproduction.

Component video systems are the ultimate realization of this objective. However, these systems are expensive to own and operate and there is always pressure for recorders that are less expensive, smaller, and easier to integrate into existing broadcast or production systems. Some of the new possibilities are discussed below.

9.19.1 Use of video compression

The discussion of HDTV digital compression in Chapter 6 made it clear that substantial reduction of data rate is possible by use of video compression. High-quality HDTV pictures can be transmitted at 19 Mb/s. However, the system including the compression and decompression is definitely not transparent. For that reason, distortions can accumulate if the compression–decompression process is cascaded. That would not be a problem if compression occurred only once at the camera and all intervening processes were done in the compressed format with a single decompression at the final display. However, it can be difficult to edit video in compressed form. An ideal solution has yet to emerge, but the potential and the market are clearly there—it's only a matter of time before compressed video will be used everywhere in production, broadcasting, and distribution.

9.19.2 ¼-inch Tape Formats

At this writing, the industry is developing new systems based on the use of ¼-inch tape for consumer and professional markets. All applications of this system will be digital, even for the consumer market. Equipment and cassettes are significantly smaller than larger-tape systems, which will be advantageous in portable systems.

The design objectives for ¼-inch tape decks and cassettes assume that a modest amount of video compression will be used, so the considerations of the preceding section apply to these systems. Professional digital formats are in the proposal stage. The ¼-inch home recorder format is discussed more in Section 10.9.

9.19.3 The general-purpose digital recorder

The digital tape recorders discussed above are specifically *video* recorders in that they are tied to one or more video signal formats. They would not necessarily handle other kinds of digital data that did not contain the specific structures of scanned video formats. These limitations are necessary because of synchronization requirements of video systems.

As indicated by the use of computer hard disks and CD-ROMs discussed above, it is possible to use general-purpose digital storage devices such as hard disks to handle video. A general-purpose digital tape device would also be usable if it had a high enough data rate. The synchronization requirements could be handled by appropriate software. This probably will lead to digital video tape recorders becoming more video format-independent and more use of general-purpose digital recorders for video.

References

1. SMPTE Standards 18M-1991, 19M-1991, and 20M-1991, SMPTE, White Plains, N.Y.

2. SMPTE Standards 249M-1991, 250M-1991, 251M-1991, and 252M-1991, SMPTE, White Plains, N.Y.

3. SMPTE Standards 224M, 225M, 226M, 227M, and 228M, SMPTE, White Plains, N.Y.

4. SMPTE Standards 249M-1991, 250M-1991, 251M-1991, and 252M-1991, SMPTE, White Plains, N.Y.

5. SMPTE Standards 263M-1993, 264M-1993, and 265M-1993, SMPTE, White Plains, N.Y.

10

Home and Semiprofessional Video Recorders

10.1 Introduction

The home video recorder is one of the great technical achievements of the last two decades. The underlying technology of home VCRs—helical scan (see Section 9.4)—was originally developed for training and education users who did not require full broadcast-quality signal performance but who did need the lower cost and easier maintenance of helical-scan systems.

Product development based on the helical-scan principle then proceeded in two directions. One path led to recorders of professional quality that surpassed the original quadruplex format (see Chapter 9). The other led to VCRs for the home—low-cost, mass-produced recorders that deliver pictures of remarkably good quality, although short of professional standards. The combination of performance and cost that is available from today's home VCRs would have been unthinkable twenty years ago. The seemingly simple VCR, operating in a home environment, employs technology that embodies greater electro-mechanical sophistication than any other consumer product.

This chapter discusses home video recorders and their close relatives in the semiprofessional market.

10.2 Comparison with Professional Recorders

Although professional recorders and VCRs employ the same basic technology, there are large differences in their design priorities, which result in wide differences in their cost, performance, and operating features (see Section 9.1).

Features that are given the highest priority for home VCRs include (in approximate order of priority):

1. Low price
2. Reliability and ease of operation
3. Recording time and tape usage
4. Picture and sound quality
5. Size and weight

Picture quality, while obviously important for home VCRs, is not as critical as for professional recorders.

Features that are given the highest priority for professional recorders include:

1. Basic picture quality—bandwidth, S/N, linearity
2. Time-base stability
3. Multigeneration performance
4. System integration features
5. Compliance with broadcast standards

The effect of these priorities on product design is shown in Table 10.1, which compares the specifications of representative home VCRs and professional recorders. These specifications are chosen to illustrate the significant differences between the two product types, and they are not based on specific designs. The following sections further discuss these considerations.

Table 10.1 Typical home and professional VCR specifications

	Home VCR	Professional
Price	$200	$30,000
Size	4 in.×14 in.×12 in.	10 in.×17 in.×21 in.
Weight	<15 lb.	60 lb.
Tape width	½ in.	½ in.
Operator skill	Same as TV receiver	Technician
Recording time	Up to 6 hours	Up to 2 hours
Bandwidth		
Luminance	2.5 MHz	4.3 MHz
Chrominance	0.4 MHz	1.5 MHz
S/N (composite)	45 dB	49 dB
Differential phase	10°	3°
Time-base instability	>2 μsec.	<2 nsec.
Number of generations	2*	4
System integration	Poor	Good
Meets broadcast standards	No	Yes

*Marginal performance

10.2.1 Price, size, and weight

Price is the most critical specification for recorders intended for the consumer market. If the product is not affordable by consumers, it will not even be considered. However, other attributes cannot be sacrificed in favor of price and that is where the consumer VCR was initially such a challenge. At the time that home VCRs were introduced to the market at prices under $1,000, professional video recorders were selling for upwards of $100,000, yet the home VCR contained nearly equivalent technology.

The size and weight of rack or cabinet-mounted professional recorders are inappropriate for a home environment. Home recorders are typically of a size that can be placed on top of a TV set, on a bookcase, or on a small table and they weigh less than 15 pounds. Remarkably, the achievement of those goals in home recorders has led to professional recorders that are just as small. This is an example of how the technological advances in one part of the recording market quickly appear in all markets.

10.2.2 Reliability and operational simplicity

Although operational simplicity is an important feature of both professional recorders and home VCRs, it can reasonably be assumed that professional products will be operated by trained technicians. Home VCRs, on the other hand, must perform satisfactorily when operated by completely unskilled individuals. This establishes a requirement for a minimum of adjustments and fully automatic operation.

10.2.3 Playing time and tape storage

The *VHS* home VCR format won a clear victory over the *Beta* format (see Section 10.4) in the marketplace primarily because of its longer playing time. This is important to home users because of the need to record long programs unattended. That happens when the consumer sets up a VCR to automatically record a program for later viewing.

The VHS system offers three choices of picture quality vs. playing time so the user can choose the one that best suits the situation. The playing time of most home VCRs is at least 4 hours; the playing time of professional recorders can be less because programs seldom run longer than 1 hour without interruption—typically 2 hours for reel-to-reel models and 1 to 1.5 hours for cassettes.

Cassette loading is mandatory in the home market because handling and threading of tape would be too difficult. They also protect the tape for easy storage.

10.2.4 Bandwidth, signal-to-noise, and linearity

For direct recording of composite NTSC signals, the bandwidth of the record-playback process must be sufficient to transmit the 3.58 MHz color subcarrier and its sidebands (see Section 5.7.4). The nominal bandwidth requirement is 4.25 MHz, but to record that as-is would require too high a head-to-tape speed for a home VCR. Therefore, the signals recorded on the tape by home VCRs are not NTSC but are another format, known as *color-under* (see Section 10.7). With color-under, the chroma subcarrier frequency is converted to about 700 kHz and it is recorded below the lower sidebands of the FM luminance signal.

The color-under signal is reconverted to the NTSC format during playback, but its chroma bandwidth becomes limited by the low frequency of the color-under subcarrier. The narrow chroma bandwidth is adequate for most picture content, but it can cause color smearing of images with high color saturation and sharp hue discontinuities. Improvement of this parameter is an objective of some of the newer home VCR formats.

The bandwidth of the luminance channel is a design choice based on a cost/performance tradeoff—one of the many that must be made in home VCR design. The lower limit of the bandwidth range shown on Table 10.1 (2.5 MHz—a common choice) results in a horizontal limiting resolution of slightly more than 200 lines. This is not a high-quality picture but one that is satisfactory to most home viewers.

The signal-to-noise ratio and the linearity (as indicated by differential phase in Table 10.1) of VCRs also involve cost/performance trade-offs. The color-under format may result in coarse, low-frequency noise in areas of high saturation. This becomes especially noticeable in second-generation color-under recordings.

10.2.5 Time-base stability

As noted in Section 9.1, time-base instability is an inherent problem with helical scanning because of the long recording tracks and the elastic recording medium. Without time-base correction (TBC), the playback instability may exceed $\pm 2\,\mu$sec, three orders of magnitude greater than the ± 2 nsec stability demanded for professional recorders. Stability is restored in professional recorders with a digital TBC (see Section 9.9.4.4) but these are (so far) too costly for consumer VCRs.

Most TV receivers can synchronize successfully to the luminance channel of a home VCR signal in spite of the instability of its time base and display a stable monochrome picture. However, the instability of a directly-recorded chroma channel is too great for the color circuits of a receiver to handle and no color can be displayed. This problem is solved by the color-under approach because the color-under subcarrier on playback is up-converted back to a stable frequency (see Section 10.7.2).

Figure 10.1 Recording drum configurations (VHS)

The result of a color-under playback is that luminance and color are displayed, but they are not synchronized with each other (the receiver is synchronizing separately to each of them). That means that the subcarrier interlacing features of the NTSC system cannot work properly, and there are spurious effects that show—particularly on edges of highly-colored areas in the picture. This is tolerated by consumers, but it is another reason for future systems to get away from color-under.

Because of the lack of luminance-chrominance synchronization in home VCRs, it is impractical to integrate them in systems that require the mixing of multiple signal sources.

10.2.6 Compliance with FCC broadcast standards

Home VCR signals generally do not comply with FCC broadcast standards. Their greatest defect is their timing instability, but they may fail to meet FCC standards in other respects such as sync and blanking level format, and signal and blanking levels. They can usually be upgraded, however, to meet FCC standards by the use of digital time-base correctors and other signal processing equipment (which can easily cost more than the VCR itself).

10.3 Recording Drum Configurations

Two recording drum configurations are commonly used in home VCR tape transport systems as illustrated in Figure 10.1, the essential difference being their diameters. The diameter of the larger drum is fifty percent greater than the smaller, and it has two heads as compared with four. With these diameters, the length of a 270° track on the smaller drum is equal to the length of a 180° track on the larger. The rotational speeds are controlled so that the smaller drum rotates 270° and the larger 180° during one television field, thus creating record tracks of equal length for one field.

The scanning sequence of the two formats is shown in Figure 10.2. The configurations are compatible, and tapes recorded on one can be played back on the other.

10.4 Home VCR Record Formats

Helical scanning is a very versatile technology, and it has been adapted to a variety of formats to meet the widely different requirements of the consumer, audio-visual, and professional markets.

A description of all the home VCR formats and their variations (some of which are in a continuous process of change) is beyond the scope of this volume. The formats described here include the most commonly used consumer formats, including the Beta format that has largely been replaced by VHS.

Table 10.2 Record parameters

Parameter		U-Matic	Beta	VHS	8 mm
SMPTE type	C	E	G	H	
Tape width, in.	1.0	0.75	0.5	0.5	0.31
Drum diameter, in.	5.4	4.33	2.95	2.44/1.63	1.57
Number of record heads	1	2	2	2/4	2
Track width, mils	5.0	3.35	2.3/1.15/0.77	2.3/1.15/0.76	0.81/0.4
Tape speed, in./sec.	9.6	3.75	1.57/0.79/0.52	1.3/0.66/0.43	0.56/0.28
Recording time, hours		1	1/2/3	2/4/6	2/4
Head-tape speed, in./sec	1,000	410	276	229	148
FM carrier, MHz					
Sync tip	7.06	3.8	3.5	3.4	4.2
Peak white	10.0	5.4	4.8	4.4	5.4
Minimum wavelength (μin)	85	70	56	46	25
Color-under subcarrier, kHz	N.A.	688	688	629	743

Figure 10.2 Recording track sequence

10.4.1 Format parameters

The parameters of the most commonly used consumer and audio-visual formats are tabulated in Table 10.2.

The record parameters of the SMPTE Type C professional reel-to-reel format are shown for comparison. As indicated by their generic title, all consumer formats employ cassettes, and the trend is toward cassettes in professional recorders as well (see Chapter 9).

The U-Matic was the first helical-scan format to achieve commercial success. It was an audio-visual and industrial product with the characteristics described in Section 10.1, and it is still used in these markets.

U-Matic recorders were followed by a series of lower-priced models designed specifically for the consumer market:

Beta   1976

VHS   1978

8 mm   1983

In spite of its earlier introduction and excellent performance, the Beta format lost the competitive battle with VHS in the consumer market, largely because of the latter's longer playing time.

VHS (including its special purpose variants for special purposes) is now the most widely used VCR format, but it, in turn, is receiving competition

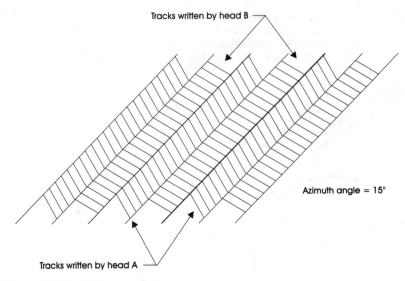

Tracks written by head B

Azimuth angle = 15°

Tracks written by head A

Figure 10.3 Azimuth recording

from 8 mm, particularly for camcorders where its small size is an important advantage.

The Beta, VHS, and 8-mm formats offer a choice of tape speeds, usually described as standard, LP (long play) and SLP (super long play). Note especially that the drum rotational speed remains constant, meaning that tape-to-head speed remains constant for different tape speeds. Therefore, changes of longitudinal tape speed result in changes of track width.

Upgraded (and more costly) models of the Beta and VHS formats have been developed—Betacam SP or Super-Beta and Super VHS or SVHS—with performance specifications that are adequate for some professional applications. These are described in Section 10.8.

10.4.2 Azimuth recording

Azimuth recording is incorporated in almost all VCR record formats because of its effectiveness in minimizing the effects of mistracking, reducing cross talk between adjacent tracks, and improving the S/N. It is not used with the SMPTE Type C professional format, but it is used in the professional digital formats.

The record pattern for azimuth recording is shown in Figure 10.3. The gap in the record head is tilted from its normal 90° orientation with respect to the head tracks, and the magnetic lines of force in the record are not parallel with it. The gap azimuth angle, the direction of tilt, is reversed on alternate scans, and on playback the signal from adjacent scans is attenuated by the difference between the head gap and magnetic record azimuths (see Section 9.7.1).

Figure 10.4 Audio head locations

The azimuth angle is normally set at about ±10° or ±15° from a 90° angle with the head path. Once chosen, azimuth must be held to tight tolerances between machines to guarantee interchangeability.

Since signal pickup from adjacent tracks is attenuated by the azimuth mismatch, it is unnecessary to provide guard bands on either side of the recorded tracks. This permits the use of wider tracks and heads that produce a higher S/N and are less sensitive to tracking errors. It also allows the tape speed and track widths to be changed while using the same physical head width.

The head width is wide enough for the widest track width needed at the highest tape speed. When the tape speed is lower, each head pass over-writes some of the track from the previous head pass, automatically resulting in narrower recorded tracks.

10.4.3 Audio recording

Audio signals can be recorded conventionally on longitudinal tracks or by a 1.5 MHz (1.4 and 1.6 MHz for stereo) FM carrier recorded with rotating heads mounted on the same drum as the video (see Figure 10.4). The performance of longitudinal-track audio is limited on home video recorders because there is not room on the narrow tape to provide wide enough audio tracks and because of magnetic cross talk from the video heads that pass over the audio tracks. The FM audio gives higher quality and is usually referred to as "hi-fi" audio. However, the FM audio shares the same tape space as the video, which precludes editing the audio once the video has been recorded.

Figure 10.5 8-mm record format. (*From Koichi Sadashige*, Developmental Trends for Future Consumer VCRs, *J. SMPTE, December 1984.*)

In the rotating head configuration, the audio head gaps are located 60° around the periphery of the drum from the video, and their azimuths are tilted ±30° from normal.

The video and audio records occupy approximately the same area on the tape. The signals are kept separate by:

1. The difference in carrier frequencies

2. The difference in the head azimuths

3. The difference in the strength of the record magnetizing force. The audio heads have a longer gap length and a longer recorded wavelength, which causes the audio recording field to penetrate more deeply into the tape. The audio signal is recorded first and the video record is overlaid on it. The video does not penetrate so deeply into the tape because of its shorter wavelength, so very little of the audio recording is erased by the video.

10.4.4 Format patterns

The format patterns for home VCRs are similar in principle to those shown in Figure 9.5 for professional recorders. The diagonal video and audio tracks occupy the center portion of the tape, one field per track, and the fixed-head audio and control tracks are recorded longitudinally along the edges. The 8-mm format (Figure 10.5) shows the essentials features of home VCR record patterns although they have many variations.

Figure 10.6 Record frequency spectrum, color-under system

10.4.5 Record frequency spectrum

The frequency spectrum of an 8-mm record is shown in Figure 10.6. The frequency-modulated luminance channel occupies the region from 4.28 to 6.8 MHz. The audio signal occupies a band around 1.5 MHz, and the color-under chroma carrier occupies a band around 750 MHz. There may be, in addition, four low-frequency pilot tones that are used for automatic scan tracking.

10.5 Automatic Scan Tracking

Automatic scan tracking (AST) systems that employ bimorph mounting of the heads as used in professional recorders were described in Section 9.9.4.2. Home VCRs may use different systems, one of which is based on the use of low-frequency pilot tones and incremental adjustments of the drum rotational speed.

The pilot tones are recorded on either side of the video tracks, the tone on one side having a higher frequency than the tone on the other. On playback, the head is wide enough to pick up both tones, and if it is properly centered the amplitude of both will be equal. If the tracking is not perfect, the amplitude of one of the tones will be greater, and the capstan servo that determines the tape speed is adjusted by a feed-back circuit to restore the tracking.

10.6 Record Information Density

10.6.1 Record density requirements

The home VCR marketplace has imposed two conflicting requirements on VCR designs—small size and long playing time. These needs can be satisfied simultaneously only by increasing the recording density (see Section 9.7.3), i.e., the amount of information recorded per unit area of tape. This can be calculated as the amount of tape area required to record one frame of the TV signal:

$$\text{tape area / frame} = \frac{\text{tape speed} \times \text{tape width}}{\text{frame rate}}$$

The increase in this parameter in the more recently introduced formats shown in Table 10.2 has been dramatic. The area required to record one frame of an NTSC signal with these formats as compared with the SMPTE Type C format is shown in Table 10.3.

Some of the area reduction is possible because of differences in the bandwidths of the recorded signals. For example, the highest frequency component of the 8-mm RF signal is about 7 MHz as compared with about 12 MHz for the Type C signal. But most of the reduction of the area per frame is due to higher density in the record pattern.

10.6.2 High-density technologies

The information density of the recorded signal is determined by the wavelength of its highest frequency component and the *pitch* or spacing of the recording tracks (equal to the track width divided by the sine of the track angle for azimuth recording). The track width and minimum wavelength of the common VCR formats were given in Table 10.2.

Reducing the track pitch and minimum wavelength from the values used in the SMPTE Type C format to the values used in the Beta, VHS, and 8-mm

Table 10.3 Tape area per frame for different formats

Format	Area/frame, in$^2 \times 100$
SMPTE Type C	32
U-Matic	6.25
Beta	2.62
VHS	2.16
8 mm	0.58

standards was a major engineering challenge, and it required the addition or improvement of a number of technologies. The problems included:

1. Reduction in bandwidth and a lower S/N because of the necessity of narrower recording heads.
2. Tracking difficulties because of the narrower tracks.
3. Cross talk between adjacent tracks.

   The solutions included:

1. Improvements in tape/head performance (see Section 9.7). The track pitch and the wavelengths in the 8-mm format are so small that metal-particle or evaporated-metal tape is almost a necessity.
2. Azimuth recording (see Section 10.4.2). This reduces the effects of mistracking, permits the use of wider recording heads for improved S/N, and reduces cross talk between tracks.
3. Automatic scan tracking (AST) (see Section 10.5).

10.7 Color-Under Signals

10.7.1 Record frequency spectrum and electronics

The record frequency spectrum of the color-under signal format is shown in Figure 10.6. The essential elements of the record electronics are shown in Figure 10.7.

The chroma component (the color subcarrier) of the incoming signal is separated by a band-pass filter, and the NTSC chroma and luminance components are processed separately. The frequency of the chroma component is downconverted to the 650-750 kHz region by heterodyning and its

Figure 10.7 Record electronics, color-under system

bandwidth reduced (see Table 10.2) to form the color-under component of the recorded signal. This places the color subcarrier and its sidebands below the band occupied by the FM video signal rather than being interleaved with it as with NTSC and PAL; hence the term "color-under." Since the bandwidth available for the color-under signal is limited, it is necessary to pass it through a band limiting filter.

The luminance component passes through a low-pass filter with a bandwidth of about 2.5 MHz (the luminance component has a wider bandwidth in some "high-band" models—see Section 10.8) and it frequency modulates the video carrier between the limits shown on Table 10.2.

The audio signal is recorded on a track along the edge of the tape, or alternatively it is multiplexed with the video record (see Section 10.4.3).

If pilot tones are used for automatic scan tracking, they are recorded at the bottom of the spectrum as shown in Figure 10.6

10.7.2 Playback electronics

Unlike professional recorders, which require a time-base stability of about ±2 nsec on playback so that the signal can be mixed and edited with other signals (see Chapter 11), home VCRs are not used in complex editing systems and can tolerate a considerably higher degree of instability. The use of time-base correctors is an option that can be included, however; see below.

Figure 10.8 is a functional diagram of the electronics of a typical playback system. The chroma and FM luminance signals are separated, and the luminance signal is demodulated. As noted in Section 10.2.5, the time-base instability of the chroma signal is too great to operate satisfactorily with a fixed reference subcarrier, and the reference subcarrier is generated by a burst-controlled oscillator whose phase varies with the timing errors on the

Figure 10.8 Playback electronics, color-under system

Figure 10.9 Home VCR chroma time-base corrector

tape. This retains the phase relationship between the color and reference subcarriers to a reasonably satisfactory degree.

Improved performance can be obtained with a relatively inexpensive analog time-base corrector for the chroma signal as shown in Figure 10.9. The playback video is separated by filters into its chroma and luminance components. The luminance component, its bandwidth limited to about 2.5 MHz, passes directly to the output. A burst-controlled subcarrier is generated, as in the playback electronics in Figure 10.8. This subcarrier is then used to demodulate the I and Q or color difference signals. These, in turn, remodulate a stable subcarrier, which is combined with the uncorrected luminance signal. This gives a stable chroma signal combined with an unstable luminance, a result which is satisfactory for many applications.

10.8 S-VHS and Hi-8

Tape and head performance has improved substantially since the VHS and 8-mm home VCR systems were introduced. Exploiting these advances, higher-performing standards have been developed for both of these systems. The new standards are called Super VHS (S-VHS) and Hi-8; they are based on the same basic tape transports, scanners, and cassettes as the original VHS and 8-mm systems. Specific parameters are shown in Table 10.4.

Because of the higher performance, these standards are very suitable for semiprofessional applications and product lines of recorders and camcorders have been developed for this market. Prices are slightly higher than home VCRs but far less than broadcast-level products.

10.8.1 Control interfaces

A significant feature of higher-priced home and semiprofessional recorders is a control interface to allow the units to be used with an edit controller. This requires that tape transport controls for play, shuttle, and recording be

Table 10.4 Record parameters of S-VHS and Hi-8

Parameter	S-VHS	Hi-8
Tape width, in.	0.5	0.31
Drum diameter, in.	2.44/1.63	1.57
Number of record heads	2/4	2
Track width, mils	2.3/1/15/0.76	0.81/0.4
Tape speed, in./sec.	1.3/0.66/0.43	0.56/0.28
Recording time, hours	2/4/6	2/4
Head-tape speed, in./sec.	229	148
FM carrier, MHz		
Sync tip	5.4	5.7
Peak white	7.0	7.7
Minimum wavelength (μin)	33	19
Color-under subcarrier, kHz	629	743

brought out to a connector for remote control. Two standards for that purpose are called Control-L and Control-M.

10.9 Digital Home Recorders

Digital video recording is now highly developed in the broadcast and professional fields (see Chapter 9). However, it is considered high-end (meaning expensive) and would seem out of place in the home market. However, digital technology continues to become less expensive and it is just a matter of time before it will be fully competitive with analog technology in nearly all fields. With that in mind, a group of companies has developed a digital home video recording system based on ¼-inch tape. This is called the *Digital Video Cassette* (DVC) system and several camcorder models are already on the market.

Video compression is the key feature that makes a digital home video recorder possible and at the same time distinguishes it from professional digital recorders. Compression is generally not used in professional recorders because it is not transparent and therefore could cause accumulating distortions in multigenerational work. However, that is not as important an issue for the home market because home users will seldom go beyond two generations of recording.

The DVC system has a basic data rate of 25 Mb/second. That is high enough to record the digital HDTV broadcast format as well as compressed versions of NTSC or PAL signals. With that data rate, performance is excellent as long as the compression-decompression process is only performed once (implying that copying must be done digitally, without going to analog between copies). A professional version of DVC is also contemplated by doubling the tape speed to deliver a data rate of 50 Mb/second. This will still require the use of compression.

Table 10.5 DVC specifications

	NTSC	PAL
Tape width, in.	0.25	0.25
Tape speed, in./sec.	1.48	1.48
Recording time, hours	4.5/1.0	4.5/1.0
Drum diameter, in.	0.85	0.85
Head-tape speed, in./sec.	390	390
Number of record heads	2	2
Track width, mils	0.25	0.25
Video sampling rates, MHz (8 bits/sample)		
Luminance	13.5	13.5
Chrominance (2)	3.375	6.75
Video compression factor	6.5:1	8.5:1
Recorded data rate, Mb/sec.	25	25
Audio recording method	PCM, stereo	PCM, stereo
Audio bits/sample	12 or 16	12 or 16
No. of stereo channels	2 or 1	2 or 1

The specifications of DVC are listed in Table 10.5.

10.10 Camcorders

Shooting video requires at least two types of equipment—a camera and a recorder. As equipment became smaller, it was natural to make a combination of these two. Thus, the *camcorder* was born. Camcorders were first developed in the broadcast market, where they were used for a new class of operation called *electronic news gathering* (ENG.) The first ENG camcorders were large and heavy, but the single unit was an advantage in that it eliminated cables and in-field hook-up. An ENG operator could begin shooting as soon as he/she took the camcorder out of its carrying case.

Later, as the home VCR market matured, home camcorders were introduced. Since home VCRs were inherently smaller to begin with, home camcorders were smaller, lighter, and more convenient.

10.10.1 Camcorder configurations

Because of their weight, early portable cameras and camcorders were designed to be carried and used on the operator's shoulder. This allowed the operator to use the stability and inertia of his/her upper body to achieve smooth camera movement. Shoulder mounting inherently requires some large dimensions because the camera viewfinder eyepiece must be in front of the eye while the center of gravity of the unit as a whole must be back at the shoulder. This is shown in Figure 10.10 (a).

(a)   (b)   (c)

Figure 10.10 Camcorder configurations

As it became possible to make a smaller unit, home camcorders that could be held in one hand were introduced. These are sometimes called *palm* camcorders, and they are held up in front of the eye for use as shown in Figure 10.10 (b). Since there is no limiting large dimension for this type of support, the unit could be as small and light as possible. Most such units weigh less than two pounds with battery and cassette. They are easy to carry and to use, but the lighter weight and the "out-front" holding makes it more difficult to keep them steady for stable pictures. This has further led to the development of *image stabilization* (see Section 10.10.4).

A third configuration depends on the use of an LCD viewfinder that is large enough for the operator to view without an eyepiece. This configuration, shown in Figure 10.10 (c), allows the camcorder to be held anywhere within the operator's reach. The viewfinder swivels so that it can be positioned for viewing from any angle.

10.10.2 Camcorder system

The system diagram for a camcorder is not very different from that of individual cameras and recorders except that some common functions are combined and the signal interface between camera and recorder does not have to be standard NTSC or PAL.

The camcorder's camera can generate the color-under format directly, which eliminates the composite encoder and any degradation that may cause. However, most camcorders have a separate composite output signal for driving a monitor, computer, or another recorder, so most units still contain the encoder.

CCD

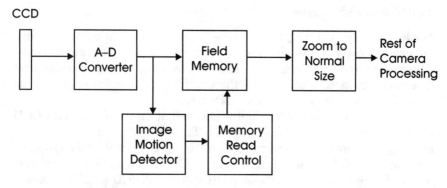

Figure 10.11 Electronic image stabilization

Sync and power-conditioning circuits for the camera and recorder are combined in a camcorder. This is one case of duplication that is eliminated by the single-unit configuration.

The camera in a camcorder is always a CCD, usually only one, although a few three-CCD camcorders are on the market for semiprofessional use (see Section 10.11). Tape transports for camcorders are a special design, optimized for small size and light weight. In the VHS format, a special small cassette (VHS-C) is available for camcorders. VHS-C cassettes are playable in any standard VHS players by using an adaptor cassette.

One might think that a camcorder would not need a playback capability. However, a major advantage of tape (compared to motion picture film) is that the result of shooting is viewable immediately, while still in the field. Therefore, playback to the viewfinder is an essential feature. Playback is also needed because most camcorders have some form of in-camera editing, such as assemble editing (see Section 11.2.2.1). Playback is essential to determine edit points.

Another purpose of playback in the camera for home use is to allow the camcorder to have a different tape format from the bulk of home VCRs, which are VHS. This has allowed 8-mm camcorders to proliferate even though not many homes have 8-mm players. A video output from the camcorder connects to a TV receiver for playback.

10.10.3 Viewfinders

Another advantage of video shooting is that an electronic viewfinder shows exactly the picture being recorded. For this reason, all camcorders have electronic viewfinders, using either CRT or LCD panel displays. The trend is to color LCD displays, and as mentioned above, a large LCD panel display gives added flexibility compared to a viewfinder that requires an eyepiece. Viewfinders, even those with eyepieces, usually have very flexible mounting so that they can be adjusted to the operator's preference.

10.10.4 Image stabilization

Holding a small, lightweight camcorder steady enough to deliver a stable picture, especially with a telephoto zoom, is not easy. Many camcorders have *image stabilization* to help with this. There are electronic, optical, and mechanical systems available, but the trend seems to be toward electronic methods. Figure 10.11 shows how this works.

The operation is similar to a digital TBC in that the picture from the CCD is digitized and stored in a field memory. The read-out from the memory is controlled by a means for sensing motion. The output signal is taken from an area that is slightly smaller than the full CCD area and motion correction is obtained by moving the output area relative to the full CCD area. The location of the output area moves according to the motion sensing to essentially cancel the movement of the picture. The output picture must be zoomed up slightly to achieve a full size picture—this can be done in the memory read-out process.

The trick of this is in the motion sensing. One approach[1] divides the image into four areas and performs frame-to-frame comparison separately for each area to develop four motion vectors. The four vectors are processed by a fuzzy logic system to identify the component of motion that is common to all areas—this is assumed to be caused by camera movement. This motion component is used to control the location in the field memory that is read out. There's more to it, because the system must deal with motion in the scene itself and it also should not try to correct if the operator is smoothly panning or tilting the camera.

Another approach[2] (not shown in the figure) uses the CCD itself instead of a field memory and applies vertical motion control signals to the CCD scanning. That can correct for vertical instability; a line memory is used to provide horizontal correction.

10.11 Semiprofessional and Professional Camcorders

The camcorder is not just a home product. Upgraded home-market camcorders are available for the semiprofessional market, and full broadcast camcorders are available for the professional markets.

10.11.1 Professional features

Semiprofessional and professional camcorders have better cameras, with more elements in a single CCD or three CCDs. Recorders are compatible with one of the semiprofessional or professional formats, such as Hi-8, S-VHS, Betacam, or M-II. Professional camcorders may have a feature known as *docking*, where the camera and recorder are separate units so the "cam" and "corder" can be taken apart. This is useful in that either part could be upgraded in the future without obsoleting the investment in the other part.

It also may be convenient at times to use the camera by itself or with a separate recorder. For this purpose, an adapter module plugs in to the docking interface of the camera. The adapter module provides all the usual system interface features that a stand-alone studio camera would have. Thus, the camcorder's camera could be used in the studio when it was not needed in the field.

References

1. Egusa, Y., Akahori, H., Morimura, A., and Wakami, N., "An Application of Fuzzy Set Theory for an Electronic Video Camera Image Stabilizer," *IEEE Transactions on Fuzzy Systems*, vol. 3, no. 3, August 1995.

2. Kubota, Y., Nishi, Y., Shintani, K., Urabe, T., Shimada, K., and Katsumoto, T., "Latest Advances in Camcorder Technology," *Proceedings of the IEEE*, vol. 82, no. 4, April 1994.

11

Video Postproduction Systems

11.1 Introduction

Before the development of electronic video recording, video programs were produced live with everything occurring in real time, or they were created on motion picture film. Live production causes serious limitations in that every element of the program has to occur on schedule when it is called for and there is no opportunity to correct mistakes or to integrate material shot at different times or locations.

Film production inherently separates the tasks of shooting and program assembly because of the need to develop the film. Thus, it avoids the limitations of live television production but it has its own difficulties. The most serious of these is that the actual success of shooting is not known until the film is developed—hours or days later. Film also requires an *editing* process to assemble the individual shots into a program. Although this is a difficult process that calls for a lot of skill, it actually is a major advantage because it allows the separation of the shooting and the program assembly steps. It gave rise to the part of the film industry known as *postproduction*.

The film-making process logically divides into two parts:

1. production—get it all onto film. This is the familiar activity where each piece of each scene in the film is set up in a studio or on-location and shot with movie cameras. The shots are separate and no effort is made to assemble them together at this time. Shooting is planned at the convenience of the set-builders and the talent. For example, all the shots on a particular set would be done in the same session without regard to where they will be used in the final movie.

2. postproduction—add effects, edit, and assemble. This has grown into a very complex process involving many different kinds of equipment and skills. Many specialty companies exist to support the different tasks of film postproduction.

As the film industry matured, different people became masters of the art of each of the parts of production and postproduction. A successful film required all of them. This style of operation greatly enhanced the artistic portfolio of film-making (called the *production values*) and it depended entirely on the ability to capture the elemental program material onto an editable medium. Of course, that was not available for television in the early days except for *kinescope recording*, which was recording from a television screen to film. However, the performance of kinescope recording was so poor that it was never seriously used as an editing medium.

When video recording was developed in the late 1950s, the opportunity to have all-electronic film-style production and postproduction was immediately recognized. However, videotape technology was not ready for that because of the difficulty of editing (it required literally cutting and pasting the tape). But when there's a need, video engineers will find a solution, and videotape electronic editing was quickly developed.

11.2 Electronic Editing of Videotape

In electronic editing, a tape of the complete program is created by recording the individual scenes onto a new tape in the precise sequence required by the program. If the scenes are already on tape, then the process involves copying the scene from its original tape to the new tape. This creates a *second-generation* recording, which has the problem that any distortions of the signals occurring in either generation will accumulate in the final recording. However, it works, and an entire industry for videotape editing has grown up from this idea and its enhancements. At the same time, analog videotape performance has improved to the point that electronic editing processes involving up to three or four generations are quite satisfactory.

By the use of multiple tape recorders and video switching systems, any of the transition effects (dissolves, wipes, etc.) that are possible in live television can also be used in postproduction. In fact, they can be done better because the source materials and timings can be adjusted much more precisely than is ever possible in live production.

11.2.1 Synchronization for editing

For electronic editing to work at all, there must be some degree of synchronization between the signals being edited. Synchronization of composite video signals occurs at three levels:

1. frame/field

2. line

3. color subcarrier

If the scanning of two signals does not match so that field sync intervals occur at approximately the same time within a few lines, all receivers will have to resynchronize their vertical scanning when the video feed is switched from one signal to the other. This causes a "roll" effect, where the vertical synchronization interval will briefly appear in the picture and the picture will slowly move to either the top or the bottom of the screen while it restabilizes. It is very disturbing and is not acceptable for almost any use.

A corollary effect occurs if such a nonsynchronized switch is done at the input of a video recorder. It causes the recorder to lose lock, which usually causes severe interference to the picture until lock is reestablished. Because of these annoying effects, frame/field synchronization was the first editing feature to be developed. It allowed direct switching between signals without disturbing receivers or tape recorders.

However, with frame/field sync alone, it was not possible to combine or mix signals to achieve effects such as a lap dissolve or a wipe. When the signals were combined, one could see that the sync stability was not good enough at the line level—the two signals jittered horizontally with respect to each other, often by a large percentage of the width of the picture. Line sync techniques were developed to correct this problem, which then allowed mixing of monochrome signals. But it didn't work for color signals—the color subcarrier phase still jittered between the signals, making the color of the mixed signal totally unstable. Color synchronization techniques were thus added to correct this problem.

The development of synchronization techniques for tape editing took a number of years, but it eventually became possible to perform fancy effects and editing without concern for synchronization. Today, most semiprofessional video recorders and all professional recorders are capable of full color synchronization. However, most home recorders have only frame/field synchronization, which limits their use in editing systems.

Another aspect of synchronization is maintaining exact frame locations so that corresponding points in different video or audio clips can be identified. This is best accomplished through the use of *time code*, which is a system that gives each frame a number. It is covered in Section 11.4.

11.2.2 Tape editing techniques

There are three basic methods of videotape editing:

1. Assemble editing

2. Insert editing

3. A-B roll editing

The first two methods require one or two tape machines and the third process takes at least three machines. In each case, special control features are required in the system to allow precise positioning of the tape and accurate synchronization between machines when they are run. The degree of sophistication and performance of these features varies widely between professional, semiprofessional, and home recorders, but they exist to some extent in all video recorders today.

11.2.2.1 Assemble editing Assemble editing is often called *add-on* editing because it is a process of building the edited tape by adding one scene after another. Figure 11.1 shows an assemble editing setup. A video source is set up to deliver the signal to be added on, and the recorder making the new tape (called the *edited master*) is positioned at a point ahead of where the actual edit is going to occur. This is called setting up for *pre-roll*. When the edit is made, both the source and the recorder are started up and at the predetermined edit point, the recorder is switched from playback to recording to record the input. The recording then is made up to a point slightly beyond where the current scene is expected to end. At that point, everything is stopped.

In assemble editing, the source can either be a recorder or a live camera. If it is a recorder, that unit must also be set up for preroll by appropriately positioning its tape ahead of the edit point. Assuming that the source tape is an original (first-generation) recording, the resulting edited master will be a second-generation recording. Assemble editing requires that the shots be added in the order they will appear and the transitions between scenes are usually limited to cuts only.

For simple scenes, assemble editing may be done live by using a camera and one recorder. The process is the same: set up the edit point and the preroll,

Figure 11.1 Equipment setup for assemble or insert editing

run the tape, and cue the source talent to start doing their thing at the edit point. A first-generation tape is produced. This works well for low-budget productions, but it offers no opportunity for correcting mistakes once editing has gone beyond a particular shot (unless all shots after the mistake are redone).

For this kind of one-camera live-edited production, it is actually better to use the insert editing technique described below, because that will allow any shot to be redone at any time as long as the duration of the shot does not change in the redo.

11.2.2.2 Insert editing The insert editing technique is an enhancement of the assemble technique that depends on having the new tape prerecorded with at least black pictures and sync. That allows edits to be done at any point on the tape at any time, so there is no need to edit in the order of the final scene. This is a better technique than assemble for any editing, but it requires somewhat more expensive features in the recorder. It works best with full color synchronization, although it is still useful with frame/field sync. It is not necessarily available in the lowest-priced recorders. The equipment setup for insert editing is the same as shown in Figure 11.1 for assemble editing.

Insert editing is operated the same as assemble editing except that the insert must be stopped at exactly the right point because it will not be overwritten by another edit.

A useful technique with insert editing is to record the audio track for the segment before inserting the video. This allows the audio to be used for the timing of the video inserts. When the insert editing is done, audio-video synchronization has also been completed

11.2.2.3 A-B roll editing By using more equipment as shown in Figure 11.2 and going to more trouble, many other possibilities are available by using *A-B roll* editing. In this technique, two source VCRs (A and B) are fed through a computer-controlled video switcher to the recording machine. Alternate shots are recorded on each of the source VCRs; these tapes are produced in a preediting session by either assemble or insert editing. When the edit is run, the A and B machines start up together and the video feed to the record machine is switched in real time between them. If the A and B recordings overlap appropriately in time, any transition type that the video switcher has can be used in the edits. This type of editing requires full color synchronization of the source machines. For more complex assembly, additional source

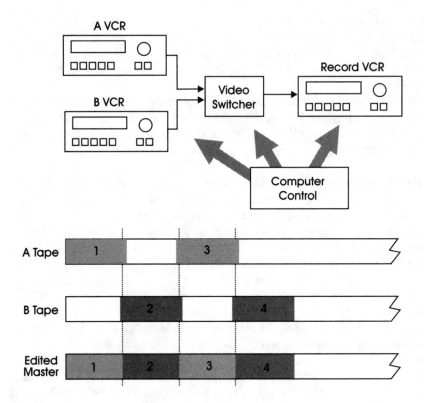

Figure 11.2 A-B roll editing setup and operation

machines can be added so that more than two signals can be combined at once.

11.2.3 Off-line editing

A professional-level editing suite with all the capabilities discussed above is expensive, it requires a highly skilled operator, and it costs hundreds of dollars per hour to operate. Since much of the time spent in editing is simply reviewing the production material and locating preliminary edit points, it is valuable to have a low-cost way to do all the preliminary work and go to the full edit suite only for the final editing. This thinking has led to the idea of *off-line* editing.

Off-line means that the bulk of the edit decision making is done in a different format (and presumably a lower-cost one) than will be used for the final editing. In film editing, there is the *Moviola* machine, which is relatively simple equipment that allows raw footage to be viewed with precise control so that someone can locate the desired edit points in each shot. A lot of

OFF-LINE WORKSTATION   EDITING SUITE

Figure 11.3 Off-line editing

development has gone into developing the electronic equivalent of the Moviola.

One easy way of doing this is to copy all the production shot tapes to a low-cost format such as VHS, and let the producer, director, and anyone else involved in the edit decision-making process take that and use home VHS players to locate edit points (see Figure 11.3). That is not as far-fetched as it sounds, because the time codes from the production tapes can be copied into the video frames themselves on the VHS tapes (this is called *burned-in time code*). Then, when a proposed edit point has been located by still-framing the home recorder, the time code value can be read from the TV screen and written into an edit list on paper. When all the preliminary decisions have been made this way, the edit list is taken into a full editing suite, entered into the computer, and preliminary edits can be immediately viewed. Of course, the result may need fine tuning, but a lot of expensive edit suite time has been saved.

The use of home-style VHS equipment does not allow the off-line decision makers to actually view the edits they have chosen. They have to come into the expensive edit suite for that. Other systems have been proposed and even developed to provide more capable off-line editing, but until the growth of digital video, they were all too expensive for the purpose. Now, with personal computers and digital video, practical systems for off-line editing are available (see Section 11.5).

11.3 Tasks of Postproduction

There is much more than editing in postproduction. Some people refer to the entire postproduction process as editing, but in this book, editing is just the process of assembling separately-produced shots into a scene or program. Postproduction can include many other tasks in the process of delivering the highest-quality, most artistic program from the material that has been produced:

video color correction or enhancement

transition effects

video effects

video titles

audio sweetening

audio mixing

audio-video integration

video editing

final format conversion

Each of these tasks could take a chapter itself; here there is only room for a summary (see the bibliography for more information).

11.3.1 Video color correction or enhancement

Many things can happen in video production to cause the resulting video to not be up to full quality. Sometimes, those problems can be corrected in postproduction. For example, if the colors do not match correctly in all of the shots to be used, a color corrector can usually fix it. Similarly, if a particular shot is noisy, the S/N can be improved by the use of noise reduction. Fixing substandard video or video errors is an important part of postproduction.

11.3.2 Video transitions

The use of video transition effects such as dissolves or wipes has become commonplace in most video production. If a program has video transition effects, it is a clue that the program was probably built in postproduction. Figure 11.4 shows some typical transition patterns. These transitions do not change the picture geometry—they simply replace one picture with another according to the motion shown by the arrows.

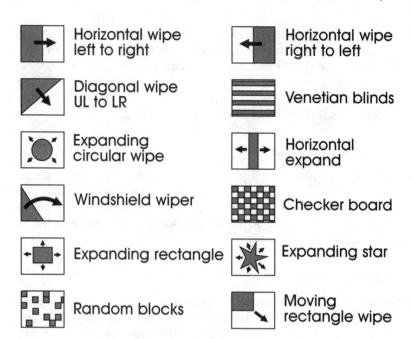

Horizontal wipe left to right

Horizontal wipe right to left

Diagonal wipe UL to LR

Venetian blinds

Expanding circular wipe

Horizontal expand

Windshield wiper

Checker board

Expanding rectangle

Expanding star

Random blocks

Moving rectangle wipe

Figure 11.4 Some typical video transition effects

11.3.3 Video effects

There are many video effects that are more than transitions. Examples are:

chroma keying—this is an analog effect where two pictures are combined by using one picture to create an outline for keying in another picture. A simple example is showing a window behind a newscaster that displays another video. The window exists as a bright blue rectangle in the background of the news set. The chroma keyer is set up to insert the other picture into any area that is bright blue. This effect can also be done digitally (see below).

split screen—this is an analog effect where different pictures appear in different areas of the screen. For example, the screen may be split between left and right or between top and bottom. It is similar to starting a wipe effect and stopping it in the center.

digital effects—a whole class of effects are referred to as "digital" because they became available in television when digital video processing was first developed. They include all effects that involve any kind of geometric manipulation of pictures, which is something that cannot be done with analog hardware. Examples are: changing the size of a picture (zooming), rotating pictures, page-turning effects, or placing pictures into odd-shaped areas, such as having pictures on the sides of a 3-D cube.

The chroma key newscaster example described above can be combined with a digital zoom so that the chroma-keyed window image zooms out to full screen for presentation of the clip that is introduced in the window.

Because digital video effects are so powerful, it makes sense in analog systems to use A-D and D-A conversion just to get the effects. Of course, in a system that is already digital, no conversion is necessary.

11.3.4 Video titles

Many programs use titles or other computer-generated overlays. Special equipment in postproduction generates titles and overlays and keys them into the video on demand. When keying computer-generated art into composite analog video signals (NTSC or PAL), special features must be included to avoid aliasing on the keyed edges.

11.3.5 Audio sweetening

The equivalent to video enhancement and color correction is *sweetening* for audio. This includes processes to make the audio sound as good as possible. Examples are: noise reduction, equalization, effects (such as reverberation), stereo balancing, and level matching.

11.3.6 Audio mixing

Mixing is the audio process for combining signals. Most audio channels for video are produced as several separate tracks that get combined in postproduction. For example, it is common to produce separate tracks for music, dialog, effects, and ambience. Each track can be optimized by itself and then the audio postproduction editor will mix these to achieve the desired final audio track.

11.3.7 Audio-video integration

It is normal to separate audio and video and operate on them differently during postproduction. This is necessary because the audio and video usually have different continuity. For example, in a dialog scene between two people, the video may be cut or dissolved between shots of the two participants or cut away to supporting material, but the dialog audio runs continuously throughout. By separating the audio, the video editor does not have to worry about the audio when doing the cuts and dissolves. Similarly, the audio mixer does not have to be concerned with the video.

Once both audio and video are completed separately, they can be combined in the final format. The only requirement is to maintain synchronization. This is facilitated by having the entire postproduction process based on time code (see Section 11.4).

11.3.8 Video editing

Techniques for video editing were described in Section 11.2.2 for tape-based editing and in Section 11.5 for computer-based editing.

11.3.9 Final format conversion

Usually, the edited master is produced in the format required for the end user. However, if additional copies are to be made from the master, it is better to perform all postproduction operations in the highest-performance format available and drop down to the distribution format only for the copies made from the edited master. This is especially important if the end product is to be a low-performance format such as VHS tape or microcomputer compressed video such as Video for Windows. In any case, the final format conversion is considered to be the last step of postproduction.

11.4 Time Code

Postproduction requires exact positioning of video and audio edit points. Points can be found by playing video and audio on machines that have shuttle (variable-speed playback) and jog (single-frame stepping) features, but once the desired frame is located, there must be a way to save it and come back to it exactly when doing the edit. That is the purpose of *time code*. All frames are identified with sequential numbers based on hours, minutes, seconds, and frames within a second.

Time codes on video or audio tape are usually placed on a separate longitudinal track that will be readable when the tape moves at any speed. This is called LTC. Variable-speed readability is important because it is desirable to run the tape as fast as possible when searching for a location and slowing down only as the time code indicates that the tape is getting close to the target point. Time code can be recorded during the shooting or it can be inserted after shooting, at the start of postproduction.

Another possibility for time code in video is to insert digital codes in the vertical blanking interval of the video. There is plenty of room for that—it is called *vertical interval time code* (VITC). However, this has limitations compared to the separate tracks because it is more difficult to make VITC readable at all tape speeds.

11.4.1 SMPTE time code

Standards for time codes have been set by the SMPTE. Both LTC and VITC versions have been defined. They are based on a bit rate of 2,400 bits/second, which provides 80 bits per frame in a 30-frames/second system and can easily be recorded on an audio track and played back at any speed from one-fifth of

Figure 11.5 SMPTE time code 80-bit data format

normal up to 20 times normal. The 80 bits per frame are allocated as shown in Figure 11.5.

Time is specified in hours, minutes, seconds, and frames; 26 bits are allocated for that. In the case of NTSC, where the frame rate is not exactly 30/second (it is 29.97), it is necessary to periodically drop a frame from the time code to keep the code in sync with clock time. One frame is dropped every 1,000 frames, which works out to 108 frames dropped per hour. A flag bit indicates when the dropped-frame mode is in use. Another flag can be used to identify one color frame of the color-frame sequence.

11.5 Digital Postproduction

As digital video technology has developed, one of the first places it was used was postproduction. The digital effects, described above, have had a particularly important impact on television postproduction. But now, with video compression technology, complete digital editing has become a reality. But the real reason for this is not simply the usual advantages of digital systems, but because digital editing allows a completely different mode of operation—*nonlinear* editing.

11.5.1 Linear and nonlinear

Videotape is a *linear* medium, meaning that it is one-dimensional in time. To get from point A to point B on a tape, one must play through all the intervening material. There is no means of fast random-access to points in a tape recording.

However, a computer hard disk is a true random-access device which provides rapid access to any material recorded on the disk by moving the read heads from track to track. Any point on a typical hard disk can be accessed in a few hundred milliseconds or less. That is fast, but not fast enough to provide seamless editing of video, which would require the access time to be less than the width of vertical blanking, about 1 millisecond for NTSC or PAL.

In a digital system this problem is easily handled by providing some buffer memory that stores enough video data to cover the seek time of the hard disk. Then, an editing system can effectively switch to any frame of video on the disk "instantaneously." It is an ideal environment for editing. Because there is no restriction of playing through material on a linear medium, hard disk editing is called *nonlinear*.

11.5.2 Other requirements of digital editing

The idea of true random access for video editing using computer hard disks and buffering has been around for many years, but there has been a large problem that has prevented much use of it—digital video was just too much data. This severely limited the amount of video that could be stored on any reasonable-sized hard disk or disks. Video compression technology that has been developed for computer video display could be used, but it did not provide full quality performance, so the use of nonlinear editing was only for low-quality services, or for an off-line medium. In that case, full editing can be done with the nonlinear system, but to obtain a full quality output, the EDL from the nonlinear system has to be taken to a full quality (analog or digital) system to perform the final editing.

However, video compression is a moving target, and performance has improved in the latest systems to the point that affordable nonlinear systems can in some cases deliver full broadcast-quality output.

A full quality nonlinear editor represents more than an order of magnitude reduction of cost for an editing system. As shown in Figure 11.6, all it takes is a powerful PC, a lot of hard disk storage, and add-in hardware for video capture, compression, and decompression. Because of the extreme competition in the markets for these systems, costs are dropping rapidly and performance is improving at the same time. It is the editing system of the future.

Figure 11.6 Hardware for a nonlinear editing system

11.5.3 Typical nonlinear editor

There are many products on the market for nonlinear editing at different quality levels. Because of that diversity, this discussion is based on a fictional system that embodies the principal features of the best real products. The main control screen is shown in Figure 11.7.

At the top of the screen, an area of "thumbnail" images shows the start frame of each shot selected for the edit. At the top right there is a preview window where actual edits (or the whole scene) can be viewed. Below that is a "timeline" view. It shows the edits on a time scale using the thumbnail names to identify them and indicates the timing and duration of each. The time scale can be expanded or compressed and this part of the display can scroll horizontally to show different parts of the scene in as much or as little detail as desired. The timelines also include audio tracks. Below the timelines there is a window that displays the EDL in actual timecode (this is for the purist who wants to fine tune the numbers) and a set of control buttons for the system.

Operation of the system is as follows:

First, all source video is copied from original tape to the hard drive. The copying includes the time code (or adds it if it was not on the tape) and video compression.

Figure 11.7 Control screen for a hypothetical nonlinear editor

Once all the video has been copied, the operator uses the menu to select a source video for *logging* of shots. (Logging is the process of specifying exact frame numbers for the start and end of each shot.)

The first frame of the video appears in the first thumbnail window. By using the control buttons, the operator can shuttle and jog the video to find an exact frame for starting the shot. By clicking the "Start" button, time code for this frame is logged in the computer, the frame is shown in the first thumbnail, and the source tape control automatically jumps to the next thumbnail space.

If the operator selects the first thumbnail again and clicks the "Duration" button, he or she can shuttle or jog the picture in the thumbnail to log the edit end point. The "End" button is clicked to confirm this.

All shots for the session are chosen by repeating the above procedure. Any points can be revised by selecting the thumbnail, shifting the frame with shuttle or jog, and clicking either the "Start" or "End" buttons. A dialog box is available for each thumbnail by double-clicking on it. Using this, the operator can type in a description of the shot, which then appears on its thumbnail.

Assembly of the scene on the timeline is accomplished by "dragging" from the thumbnails to the timeline view. The shot names are reproduced on the timeline at their locations and their durations are shown. The EDL is automatically created by this operation.

Transition effects are chosen by placing the timeline cursor (dotted vertical line) over the desired location and clicking the "Transitions" button, which brings up a dialog box containing transition choices and their parameters.

At any time, the editing can be viewed by clicking the "Preview" button. It can also be examined by using the jog and shuttle controls.

Nonlinear editing greatly simplifies and speeds the editing process.

11.6 Summary

The use of postproduction allows the shooting of video and audio to be unencumbered by the considerations of program assembly while the assembly process gains far more flexibility than could ever be handled during shooting. Because of these advantages, nearly all program production is done this way and postproduction equipment is available at all quality levels.

Postproduction techniques are highly developed and, with the arrival of digital video systems, new possibilities are being explored to even further enhance the process while making it easier at the same time.

Television Receivers and Video Monitors

12.1 Introduction

The performance of today's color television receivers is superior by orders of magnitude to those available in 1953 when color television broadcasting was first approved. They produce far better images in every measurable respect (see Chapter 4), they make larger pictures with smaller sets, and they are far more stable and reliable. Best of all, the improved performance has been achieved while offsetting manifold inflationary cost increases, and the average 1992 receiver is lower priced than in 1954, even in current dollars. The engineering resources of many countries have contributed to this progress, and it has truly been a superb worldwide achievement.

A remarkable feature of this performance is that it has been achieved without any revolutionary technical breakthroughs. The same superheterodyne principle continues to be used for the radio-frequency section of the receiver. The original three-gun shadow-mask principle continues to be used in most receivers. The transmission format established by the FCC in 1953 (and by the EBU in Europe) is essentially unchanged. As is often the case in war and technology, success lies in the details.

12.2 Receiver Configuration

A functional diagram of a typical television receiver is shown in Figure 12.1. (A similar diagram of a receiver designed for digital HDTV signals is shown in Figure 13.6.) Its layout is conventional—a tunable RF stage for channel selection including a local oscillator and first detector, and an IF amplifier and second detector for video and audio. The video channel employs a diode

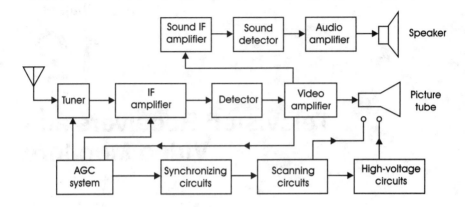

Figure 12.1 Typical receiver configuration

detector or synchronous detector. The audio channel employs a discriminator or other FM detector. The output of the video channel includes the signal multiplexed with the timing and color synchronizing signals in the format described in Section 5.7. The timing and color synchronizing signals control the kinescope sweep circuits and the color phase reference.

12.3 RF Amplifier and Channel Selector

Figure 12.2 is a diagram of an RF amplifier and channel selector. VHF signals are amplified before mixing with the output of the local oscillator, which operates 45.75 MHz above the picture carrier and 41.25 MHz above the sound carrier. The level of the incoming carrier signal can vary from 10 microvolts to 100 millivolts, but the AGC amplifier maintains the amplified signal level within reasonable limits. The maximum gain of the channel selector amplifier is approximately 20 dB.

Broadcast receivers can be tuned to the VHF and UHF channels in the television broadcast bands, channels 2 to 13 and 14 to 83. For reception of the additional cable channels, the receiver is tuned to channel 3 or 4 and the cable channels are heterodyned to it by means of a *set-top converter*. Alternatively, *cable-ready* receivers are capable of picking up the standard cable channels directly (see Chapter 16) without an external converter.

The difficulty in designing a tunable UHF amplifier of reasonable price and performance often leads to the design shown in Figure 12.2 where the channel selector has a loss rather than a gain. Its IF output is fed to the mixer of the VHF section, which operates as an amplifier under these conditions.

In earlier sets, VHF tuning was accomplished by variable capacitors with mechanical detents, while UHF tuning required the more difficult and time-consuming method of continuous tuning. The FCC then decreed that

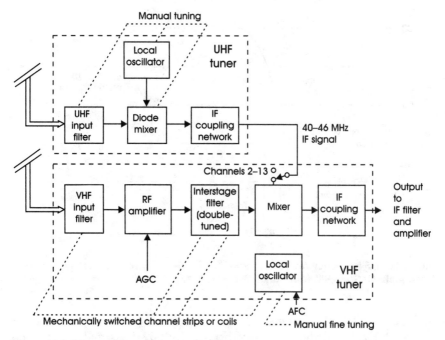

Figure 12.2 RF amplifier and channel selector

VHF and UHF receivers must have the same ease of tuning, and this requirement has been met by the use of electronically controlled systems that can have preset values for the desired channels. *Varactor tuners* that make use of the change in capacitance of a reverse-biased solid-state diode are often used as the variable tuning elements in both the VHF and UHF local oscillators. The most advanced receiver designs employ a microprocessor that permits easy preset and readout of selected channels and a *frequency synthesis* system to generate a precise local oscillator frequency for any channel.

12.4 IF Amplifier

12.4.1 Video channel frequency response

The frequency response of the video channel of the IF amplifier must meet rather exacting requirements (see Figure 12.3). It must provide high rejection of the adjacent channel picture carrier at 39.75 MHz and the adjacent channel sound carrier at 47.25 MHz. The treatment of the channel sound carrier differs from system to system, but it must be highly attenuated at the input of the video detector. The picture carrier at 45.75 MHz should be located at the 50 percent (–6 dB) point on the upper slope of the response curve. This position equalizes the sum of the responses of the sidebands (at 100 percent

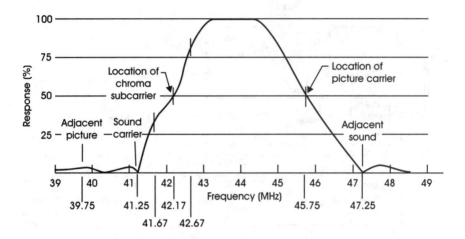

Figure 12.3 IF frequency response. (*From K. Blair Benson,* Television Engineering Handbook, *McGraw-Hill, New York, 1986, Fig. 13.17.*)

in Figure 12.3) in the vestigial sideband area for baseband frequencies from 0 to 1.5 MHz. For modulating frequencies above 1.5 MHz, equalization of the video channel must be provided when the response of the IF falls below 100 percent. In the example in Figure 12.3, this occurs at sideband frequencies below 43.25 MHz or baseband frequencies above 2.5 MHz. Equalization is particularly important in the vicinity of the chroma subcarrier in the range 41.67 to 42.47 MHz. Figure 12.4 shows the details of the equalization required.

The presence of the subcarrier imposes requirements on the performance of the IF amplifier in a color receiver that are not present with monochrome. The bandwidth must be greater to accommodate the subcarrier and its sidebands. The tuning must be stable so that the correct ratio of luminance to chrominance signals is maintained. Linearity must be sufficient to avoid intermodulation between subcarrier and carrier. The envelope delay as well as the frequency response must be equalized over the channel pass band.

The frequency response is often established by a SAW (surface acoustic wave) filter (see Section 15.8.1.6).

12.4.2 Sound IF channel

The sound IF subcarrier operates at 41.25 MHz, 4.5 MHz below the video. It can be handled in a number of configurations. In one, it is carried through the IF stages at a level 10 to 20 dB below the picture carrier. The 4.5 MHz difference signal, including its frequency modulation then appears at the output of the detector. In another, the sound IF is separated from the video

Figure 12.4 Color subcarrier equalization. (*From K. Blair Benson,* Television Engineering Handbook, *McGraw-Hill, New York, 1986, Fig. 13.18.*)

at the output of the 2nd IF stage and is then processed by a separate sound IF amplifier.

In both cases, the final difference between the picture and sound IF frequencies is 4.5 MHz as established at the transmitter; this is known as *intercarrier sound*. It significantly simplifies a receiver but it does have the problem that overmodulation of the video carrier can cause buzz interference in the received audio. That occurs because the video carrier level can go to zero during bright areas of the picture, causing momentary loss of the intercarrier sound carrier.

12.4.3 Gain

The IF amplifier gain is typically 90 dB, which is offset by 20 dB loss in the input and detector stages:

Stage	Gain (dB)
Input coupling	−10 dB
1st IF stage	+30 dB
2nd IF stage	+30 dB
3d IF stage	+30 dB
Detector	−10 dB
Total gain	+70 dB

12.4.4 AGC and AFC

Automatic gain control, AGC, and automatic frequency control are standard features of most receivers.

Automatic gain control is often provided in both the RF and IF stages of the receiver. For low and medium level signals, gain control is applied only in the IF section. Additional gain reduction is applied in the RF section for very strong signals.

Automatic frequency control is applied to the local oscillator and maintains the IF carrier frequency at 45.75 MHz.

12.5 Detection

Detectors can either be a diode or the more advanced synchronous and balanced type. The latter is more immune to cross-modulation between the video carrier, the color subcarrier, and the sound carrier.

12.6 Sync Separation and Scanning Generation

The vertical and horizontal sync pulses are transmitted at a higher modulation level than the video signal and are easily removed for driving the scanning generators.

The vertical scanning generator operates at the relatively low frequency of 60 Hz and is usually a multivibrator-type sawtooth generator that is synchronized by the vertical sync pulses as modified by the equalizing pulses (see Section 1.8.2).

The horizontal scanning generator operates at the much higher line rate, and the retrace time is much shorter. As a result, the voltage, $L(di/dt)$, developed across the deflection transformer and/or deflection yokes is very high. This voltage can be rectified and used to provide the anode voltage for the kinescope.

12.7 Subcarrier Demodulation and Video Amplification

The color subcarrier is removed from the composite video signal following the second detector and is demodulated to recover the I and Q signals (see Section 5.5.2). The I and Q signals are matrixed with the Y luminance signals to generate first the R–Y and B–Y and then the R,G,B primary signals that are applied to the three kinescope electron guns. The video amplifiers at the output of the matrix circuits must be capable of amplifying the level of the video signal from 1 or 2 volts to 20 volts or more. Processing circuits for image enhancements may be added at this point.

The level of the subcarrier is adjustable at the input of the demodulator by the "color" control to establish the saturation of the image.

The phase of the subcarrier with respect to the color burst can be adjusted to control the hue, and this is the "hue" or "tint" (sometimes confusingly called the "color") control.

12.8 Color Display Devices

The display device, the component that produces the visual image, is the end point and the final key to the performance of color television systems. No matter how perfect the performance of the signal generating and transmission equipment, the image quality can be no better than that of the display device. Improvements in these devices, probably more than of any other single component, have been the greatest contributor to the dramatic advances in television receiver performance described in Chapter 4 and at the beginning of this chapter.

12.8.1 Shadow-mask kinescopes

Shadow-mask kinescopes were used as the display devices in the original color television receivers. This principle has been remarkably durable, and it is still in wide use 40 years later. It is produced in a number of formats, two of which are shown in Figures 12.5 and 12.6.

The format in Figure 12.5 has round holes in the shadow mask arranged in triads and a "delta gun" structure with the electron guns arranged in a triangle.

The format in Figure 12.6 has an "in-line" gun structure with the guns mounted side-by-side rather than in a triangle and with slots rather than round holes in the shadow mask. An advantage of this configuration is that there is no color registration problem in the vertical direction.

As a result of the relative positions of the electron guns, shadow-mask holes, and tri-color phosphors, the electrons from each gun impinge only on phosphors of one color. By controlling the amplitude of the signal current from each gun in accordance with the brightness of the primary component of the scene at each point, the amplitude of the image brightness for each primary color will correspond to the scene brightness for that color.

In addition to the formats shown in Figures 12.5 and 12.6, the shadow-mask principle is applicable to a variety of additional formats in the shape and location of the holes in the masks and the arrangement of the guns.

A black deposit around each phosphor dot (the surround) is an especially important feature of more recent shadow-mask designs. It absorbs the ambient light striking the faceplate so that blacks appear to be really black, thus increasing the contrast ratio (see Section 4.14). Black portions of the image can be no blacker than when the set is turned off, and this was a light gray in early TV receivers because of reflection of ambient light. The production of black blacks has been one of the major contributors to the improvement in image quality in recent years.

A trend toward larger pictures has been another important development, and kinescopes with diagonals up to 36^2 are now standard. The performance of larger pictures is now satisfactory because of the significant improvements in the technical quality of broadcast signals (see Chapter 4), which reduce the

Figure 12.5 Shadow-mask kinescope, delta gun, round hole. (*From K. Blair Benson, Television Engineering Handbook, McGraw-Hill, New York, 1986, Fig. 12.4.*)

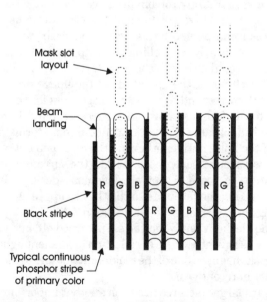

Figure 12.6 Shadow-mask kinescope, in-line gun, stripes. (*From K. Blair Benson, Television Engineering Handbook, McGraw-Hill, New York, 1986, Fig. 12.4.*)

defects that become more visible with larger images and lower viewing ratios. Some homes are now designed with one room devoted to TV viewing—a "home theater."

The minimum screen sizes recommended by one leading manufacturer for various viewing distances are:

Viewing Distance (ft)	Screen Size (in.)
<7	27
7–10	31
10–11	35
11–15	46
>15	52

These recommendations provide viewing ratios (see Section 4.4.2) in the range of approximately 5:1 to 6:1.

The use of large screens will be accelerated further by the introduction of HDTV—in fact HDTV will require large screens to be fully effective.

The transition to larger screens has been made practical by advances in design and production techniques that have steadily improved the quality and reduced the cost of shadow-mask kinescopes.

12.8.2 Projection kinescopes

Until recently, direct-view shadow-mask kinescopes have been overwhelmingly preferred by consumers for television display devices as compared with their competitor, projection kinescopes. The latter have suffered from inferior brightness, contrast ratio, resolution, and high costs. Recently, however, the same developments that have increased the demand for larger direct-view tubes have created a growing desire for screens larger than 36"—the approximate practical limit for direct-view tubes. This has caused a modest but significant increase in the demand for projection sets, which can produce far larger images, typically with 46" to 52" diagonals and greater. The demand for projection sets has also been increased by impressive improvement in their performance—particularly their brightness, contrast ratio, and resolution.

Projection sets have three small but very bright tubes, one for each primary color. The images on the tubes are enlarged, converged, and focused on a screen by means of an optical system that may employ mirrors, lenses, or both.

12.8.3 Flat-panel displays

Flat panels are attractive since they would make "picture on the wall" displays possible. They are widely used for portable computer displays (see Section 13.6), but they are not presently economic for the large sizes required for broadcast receivers. The design and manufacturing problems for large luminescent screens are formidable, and this must be classified as a future development.

12.9 Special Receiver Features

A number of special features are available in commercial television receivers; four of these—video inputs, closed captioning, remote control with read-out status menus, and "picture-in-picture" are described below.

12.9.1 Video inputs

VCRs typically have an NTSC RF output signal, usually on channel 3 or 4, to which the receiver is tuned. This arrangement means that the VCR must modulate its video on a carrier and the receiver must demodulate that back to NTSC. These extra steps of processing cause significant signal degradation, which can be eliminated by providing the VCR with a baseband video output and the receiver with a baseband video input. Most VCRs and many receivers now have this capability, which is called the *video monitor* feature. This is also useful for interfacing a TV receiver to a video game or a computer.

12.9.2 Closed captioning

Closed captioning is a system for displaying lines of alphanumeric text on the screen for the deaf and the hearing-impaired. This capability is now required by law in all television receivers. The text can include titles, the program dialogue, or other explanatory material. The text characters are transmitted in line 21 of odd-numbered fields by digitally encoded symbols. They are decoded at the receiver for display on the screen.

The standard format for line 21 is shown in Figure 12.7. The digital text symbols are transmitted in a $33.76\,\mu$sec segment of the line at a rate of 960 bits/sec. The data bit stream is synchronized with the caption generator by means of 7 cycles of a 0.503 MHz clock signal that is transmitted at the beginning of line 21. Each data block consists of two 7-bit characters coded in a *nonreturn-to-zero* (NRZ) ASCII format with an 8th bit added for parity. This gives a data rate of 120 characters per second, which is high enough for any captioning purpose.

Closed captioning is one example of the use of the vertical blanking interval of an analog television system for the transmission of digital data. There are a number of proposals for other uses of this capability to deliver new services such as program schedules or multilingual features.

12.9.3 Remote control and read-out menus

Nearly all modern receivers are equipped with remote control facilities, usually employing infrared communication between a hand-held controller

Figure 12.7 Closed caption signal format. (*From Andrew F. Inglis (ed.)*, Electronic Communications Handbook, *McGraw-Hill, New York, 1988, Fig. 17.12.*)

and the receiver. As a minimum the remote unit provides the capability of selecting the video channel and controlling the volume. More elaborate systems provide tone and stereo balance controls for the sound and a complete set of video controls including color, hue, brightness, contrast, and sharpness. The control signals are usually transmitted from the control unit by means of an infrared beam.

To make remote-controlled sets more user friendly, a menu is shown on the screen that indicates the availability and status of the controls. The format of the menu is similar to that shown on computers but it is simplified to be consistent with the capabilities of home users. When the menu is not in use, it automatically goes away.

12.9.4 Picture-in-Picture

The picture-in-picture feature is one of the most interesting and popular in modern television receivers. It enables the viewer to display a reduced-size version of one picture while viewing a different program at full-screen. The reduced picture is displayed in a small aperture on the screen. The size and position of the aperture are adjustable by the user (see Figure 12.8).

Since the scanning standards of the two pictures may not be identical and synchronized, it is necessary to convert and synchronize the scanning of the inserted picture with the scanning of the base picture. This is accomplished with a digital standards converter (see Section 3.6.2). Geometric manipulations, such as varying the size, position, and shape of the picture, can also be introduced at this point (see Section 11.3.3). The signal is then reconverted to the analog format and inserted in the base picture by a video effects technique (see Section 11.3.3).

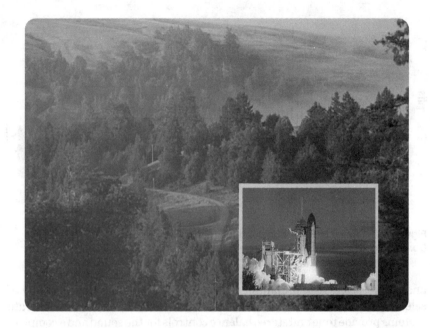

Figure 12.8 Picture-in-picture display

12.10 Receiver Performance Criteria

The performance criteria of television receivers can be divided into two categories: the performance of its signal amplification sections, and the performance of its image display device.

The primary criteria of the signal amplification sections are its noise figure and its selectivity (this presumes that they conform to industry practice with respect to frequency response and linearity). These criteria determine its performance with weak signals and in the presence of adjacent channel interference. The primary criteria of the display device, in addition to its size, are its resolution (including the convergence of the three primary colors), its brightness, and its contrast ratio.

12.10.1 Noise figure

The noise figure criterion was originally developed for specifying the performance of radar receivers. It is defined as the signal-to-noise ratio at the output of a receiver to the signal-to-noise ratio at its input when the input is at room temperature, T_O (assumed to be 290°K or 17°C).

Since the receiver inevitably adds noise, the noise figure expressed as a fraction will be less than unity and expressed in dB will be negative. To avoid this, it is conventional to express the noise figure in dB as a positive number as shown in Eq. (12.1):

$$N_F = -10\log \frac{\dfrac{S}{N}\text{(output)}}{\dfrac{S}{N}\text{(input)}} \qquad (12.1)$$

where: $T_{INPUT} = T_O = 290°K$

The receiver's noise figure is a critical performance criterion for receivers located in areas of weak local signals.

Typical noise figures are 4 to 8 dB on VHF channels and 7 to 12 dB on UHF channels.

12.10.2 Selectivity

The selectivity of a receiver is a measure of its ability to reject off-channel signals. Interfering off-channel signals can be adjacent channels that are not sufficiently attenuated by the selectivity of the receiver or spurious signals such as harmonics or heterodyned cross modulation products.

The receivers's resistance to adjacent channel signals is largely determined by the response of the IF system (see Figure 12.3). Interference from cross modulation products is minimized by the FCC's assignment plan, particularly for the UHF channels, that avoids the assignment of image frequencies and other potential off-channel sources of interference in the same area (see Section 15.2.2).

12.10.3 Picture tube resolution

The resolution of a shadow-mask picture tube is determined by the pitch or spacing between the rows or columns of dot triads. The choice of the pitch is influenced by the picture tube size, larger pitches being used for larger tubes. In part this is an economic decision and in part an assumption that a larger tube does not need any more triads per picture height. A typical pitch for a 19-inch picture tube is 0.60 mm which gives approximately 500 rows per picture height.

12.10.4 Picture tube brightness and contrast ratio

The typical brightness and contrast ratios for home receivers are tabulated in Table 4.2.

12.11 Professional Picture Monitors

Television production facilities require high-quality professional picture monitors to control and maintain the quality of the images that are produced, recorded, or transmitted by the facility. The purpose of these monitors is quite different from that of continuity monitors, i.e., to observe and control picture quality rather than content. In effect, they are test instruments.

The technical specifications of professional monitors differ from receivers and continuity monitors in the following respects:

- Resolution
- Brightness and contrast ratio
- Component and composite inputs
- Calibration adjustments
- Stability

The resolution of shadow-mask kinescopes is determined by the pitch of the dot triads as compared with the picture height. As noted above, typical 19-inch picture tubes have a pitch of 0.6 mm. The range for professional monitors of this size is approximately 0.3 to 0.4 mm giving 750 to 1,000 rows of triads per picture height.

The brightness and contrast ratio of typical professional studio monitors are tabulated in Table 4.2.

Professional monitors are frequently used in system configurations where the signals are in a component format (E_R, E_G, and E_B or E_Y, E_R–E_Y, and E_B–E_Y). Inputs must be provided for both of these formats.

Since the monitor is used as a test device, means must be provided to adjust and calibrate contrast, brightness, and color balance by using external test signals. It is not desirable to make these adjustments constantly, and the monitor circuitry should be sufficiently stable to hold adjustment for considerable periods of time.

13

Digital Video Display Systems

13.1 Introduction

The task of displaying an analog video signal is simplified by the fact that the incoming signal contains all the information in the order to be displayed, with constant updating. Analog display systems only have to synchronize their scanning with the signal and route the video in real time into the display device(s). Digital video display systems, used in computers and HDTV, have to do these same things, but they may need to do much more to deal with other features of the digital video signal such as packetizing, compression, scan-rate conversion, embedded commands, multiple choices, and others. These are explained in this chapter.

13.2 Computer Video Subsystem

Many of the concepts of digital video display can be introduced by describing how the video display part of a personal computer is built. Personal computer video requirements are so varied that most systems provide for them as a separate module slot that can accept a number of different devices depending on the user's needs. Even when a PC includes a built-in video subsystem, it usually can be disabled in favor of a separate plug-in device. These devices are called *video adaptors* and they provide one or more of these features:

1. Display refresh
2. Graphics acceleration
3. Motion video decompression
4. Analog video output

5. Video digitizing

6. Video compression

Each of these is discussed below.

13.2.1 Display refresh

Most video display devices, such as CRTs, must receive video data continuously to keep the picture on the screen. The process of doing this is called *display refresh*, and even a digital system that stores its picture in a memory must continuously read that memory out to the display.

Although it would be possible to build a display monitor as a standard PC bus device (see Section 7.2.1.1) and have the CPU send data to its address to get a picture on the screen, the amount of data transfer required would totally overwhelm the capability of most machines. The CPU wouldn't be able to do anything else. Since most of the data that shows on a computer screen remains the same for long periods of time, it makes sense to buffer the screen in separate RAM on the display adaptor and have the adaptor completely handle the screen refreshing task. That way, the CPU and main bus would be responsible only when screen data has to be changed. A typical hardware architecture for that is shown in Figure 13.1.

Some of the numbers for computer screens were given in Chapter 1, Table 1.2. A typical example is the Super-VGA 640 × 480 resolution format displaying RGB-24 information and refreshing a noninterlaced screen at 70 Hz. Table 1.2 shows that requires data at 64.5 MB/second. The amount of RAM needed to store one screen on the display adaptor can also be calculated from these numbers by multiplying the horizontal and vertical resolutions by each other and by 3 (for 3 bytes per pixel). This gives 921,600 bytes. Thus, a 1-MB memory will do the job for this screen. Most SVGA display adaptors provide 1 MB of memory—some even have more than that to support higher resolution modes (see Table 13.1).

All of the resolution combinations in Table 13.1 except one provide *square pixels*, that is the horizontal and vertical pixel counts are in the same ratio as

Figure 13.1 Block diagram of a video display adaptor

Table 13.1 Maximum SVGA resolution mode vs. color mode and display memory size

Display RAM Size	Color Mode			
	4 bpp	8 bpp	16 bpp	24 bpp
1 MB	1600 × 1200	1024 × 768	800 × 600	640 × 480
2 MB	1600 × 1200	1280 × 1024	1024 × 768	800 × 600
4 MB	1600 × 1200	1600 × 1200	1600 × 1200	1280 × 1024

the aspect ratio (4:3 for computer displays.) The 1280 × 1024 combination is not an exact 4:3 ratio, but it is close. Square pixels are advantageous on computer displays because they simplify the arithmetic involved in computing vector graphics (see Section 14.6.1.5), but any resolution combination can be accomplished with a little more software complexity.

Figure 13.1 shows that the display RAM needs two interfaces—to the PC main bus (and acceleration processor, when used) and to the display refresh circuits. Conventional DRAM as used in main memory has only one interface, used for both reading and writing. To use DRAM in a display adaptor, the RAM interface must be time-shared between bus I/O and refresh tasks. That is not difficult to do, but it limits the rate at which communication with the bus can occur. Massive screen changes are therefore limited in speed. Another architecture uses special *video RAM* (VRAM) that has two ports directly on the RAM chips as shown in Figure 13.2. The shift register is loaded during

Figure 13.2 Block diagram of a VRAM memory chip

horizontal blanking with a full row from the memory array. This only takes one memory cycle, and all other memory cycles are available for use by the CPU. Screen refresh is done by clocking the shift register, which is a completely independent operation from the memory cycles. VRAM is more expensive, but it speeds up screen changes. This can be important to full screen motion video applications.

13.2.2 Graphics acceleration

Screen operations that involve many pixels are often simply repeating the same operation on each pixel. Examples are: filling an area of the screen with a color, moving a block of pixels from one place to another, drawing lines or other graphic objects, and so on. If these things are done by the CPU, every pixel involved must be sent over the bus to the display adaptor. This slows the operation and ties up the bus and CPU. Many display adaptors now have their own dedicated processor directly connected to the display memory. This processor knows how to do the special graphic operations, so the system CPU only has to send commands to the display processor and it carries out the operations in parallel with whatever else the system is doing. This is called *graphics acceleration* and it vastly speeds up the drawing of graphic screens.

Most video display adaptors today include some form of graphics accelerator that usually is a custom IC designed specifically to work in the environment of that adaptor. Since the accelerator capabilities to a large degree will determine the performance of the adaptor, this has become a major element of competition between adaptor manufacturers. Each manufacturer has his own proprietary accelerator design and there is no standardization between manufacturers. This is not a problem because adaptor manufacturers must supply software drivers anyway to connect their adaptors to standard operating systems such as DOS/Windows or OS/2. The proprietary accelerator design is hidden inside of the driver software; the rest of the system only sees the standard operating system interface.

13.2.3 Digital signal processors

Another way to accomplish graphics acceleration is to use a general-purpose *digital signal processor* (DSP) chip in the display adaptor. A DSP is a special microprocessor that is optimized for speed in doing signal processing operations. It uses a stored program just like the system CPU, so it can be very flexible in what it does. This is a more expensive approach than a hard-wired custom accelerator, but it can offer additional features and of course its features can be changed or updated by changing software. The DSP software is usually handled as a part of the display adaptor driver, so changes would be made by updating that driver.

13.2.4 Motion video decompression

The same kind of acceleration concepts can be applied to video processes such as decompression and new video adaptors are adding this functionality, too. With a video decompression processor or DSP on the adaptor, the CPU only has to manage the transfer of compressed data to the adaptor and all the rest of the video work occurs on the adaptor.

13.2.5 Analog video output

The output that goes from the display adaptor to the monitor is analog RGB component video. However, there is sometimes a need to deliver television-format video in either a composite or component format. This is useful, for example, if one wishes to record the computer's video on tape. It also may be used to feed a large-screen television display. (There are large-screen computer displays available, but they are more expensive than TV displays.)

Converting computer video to analog TV formats usually will introduce degradation because of interlaced scanning and composite color encoding. It is satisfactory only if the video material is chosen to avoid fine detail in bright colors and to not use single-pixel horizontal lines. That means text must be significantly larger than would be used with the computer's own display.

Since the scanning parameters of computer display adaptors are usually programmable, setting up the TV scanning is easily accomplished. However, if composite encoding is required, it takes added circuitry. Fortunately, there are ICs available for TV encoding, so it still isn't expensive. The real problem is that changing the adaptor scanning for this purpose means that the computer display also must change to TV scan. That may not be desirable and some computer monitors won't even handle it.

The solution is to leave the computer display adaptor alone and intercept the SVGA RGB signals in a separate module to perform the conversion. Although that is somewhat more expensive, it is then possible to have both computer and TV-scan displays at the same time.

As HDTV comes into use, there will be a need to output that from computers, too. HDTV is more compatible with the normal contents of computer video, so there will not be the quality problems discussed above. However, there will be the problem of how to handle the different aspect ratio of HDTV. Again, the computer can be programmed to operate at HDTV scanning, but then computer displays won't work—at least not until wide-format computer displays become available. (That will probably happen as HDTV display components become readily available and users become familiar with HDTV in their homes.) Until 16:9 aspect ratio computer displays are produced, the best choice for HDTV output from computers will probably be the separate converter module approach.

13.2.6 Video capture and compression

To bring analog video into a computer so that is can be compressed, edited, or displayed, requires digitizing and storing the digital data on the computer's hard disk. This can be done with a separate adaptor or it can be added to a display adaptor. Architectures for this are discussed in Section 7.4.3. The issues of video compression are also discussed in that section and in Sections 3.10 to 3.12.

13.3 Analog Video Overlay

Effective audio–video–computer applications have been built for many years using a technique known as *video overlay*. The output from a computer-controllable analog video source such as a laser video disc is displayed on the video monitor along with graphic output from the computer. The analog video is mixed with the computer graphics, but there is no attempt to input the video data into the rest of the computer. This technology has been widely used for education and training applications and it is very effective.

Video compression is not required for video overlay because the video never has to pass through the computer bus or be stored in mass storage. It is directly displayed as it is played from the laser disc. Thus, the system is simpler than a true computer digital video system. However, the overlay approach has significant disadvantages:

The laser disc player is a separate unit from the computer that must be connected with cables. That makes a complex and inconvenient setup. It also is more expensive than a computer that plays digital video.

Figure 13.3 Block diagram of the use of an analog video overlay board

The program (computer disk) and the video data (video disc) are stored on separate media that must be coordinated for proper use.

The video disc is not recordable except through mastering (see Section 9.18) so it is difficult to make updates.

Since the video never actually enters the computer, the computer cannot manipulate the video in any way. All the video for the application must be on the video disc in exactly the format that it will appear.

Scanning synchronization is required between the computer and the video disc. This is usually accomplished as shown in Figure 13.3 by using TV scanning on the display and synchronizing the computer graphics with the TV scan. This limits the system to interlaced scanning, which does not do a good job on the graphics. (There are more complex overlay approaches that digitize the video disc signal and display it on a computer monitor.)

Modern video overlay boards combine overlay with a full video digitizer and display board. This is a useful feature for video capture setups because the analog overlay can be used for monitoring during capture without using any of the computer's resources. Such a combination video board can serve many applications:

Frame grabbing

Movie grabbing

Live overlay

Hard disk editing

Tape editing

Analog video output

TV tuner

Figure 13.4 is a block diagram of such a combination board. It differs from Figure 13.3 by performing the entire overlay process digitally. This means there must be decoding and A/D conversion for analog video input and a D/A converter on the output to the display. A frame memory is included with separate write and read clocks so the analog video is resynchronized to the computer's standard video. Therefore, the display scanning can be progressive as required for a high-quality computer display. By giving the computer access to the overlay board's frame memory, still images or video can be compressed and stored to the computer's hard disk using the processing resources of the computer.

13.4 Digital Television Displays

The display refresh portion of a digital television display is similar to a computer display adaptor. There must be enough RAM to store the current screen and the means to continuously read that to the display at the proper frame rate. The rest of a digital TV display is concerned with processing the input signal format to place it into the RAM in the correct format for display.

Note that the input processing may result in scan conversion—it is not really necessary that the display refresh scanning run at the same rate as the scanning that is implicit in the input signal. This is one of the major advantages of being digital. An example of the digital display processing for the Grand Alliance HDTV system is given in Section 13.7.

13.5 Computer Display Monitors

Computer display monitors differ from analog television displays in several respects:

1. A computer display is viewed at much closer range than an equal-size TV display (see Section 6.3). That calls for significantly higher CRT resolution, higher vertical-scan rates, and a flatter screen. These features also mean that computer monitors are significantly more expensive than TV displays.

2. Interlaced scanning is not appropriate for computer display (see Section 4.16.4), so all computer displays should use progressive scanning.

Figure 13.4 Block diagram of an all-digital video capture and overlay board

3. Most computer display adaptors are capable of multiple scan rates. A display monitor must accommodate this, preferably automatically. This gives rise to such terms as "multisync" or "multiscan" when describing computer monitors.

4. Because computer display formats use progressive scanning and usually have more horizontal lines than analog television, horizontal scanning frequencies are significantly higher. That is another higher-cost factor.

CRT resolution is specified in terms of *dot pitch*, which is the distance on the screen from the center of one phosphor color dot (or line) to the next adjacent dot of the same color. Since the CRT dots are a form of sampling, there should be more than one dot for each pixel to avoid loss of resolution. Unlike television, the viewer will usually sit the same distance away from a computer monitor regardless of its size. A larger monitor is expected to display more information at once. Thus, the dot pitch should be the same independent of display size.

For example, a 14" (diagonal) display has a useful screen width of about 10 inches. With a dot pitch of 0.28 mm (which is typical for a good 14" monitor), there will be 90 color dots per inch or 900 dots across the screen width. This will perform well with 640 pixels horizontally, it will also work with 800 pixels across, but at 1024 or more horizontal pixels, the dot size will severely limit display quality. For the higher resolutions, larger displays that have more pixels across the screen are necessary.

13.6 Flat-Panel Displays

The "picture on the wall" display has long been a goal of television display research. Numerous technologies have been developed but none has been commercially significant in the TV field except for small portable receivers where CRT displays are impractical. In general, flat panels do not compete with CRTs except in places where the size and power needs of a CRT display are prohibitive.

The portable computer market fits this model. It is impractical to make a portable computer with a CRT display, so this is a market which will pay the price necessary to have small, light-weight, and low-power displays. The technologies of flat-panel displays are now highly developed as a result of the impetus of this market. Compared to TV displays, portable computer displays are small—10" diagonal displays are the largest.

Oddly enough, the entre for flat panels into the TV market may be HDTV, where large high-resolution displays are required. CRTs for this market are going to be expensive monsters in the living room and it would be an opportunity for flat panels to get into the market if they can achieve the screen sizes at anywhere near reasonable costs. Many companies are working on it.

There are two technologies for flat panels that have reached mass marketability, described below.

13.6.1 Liquid-crystal displays

Nearly all displays used in portable computers employ *liquid-crystal display* (LCD) technology. In these displays, the controllable element is a liquid-crystal (LC) material sandwiched between two glass plates. The LC material has the property that its oblong molecules can be aligned by grooving the glass plates enclosing the material but when an electric field is applied to the panel, the molecules will realign themselves with the field. By building a control matrix behind the glass panel that allows the electric field to be controlled locally for each pixel, a display screen can be built. This is diagrammed in Figure 13.5 for a panel that contains its own light source. (Reflective panels that use ambient light as the source are also used, but the self-illuminated panels are more popular for computer or TV displays.)

The LC material is contained between two glass plates; each plate has microscopic grooves on its inner surface, but the grooves are aligned at 90° to each other. This causes the molecule alignment to spiral by 90° between the bottom and top of the panel, which explains the name "twisted." To obtain control of light transmission in the panel, two polarizers are placed behind and above the panel, with their plane of polarization aligned with the grooves in the glass plates. As the light passes through the LC material, its polarization is rotated as the molecule alignment twists, so that full light transmission occurs even though the polarizers are at 90° to each other.

When an electric field is applied to a location on the panel by applying voltage to the control electrodes, the LC molecules begin to align with the field and the 90° twist is disrupted. The result is that light transmission is reduced—the panel gets darker where the field is applied.

Because of their use of polarized light, LCD panels in general have a limited angle of view, usually only about 30°. This is acceptable for the single-viewer application of portable computers, but it is not suitable for TV-style viewing by a group. This is one of the challenges in applying LCD panels to TV.

There are two ways to achieve control of the electric field: passive-matrix and active-matrix, explained below.

13.6.1.1 Passive-matrix LCDs

A passive-matrix panel is built by placing an array of row and column wires behind the panel as shown in Figure 13.5. The row and column wires are pulsed in sequence to provide a scanning action. By controlling the pulse amplitudes, the light transmission may be set for each pixel. Because the LC material changes back to its unexcited state slowly after the voltage is removed, the panel appears to be showing a continuous image that does not flicker.

Color is obtained by using a color filter pattern over the display, much like that used in single-device color cameras. Thus, each color pixel requires three LCD pixels in the matrix; they are usually arranged horizontally, so for example, a 640 × 480 panel would actually have 1920 × 480 LC cells in its matrix.

Figure 13.5 Cross section of a twisted nematic LCD panel, transmission mode

Because of the short pulsing of each pixel and the slow decay of the LC behavior, a passive-matrix display has slow response and low contrast ratio. However, it is the least expensive form of LCD and it is used in low-end portable computers despite its shortcomings. Even so, passive-matrix LCDs are four or five times more expensive than CRT displays of similar size.

13.6.1.2 Active-matrix LCDs The leading technology for portable computer displays is the active-matrix liquid-crystal display (AMLCD) panel. It provides the best combination of cost and performance with display resolutions of 640 × 480 in sizes up to about 11" diagonal. It uses a liquid-crystal panel that has a control transistor associated with each pixel (or subpixel for color). The transistors are integrated directly onto the bottom glass of the panel. This is why it is called "active-matrix." The control transistors set the voltage on each pixel that causes the liquid-crystal molecules to scatter or transmit light. Because the control transistors are individually addressed and are much faster, the panel response time and contrast ratio are much improved. However, the addition of the control transistor array nearly doubles the cost of these panels compared to passive-matrix. Many buyers of portable computers consider that cost to be worth it.

Larger AMLCD panels have been made in sizes up to 21" diagonal, but the production yield drops rapidly with size, making such large panels prohibitively expensive. The technique of *tiling* is being explored to assemble large panels from smaller ones. If this technique can be mastered, it promises to lower the cost of large AMLCD panels and allow HDTV-sized units to be marketed.

13.6.1.3 LCD improvements There are many possibilities for improving LCD panel performance and cost that apply to their use in portable computers and also the HDTV market. For example, researchers are building AMLCDs using single-crystal silicon (x-Si) for the substrate of the transistor array. This allows much higher resolutions (pixels per inch) and, because high-performance circuits can be built on x-Si (it is the substrate for nearly all integrated circuits), much of the drive hardware can be integrated onto the display panel.

Panels using an x-Si substrate are limited in size to the dimensions of integrated-circuit wafers (8" diameter or so), but the pixel density can be so high that the HDTV resolution of 1920 × 1080 pixels can be achieved on a device only about 1½" across. This is ideal for a large screen projection display, and that application has a lot of potential.

13.6.2 Other flat panel technologies

Many other technologies exist in the laboratory but so far they are not ready to compete with the ones now on the market. However, the continued growth of the portable computer market and the emerging demand for HDTV displays will keep the laboratories working at high pitch. Several of these contenders are described below.

13.6.2.1 Field-emission displays The field-emission display is a form of flat CRT except that there is a separate cold-cathode electron emitter for each pixel. Colored phosphors are deposited on a transparent substrate in front of the array of emitters. This is a promising technology, but it is still in the laboratory.

13.6.2.2 Micromirror displays Another promising technology involves an array of mirrors on an IC chip. This is called a *digital mirror device* (DMD). The mirrors can be physically moved electronically to control the displayed brightness by reflecting a high-intensity light source. Panels up to 1.6 million pixels projected on a 12-foot screen have been demonstrated.

13.6.2.3 Plasma displays For large-size, bright, direct-view displays such as are required for HDTV, the plasma display is an important contender. These panels produce light from a gas discharge in a vacuum. In order to produce colors, each pixel produces three controlled ultra-violet discharges that excite photoluminescent color phosphors that deliver the visible light. The design of the plasma panel includes a memory unit for each subpixel so that the pixels stay on until they are turned off. Thus, the display does not have to be continuously scanned to produce a stationary picture. Plasma panels have been demonstrated up to 40" diagonal in HDTV resolutions.

13.6.2.4 Electroluminescent displays These displays are an extension of the familiar red-light-emitting diode displays (LEDs) used for numerical display. EL displays were used in early portable computers, but were more expensive than LCDs once that technology was developed. EL panels are now being developed for color display.

13.7 Digital HDTV Display Processing

As can be deduced from the transmission processing discussion in Chapter 6, a digital HDTV receiver has to perform a lot of processing. It is expected that custom integrated circuits will be designed for most of this processing, so it will not be expensive in the long run. Figure 13.6 is a simplified block diagram of a typical digital HDTV receiver (actually it is the GA prototype receiver). Since only the transmission format will be standardized, there will likely be many variations in receiver design.

13.7.1 Tuner, IF, and detector

The first step of reception is to tune the desired channel and down-convert to an intermediate frequency suitable for detection of the VSB amplitude modulation. This is done by a double-conversion heterodyne system in the GA prototype to obtain the desired selectivity, ending at an IF of 44 MHz. A frequency and phase-lock loop (FPLL) synchronizes to the pilot carrier of the signal to drive a synchronous detector that converts the 44 MHz VSB amplitude-modulated IF signal to the baseband 8-level symbol sequence.

A comb filter is available in the signal path for rejection of NTSC co-channel interference when it exists. (Because this filter slightly degrades the system noise performance, it is automatically switched out of the signal path if no NTSC interference is present.) This filter provides seven comb nulls in the 6 MHz channel; three of which reduce the energy at the NTSC visual carrier, the NTSC color subcarrier, and the NTSC aural carrier frequencies. The comb nulls and digital processing rates are coordinated so that the NTSC rejection filter has no effect on the digital data.

13.7.2 Segment sync and symbol clock recovery

The 10.76-MHz symbol clock is recovered by a phase-lock loop that locks to the random symbol pattern by looking for the 4-symbol segment sync intervals. This simultaneously synchronizes the symbol clock and the seg-

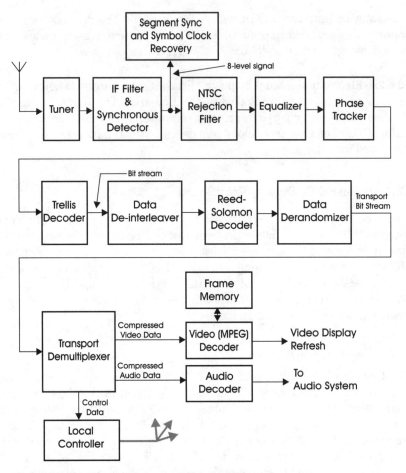

Figure 13.6 Block diagram of a typical digital HDTV receiver

ment sync. Data field synchronization is also obtained by looking for the data field sync pattern in the 8-level data.

13.7.3 Equalizer and phase tracking

The channel equalizer compensates for linear frequency response distortions and ghosts due to multiple reflections in transmission. It contains a 64-tap transversal filter that is adjusted by examining the training signal that is contained in the field sync pattern.

The phase tracking loop is an additional feedback loop that further reduces phase noise that may not have been removed by the IF FPLL operating from the pilot signal. It operates on the 8-level data signal to identify and track out residual phase errors.

13.7.4 Trellis decoder

Because of the 12-symbol interleaving at the transmitter (see Section 6.7.3.3), the receiver contains 12 trellis decoders in parallel, where each decoder sees every 12th symbol. The 8-level symbols are converted back to the original 2-bits per symbol. The output of this is a serial bit stream in the same order it existed at the transmitter before trellis encoding.

When the NTSC rejection filter is used, the trellis decoder is automatically modified to provide optimum performance with the different system response caused by the NTSC filter.

13.7.5 De-interleaving, Reed-Solomon decoding, and derandomizing

The data interleaving applied at the transmitter must be removed before Reed-Solomon decoding. This ensures that long burst errors will not overload the R-S decoder. The R-S decoder is capable of correcting up to 10 bytes of errors per data segment. The result of the combination of this and the data interleaving is that bursts of noise or interference up to 193 μsec. can be handled.

The randomizing pattern introduced at the transmitter is now removed using the same randomizing code as at the transmitter. This restores the signal to the transport packet format.

13.7.6 Transport demultiplexing

Since there is a 1:1 relationship between data segments and transport packets, the segment sync already extracted is used to decode the packet headers. Packet IDs are used to separate the packets into individual bit streams for audio, video, or data. This process results in separate bit streams feeding audio and video compression decoders. When control data is present in the packets, that is sent to the system controller (CPU) that identifies it and acts appropriately.

13.7.7 Audio and video decompression

The final step of digital processing is to decompress the audio and video bit streams. These processes are described in Section 3.12 and Section 6.5.

Interactive Video Systems

14.1 Introduction

Television viewing is ordinarily considered to be *passive*—the user sits back and watches the program but has no control except to change the program or turn it off. The other side of the coin is the personal computer style of viewing, where the user carries on a continuous dialog with the machine. This is known as interactive viewing or simply *interactivity*. Interactivity is everywhere—any product that is controlled by its user is interactive—this even includes mundane products like the faucet on the kitchen sink. Other important interactive products (see Figure 14.1) are automobiles, photographic cameras, and musical instruments.

Interactivity is creeping more and more into video systems. Even broadcast television receivers will acquire interactive features as the capabilities of digital transmission are applied to deliver richer capabilities for entertainment, information delivery, education, and play. This chapter discusses interactivity in video systems of all types.

Figure 14.1 Interactive products

Figure 14.2 The interactivity feedback loop

14.2 Ingredients of Interactivity

Any interactive situation requires three main ingredients (see Figure 14.2):

1. Two-way communication—in a personal computer, the machine talks to the user through its video display and sound capability. The user talks to the machine with keyboard, pointing device (mouse or trackball), sound (sometimes), or other more specialized controller devices. In an automobile, the driver communicates with the vehicle through the steering wheel, the pedals, and other controls. The vehicle communicates with the driver through its motion, the speedometer, and other instruments or displays. In each of these examples, a continuously-operating feedback loop is made possible by the two communication channels.

2. User interface—this is the protocol of the communication involved in interactivity. It is ordinarily something that the user has to learn, which may be easy or hard. Often, a degree of skill or dexterity is involved in operating a user interface. A good example of a difficult interface is the violin. Years of learning and practice are required to master that interface, and only a few of those who attempt it succeed. Yet the result is worth the effort and no proposal to redesign the violin has ever been taken seriously. The user interface of an automobile is much easier and can be learned by almost anyone. Most of the learning with an automobile is concerned with the rules of the road rather than the physical dexterity needed to operate the controls.

3. Computing capability—when the user issues a command, the system must know how to respond—that is, to close the loop. This is most ef-

fectively handled with a programmable computing device (an embedded processor; see Section 14.5) or an actual computer. Most video devices already contain computing capability for this and other purposes.

Each of these ingredients of interactivity will be discussed in more detail, specifically for video-based environments.

14.3 Communicating with the User

The communication devices available in video systems are the video display or sound (machine to user), and pointing devices, keyboards, touch screens, control panels, or sound (user to machine). Even more advanced devices are possible for user to machine communication based on forms of body sensors such as gloves or eye tracking.

Although in the examples given above the user operates the interface devices hands-on, interactivity can also occur remotely, where one or both communication channels involve radio-frequency, optical (infrared), or cabled communication. A simple example of this is the programming of a VCR using an infrared remote device and on-screen menus. The user sits across the room from the machine but still communicates with it for interactive control.

Good performance of the user–machine feedback loop depends on an adequate speed of response around the loop. If the loop becomes too slow, the user has to wait and he/she may perform superfluous actions that could even cause the loop to become unstable. Whenever the user performs an action, the system should respond somehow in (preferably) less than a second. If it doesn't, the user may think the action was not noticed by the system and perform it again. Even if the system action requested will take longer than a second, an acknowledgment of the request should be given in less than a second. An example of this is when a system responds with a click sound when a button is pressed. The user knows from this that the system got the message.

14.4 Video-Based User Interfaces

In a video environment, there are many ways to construct user interfaces.[1] The video screen is such a flexible device that there is usually no need to have any other means for the system to communicate with the user. The one exception to this is sound, which has the advantages that the user doesn't have to be able to read from the screen (maybe he/she doesn't always look at the screen or maybe he/she can't read) and a sound channel operates in parallel with the visual channel. The parallel nature of sound and vision means that together they can convey more information.

The communication channels provide rich capabilities that challenge designers to effectively use them. However, that same richness offers so many different ways to accomplish the same task that there can be serious problems

Figure 14.3 Visual user interface objects

in making interfaces easy to learn and consistent between different products. Standards are really valuable but in most fields they have been difficult to establish.

14.4.1 Types of user interfaces

Video-based interfaces usually employ *metaphors* that help the user remember how the interface should work. Typical metaphors are shown in Figure 14.3:

Push buttons—these are objects that the user "touches" to cause a specified action. ("Touching" means just that, if a touch-screen is employed, or it can mean moving a pointer object over the push button object and clicking a button.) Push buttons usually are labeled to indicate the action to be invoked when they are touched.

Menus—these are lists from which the user makes a selection by touching. The menu can be text or it may be actual pictures or other graphical representations of things that show the choices available in the menu. Another form of menu is the "check box" list, where the user touches one or more objects to choose several things at once. In this kind of list, the selected objects must highlight some way so the user can distinguish between checked and unchecked items.

Continuous controllers—these are knobs or scroll boxes (sliders) to set an analog parameter. They must show the setting of the parameter within its range by using position or rotation.

Entry fields—these are used when the user is required to type in something, such as a name. This ordinarily involves the use of a keyboard for typing, although in a very simple situation other kinds of controllers might be used.

Pointers—a pointer is an on-screen object that is moveable by the user. It usually functions to operate other controls or to identify specific locations or objects on the screen (called *selection*). In complex situations, the shape of a pointer can change to indicate the type of action it currently represents. For example, the normal pointer might be an arrow but if it is moved over an entry field, it changes to an I-beam shape and now indicates the text insertion point.

Metaphorical devices shown on a video screen should look and behave like the physical device they represent; this tells the user how they work. Such on-screen representations are called *icons* (see Figure 14.4) and they can be realistic or abstract. The important thing about icons is that they should be used consistently so that once a user learns the meaning of an icon and its use, it will behave the same way every time it is encountered in the interface.

Figure 14.4 Icons

Figure 14.5 A simple interface for a CD
player (from *Turtle Beach Systems*)

14.4.2 Typical interfaces

Depending on the application, video-based user interfaces range from simple
to very elaborate with multiple dimensions and levels. Examples of both
extremes are given below.

14.4.2.1 A simple interface Controlling a tape deck is a simple task; usually
no more than five buttons are needed for Play, Pause, Stop, Fast Forward,
and Rewind. A CD player is a little more complicated because of the possibility
of choosing a track. Figure 14.5 is an example of a very simple video-based
interface for a computer CD-audio player. The buttons provide all the
possibilities listed above except Pause (the only way to suspend playing is to
choose Stop, which returns to the start of the current track). However, this
is a good interface whose use is almost obvious.

14.4.2.2 An elaborate interface Microsoft Windows is a good example of a
video-based interface that can handle tasks of almost any complexity or depth.
Figure 14.6 shows the Windows video screen used for the formatting of this
book. The large window contains this page open for editing. That window is
surrounded by other windows that offer choice of special characters, page,
format tags, file selection, etc. It is clear from this screen how Windows got
its name. The screen also shows icons at the bottom for programs that are
running but their windows are not open.

14.4.2.3 TV-based interface Both examples shown above are from personal
computers. However, video-based interfaces aren't only on computers. Figure

Figure 14.6 The Microsoft Windows user interface

14.7 is a screen from a television receiver showing the menu setup used to adjust color parameters. In this case, the user is responding by means of an infrared remote unit. This actually proves awkward because the user has to learn how the buttons on the remote controller relate to what is on the screen. The computer idea of using a pointing device is really better because once the user has learned how to use the pointing device, all other instructions about the specific task at hand can be presented on the screen.

14.5 Computing Capability for Interactivity

Responses must be constructed for the user's actions in an interactive environment. This is defined here as "computing capability," although it may range all the way from a little hard-wired logic in a simple situation to a full-scale personal computer in other cases. Intermediate situations can be handled by an *embedded* microprocessor, which is a special microprocessor designed to be built-in to a product. Embedded processors are available as CPUs only, or they come in the form of a computer-on-a-chip, which contains all the functionality of a complete computer. Typically, this includes the CPU, RAM, ROM (for programs), and I/O capability. The embedded processor approach is usually applied in "smart" appliances or in electronic products such as VCRs, CD players, or tape decks.

Figure 14.7 A user interface for adjusting a TV receiver

As video products get more and more interactive features, which will require greater computing capability, the trend will be to emulate more of the features of personal computers in video products. This could include such things as loadable programs, standard keyboards, and internal mass storage.

This latter feature is especially important in order to provide rich interactivity in a situation where the video device is receiving broadcasted or cable signals. Although two-way communication with the broadcast station or cable head-end is possible to set up interactivity, the return path is likely to have slow response (seconds, or longer). That may be all right for a task such as program selection, but it certainly is too slow for exploring a database or playing a game. The problem can be solved by having enough mass storage in the local video product that data and programs for close interactivity can be downloaded from the head end. Then the user at the local product can interact with the data at his/her leisure, without the need for constant communication with the head end (see Section 14.7).

14.6 Multimedia

As personal computers have become capable of presenting realistic audio and motion video, the word "multimedia" has been used to refer to applications that use these new capabilities. New PCs that have audio, video, and CD-ROM

VIDEO

AUDIO

TEXT

As personal computers have become capable of presenting realistic audio and motion video, the word Multimedia has been used to refer to applications that use these new capabilities.

GRAPHICS

IMAGE

ANIMATION

Figure 14.8 Multimedia data types

capabilities are called multimedia PCs, or MPCs. A whole new class of applications has appeared that combine audio and video with other PC data types. These are typically distributed on CD-ROM discs because of the large amounts of data involved in audio and video.

14.6.1 Ingredients of multimedia

Computer multimedia is unique in its ability to combine many different data types into one application and under one user interface. The sections below explain each of these and their significance to multimedia. Figure 14.8 shows on-screen examples of each of the visual data types.

14.6.1.1 Motion video PC motion video employs the video compression technologies discussed in Chapter 3, and special algorithms have been developed for video playback using software without extra hardware except for an accelerated display adaptor. This performs well on the fastest PCs now available, but it is not very good on an average machine. There is also special hardware for higher-performance algorithms, such as MPEG-1, but the market still favors the software approach.

To speed up software-based video or allow larger video windows, some display adaptor vendors are adding video capabilities to their boards, which can be accessed by driver software that they provide with the boards. This will probably be the best solution until all machines are as fast as the best today. The advantage of not requiring dedicated video hardware that may quickly become obsolete is overwhelming.

14.6.1.2 Audio Multimedia audio mostly does not use compression. PC audio hardware supports PCM audio at both 8 bits/sample and 16 bits/sample, with very little load on the system CPU. Using compressed audio would increase the CPU loading and since the CPU is already heavily loaded by video (that often plays at the same time), there isn't much left for audio processing. Dedicated hardware could be used, but the arguments against it are the same as for video.

PC audio formats provide a wide range of performance choices from speech quality up to full audio CD-quality (see Table 1.3). In general, applications should use the lowest quality level that serves their purpose, because higher audio quality costs data space.

Uncompressed audio takes considerably less data than compressed video, so it tends to be more heavily used than video. A single CD-ROM disc can hold only an hour of video, but it can hold 10 or more hours of audio. The CD-ROM capacity can only be used once; space taken for audio data subtracts from that available for video. Trading between different data types based on the capacity of a CD-ROM disc is one of the challenges of designing multimedia applications.

14.6.1.3 Text The original computer interface medium was text and, even in these days of video interfaces, text is still an inescapable medium. It also is very efficient in terms of the data used. Consider that a CD-ROM disc devoted to text can hold upwards of 200,000 pages! Many existing databases are in the form of text and a lot of design effort in the multimedia field has gone into ways to use the video capabilities of computers to give easier access to text.

One very important multimedia text format is known as *hypertext*. With hypertext, words or phrases in a block of text can be made into push buttons (see Figure 14.9). If the user touches one of these words, which are highlighted so he/she can tell they are special, the program branches to a new screen or window that offers additional or related information about the subject of the word touched. For example, when reading a biography of a person, the person's birthplace may be mentioned. If the birthplace name is hypertext,

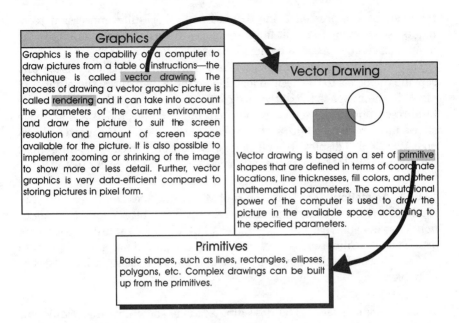

Figure 14.9 Hypertext

the user can touch it and get more information about that place. When the user is done with the birthplace information, he/she can instantly return to the biography and continue reading. It's just like what is done in a book with "see... references" except that the reader doesn't have to look up the reference—it pops up when he/she simply touches the hypertext button. This is a valuable user interface technique.

14.6.1.4 Graphics This is the capability of a computer to draw pictures from a table of instructions—the technique is called *vector* graphics. The process of drawing a vector graphic picture is called *rendering* and it can take into account the parameters of the current environment and draw the picture to suit the screen resolution and amount of screen space available for the picture. It is also possible to implement zooming or shrinking the image to show more or less detail. Further, vector graphics is very data-efficient compared to storing pictures in pixel form (see below).

The disadvantages of vector graphics is that the pictures are drawings and may not be realistic. Also with drawings, the result depends on the skill of the artist who did the original work.

14.6.1.5 Images Pictures can also be displayed on a video screen by digitizing photographs or capturing stills from motion video. Both of these produce *bit-mapped* images, where the actual pixels to display are stored (of course, they can be compressed using JPEG or another still-compression technique—see Section 3.12). Bit-mapped images are very realistic as long as the image has enough pixels and enough colors.

Displaying bit-mapped images consists of decompressing if necessary and copying the pixels to the desired location on the screen. Since the number of pixels in the image is fixed, the size of the image displayed depends on the total resolution of the screen being used. For example, a 640 × 480 image shows full-screen on a 640 × 480 display, but it will only be quarter-screen on a 1280 × 960 display. If a bit-mapped image is enlarged (zooming in), it will become blocky because the pixels become larger and easier to see.

14.6.1.6 Animation The techniques of vector graphics can be extended to render a series of frames to produce motion video. This is computer *animation*. It is very effective and data-efficient, but it has the same dependence on artistic skill. Animation is very difficult to do, an animator has to devote his/her life to it. If animation skill is available for an application, it can be very valuable.

14.6.1.7 User interface The preceding sections show how many different kinds of information and presentation techniques are available for multimedia applications. Making use of them all can present a user with a dazzling array of choices and that can mean a real user interface problem. The process of the user moving about in a multimedia application is called *navigation*. Many approaches are available for navigation and many user interfaces have been designed. The challenge is to find the one that suits the application at hand.

14.6.1.8 Operating system Multimedia also places demands on the computer's operating system (see Section 7.3.2). Lots of data has to be moved quickly from mass storage to display or sound system, sometimes with several things going on in parallel. This calls for a multitasking operating system (see Section 7.3.4.1). Multimedia is one of the things that are moving users to the more advanced operating systems.

Multimedia could, of course, be done on any equipment that had the same capabilities as a personal computer. Non-PC products for multimedia (multimedia players) have been proposed and some are even on the market. They work for their intended audience, but they offer little cost advantage compared to a standard PC and they lack the access to the vast software resources of the standard PC marketplace. They may or may not succeed in the long run.

14.6.2 Integration of multimedia applications

Upon learning about the capabilities of PC multimedia, many people in industry, education, military, or the general public get ideas of applications they would like to do. But the production of multimedia is complex, expensive, and just plain difficult. This has not been lost on the software industry and there is a large market for multimedia development *tools*—software programs that assist the multimedia development process.

The most important tools are those known as *authoring* tools. These are programs that support the assembly of all the content material into a single application and provide the user interface and navigation features needed. That, of course, is something that any computer programmer could do using his/her favorite language, but most of the people who want to create multimedia are not programmers. Some authoring tools are specifically designed to allow nonprogrammers to create multimedia; other authoring tools still require programmers, but they make the job easier. Authoring is beyond the scope of this book—there are whole books about it[2]—but a few important considerations will be covered here.

An authoring tool is always associated with a program that the user will employ to run the authored application. These are called *runtime packages*, or *players*. The authoring tool packages the application in a form that will be understood by the player, so that the end user does not get the full authoring package, only the player. Often the application content (audio, video, text, etc.) is kept in its native format and is simply accessed from mass storage by the player. However, this poses a security issue if the application developer wants to prevent others using their content material for different purposes. Some professional-level authoring systems employ data *encryption* techniques to prevent unauthorized use of the application or its materials.

The simplest and lowest-cost authoring systems may not have a player at all. The user is required to have the full authoring package to play the application. This is fine for individual users who create their own applications for their own use, but it is unacceptable for professional distribution of applications. This is just one example of why there are authoring packages available for less than $100 and other packages that cost thousands of dollars. The range of features matches or exceeds the range of prices.

14.6.3 A multimedia example

Multimedia applications are usually dynamic and it is difficult to convey them in book illustrations. There are also hundreds of different kinds of applications. Figure 14.10 is an example of a hypothetical multimedia encyclopedia. The screen shows an entry for the city of San Francisco and offers controls for navigating that entry or the rest of the encyclopedia. It shows how text, images, graphics, and video are combined on the screen to provide a rich presentation of any subject that the user chooses.

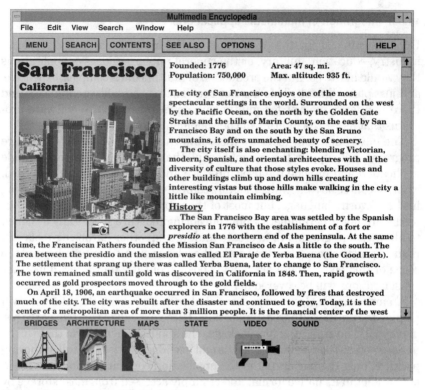

Figure 14.10 An interactive application—a multimedia encyclopedia

14.7 Interactive Television

Many experiments have been done over the years to add interactive features to television. Most of these were conducted on cable systems using reverse transmission on the cable for the user to head end communication. Generally, the interactivity was concerned with subscriber control of access to programs that were already on the cable but scrambled for limited access. By communicating with the head end, a subscriber could request access to a particular program and the head end would charge for the service. This has become known as pay-per-view. It can also be implemented using the telephone for the reverse channel and that has proven to be a better means of communication for this purpose.

With digital television on the horizon and with digital capabilities already appearing in receivers, there are some new technical possibilities for interactive television. It hasn't yet been shown that any of these will prove valuable enough to consumers for them to succeed, but that is being explored. One example was described above in Section 14.5; it will be covered more here.

Consider a television commercial for an automobile. The commercial shows glitzy attention-getting pictures of the car and an announcer describes one or two key features of it. That is all that there is time for in a 30-second or even a 1-minute commercial. However, an interested viewer would like to know a lot more about the product. This can be supported by having the digital transmitter download a database and access program to the receiver during the commercial. After the commercial is over, the consumer can activate the downloaded program and explore specifications, pictures, even prices for the car that was advertised in the commercial.

With the GA HDTV transmission system that sends data at 19.3 Mb/s, it is possible to send several megabytes of programs and data during a 30-second commercial simply by turning up the screws on the compression of the commercial to reduce the data rate below the 19.3 Mb/s. For example, downloading a 3-MB interactive package during a 30-second commercial requires a data rate of $3,000,000 \times 8/30 = 0.8$ Mb/s. This is only a little more than 4% of the channel capacity—compressing the commercial's video by another 4% to handle the downloading would have a negligible effect on picture quality.

The transmission part of this scenario seems very reasonable, but what is more questionable is the capability required at the receiver to store the data and then run the interactive program. That will take most of the functionality of a personal computer and so far, it does not seem reasonable to build that into a TV receiver because it would probably double the cost of the TV. A better solution might be to add an interface feature in the TV that would allow connection of a standard PC to the TV. Since a high percentage of homes already have PCs, the incremental cost to the consumer would be much less.

Further ideas about interactive TV are discussed in Chapter 19.

References

1. Laurel, Brenda, *The Art of Human-Computer Interface Design*, Addison-Wesley, Reading, MA, 1990.

2. Luther, Arch C., *Authoring Interactive Multimedia*, AP Professional, Cambridge, MA, 1994.

15

Terrestrial Broadcasting Systems

15.1 A Brief History and the Future Outlook

Commercial television broadcasting began in the United States in 1946. After a slow start, during which there were real questions as to its financial viability, growth accelerated, and by year-end 1950 there were 107 stations on the air.

Two serious questions about previous Federal Communication Commission regulatory policies were raised at this time:

1. What technical standards should be adopted for the transmission of color?

2. Were the technical restrictions that had been adopted for the assignment of channels adequate to avoid excessive co-channel interference?

To resolve these questions, the FCC convened a rule-making proceeding in September 1948. While the hearing was in progress, it imposed a "freeze" on the granting of construction permits for new stations. The hearing was lengthy and contentious, and it lasted nearly four years until April 1952. During this time the profitability of television broadcasting was demonstrated by the 107 on-air stations.

The FCC decision at the conclusion of the hearing was far-reaching, and it included the following provisions:

1. The system for transmitting color, the *field sequential system*, proposed by CBS was approved.

2. More stringent requirements for mileage separations between VHF co-channel stations were imposed. The result was a major reduction in the number of stations that could be assigned VHF channels.

3. To make more channels available, a large section of the UHF spectrum, 470–806 MHz, was allocated for broadcasting.

The decision for CBS color did not survive the test of time, primarily because its transmission standards were not compatible with the existing monochrome standards, i.e., color broadcasts could not be received on mono-

Table 15.1 Television broadcast channels

Channel	Frequency, MHz	Channel	Frequency, MHz
Low-band VHF		UHF (cont'd)	
2	54–60	36	602–608
3	60–66	37	608–614
4	66–72	38	614–620
5	76–82	39	620–626
6	82–88	40	626–632
		41	632–638
High-band VHF		42	638–644
7	174–180	43	644–650
8	180–186	44	650–656
9	186–192	45	656–662
10	192–198	46	662–668
11	198–204	47	668–674
12	204–210	48	674–680
13	210–216	49	680–686
		50	686–692
UHF		51	692–698
14	470–476	52	698–704
15	476–482	53	704–710
16	482–488	54	710–716
17	488–494	55	716–722
18	494–500	56	722–728
19	500–506	57	728–734
20	506–512	58	734–740
21	512–518	59	740–746
22	518–524	60	746–752
23	524–530	61	752–758
24	530–536	62	758–764
25	536–542	63	764–770
26	542–548	64	770–776
27	548–554	65	776–782
28	554–560	66	782–788
29	560–566	67	788–794
30	566–572	68	794–800
31	572–578		
32	578–584		
33	584–590		
34	590–596		
35	596–602		

chrome sets, even in black and white. In December 1953 the FCC approved a compatible color system developed by the *National Television Systems Committee,* which was an outgrowth of a system originally proposed by RCA. It has since been known as the *NTSC system.*

Except for CBS color, the standards and rules adopted by the FCC at the conclusion of the hearing, which specified *analog* video signals and *vestigial sideband* transmission, have been remarkably durable and they are largely in place 40 years later. The technical rules for the assignment of channels are virtually unchanged, and UHF channels are widely used for broadcasting, although their coverage with analog transmission is inferior to VHF. The NTSC color system and its variations have been adopted by most of the major countries of the world (France and Russia excepted).

But technology continues to progress, and a shift from analog to digital transmission is now underway, a shift that will eventually be incorporated in the FCC's Rules. The immediate impetus for this change is the increased information transmission rate required for HDTV (high-definition television). With analog transmission, HDTV requires greatly increased video bandwidth and scarce or unavailable spectrum space. But digital systems, making use of bandwidth compression technologies, can transmit HDTV signals in the same bandwidth as analog NTSC. Digital transmission also has other important advantages as described in Chapters 4 and 6.

As of this writing (1996) it appears that digital transmissions will be authorized as a short-term (15-year) supplement to analog and as a long-term replacement. The added capacity could be used (technically) not only for HDTV but also for additional NTSC signals and as a part of the "information highway" now being advertised widely.

The first sections of this chapter are devoted to a description of the existing analog transmitting system. This is followed by a description of the digital broadcasting system recommended by the Grand Alliance, an industry committee that was established under the aegis of the FCC for this purpose.

15.2 Allocations and Assignments

The cognizant regulatory agency in each country *allocates* bands of frequencies to specific services, subject to international agreement. Within these bands, specific *channels* are *assigned* to individual licensees.

Table 15.2 Minimum co-channel station separation in miles

Zone	Channels 2–13	Channels 14–69
I	170	155
II	190	175
III	220	205

Figure 15.1 Standard U.S. analog television broadcast channel

15.2.1 Current U.S. allocations

In the United States, the VHF bands 54–72, 76–88, and 174–216 MHz and the UHF band 470–806 MHz are allocated for television broadcasting. (The entire VHF portion of the spectrum extends from 30 to 300 MHz while the UHF extends from 300 to 3,000 MHz.) These bands are divided into *channels*, 6 MHz wide, for assignment. Each channel is identified by a number as shown in Table 15.1.

The VHF channels are divided into two segments, low-band — channels 2 to 6, and high-band — channels 7–13. The small gap between channels 4 and 5 is reserved for aeronautical use.

15.2.2 Current U.S. assignments

AM radio channels were assigned on the basis of demand. The FCC established co-channel and adjacent-channel interference criteria, and if the applicant for a new station could show that his proposal would meet these criteria, a Construction Permit and ultimately a license could be granted. But stations were often authorized that did *not* meet these criteria because of the intense commercial pressure for additional AM stations, and this led to overpopulation of the AM band and excessive mutual interference.

The FCC was determined not to repeat this mistake with television. Stations were authorized on a demand basis but only in accordance with a predetermined table of assignments (which was developed during the "freeze") that assigned specific channels to each city. The table could be changed but only as the result of a cumbersome rule-making procedure.

(a) Video Signal

(b) Modulated Carrier Signal

Figure 15.2 Amplitude modulation—video signal

At the end of the "freeze," a new table of assignments was issued, which was based on increased spacing between co-channel and adjacent channel stations as shown in Table 15.2. This resulted in a major reduction in the number of VHF stations, particularly in the larger cities.

Zone I in Table 15.2 encompasses the northeast and middle west states. Closer spacing is permitted in this zone in recognition of the denser population.

Zone III includes the Gulf Coast states. Greater spacing in this zone is required because of the prevalence of inversions in the troposphere and greater tropospheric propagation.

Zone II covers the remainder of the United States.

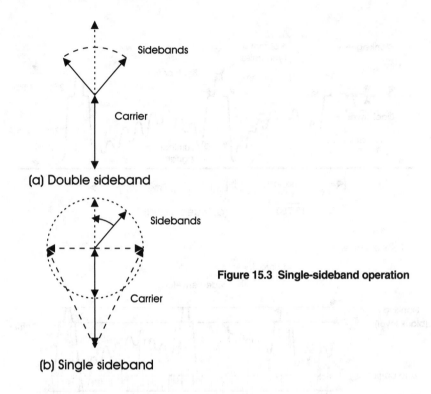

(a) Double sideband

(b) Single sideband

Figure 15.3 Single-sideband operation

The table also includes the effect of additional criteria or "taboos" for UHF channels based on the characteristics of superheterodyne receivers. For example, no channel is assigned on the image frequency of another channel in the same city.

The application of these criteria has resulted in many unassigned channels in each city, for example, channels adjacent to assigned channels. This has provided an important opportunity for the assignment of additional channels for HDTV (or ATV) (see below).

This table of assignments has generally been successful in minimizing both co-channel and adjacent channel interference.

15.2.3 Proposed HDTV (ATV) assignments

The current (1995) FCC policy anticipates the establishment of a second set of channel assignments for HDTV in the present VHF and UHF broadcast bands so that these bands will be shared by NTSC and HDTV stations. The criteria for assigning HDTV channels have not been firmly established, but they will probably include the following principles:

1. The HDTV channels will occupy the same 6 MHz bandwidth as NTSC.

2. Restrictions on the assignment of adjacent channels and taboo UHF channels will be greatly reduced for HDTV channels; adjacent NTSC and HDTV channels will be assigned in the same city.

3. Co-channel mileage spacing requirements for HDTV stations, whether to other HDTV stations or to NTSC stations, will be equally strict or perhaps stricter.

Active programs of laboratory and field testing are now underway to establish firm standards for the assignment of HDTV channels. These include tests for the *robustness* of the signals—their insensitivity to interference, and their *benignity*—their lack of interference potential to other signals. These tests are necesary because the characteristics of digital television broadcasting are radically different from analog broadcasting.

The results of HDTV assignment standards cannot be determined precisely until they have been firmly established, but an industry group, the Association for Maximum Service Television (AMSTV), presented a proposal to the FCC that assigned an HDTV channel as a companion to every NTSC channel.[1] The added channels in this proposal were overwhelmingly UHF—45 VHF and 1,645 UHF.

15.3 The Standard Analog Broadcast Television Channel

The amplitude/frequency characteristic of the 6 MHz television broadcast channel as specified by the FCC is shown in Figure 15.1. It has two carriers, the visual carrier 1.25 MHz above the lower channel edge and the aural carrier 4.5 MHz above the visual carrier.

15.4 The Visual Carrier

15.4.1 Modulation method and power rating

The visual carrier is amplitude modulated by the video signal (see Figure 15.2). Unlike the audio signal, the video signal has a dc component. It is transmitted by *clamping* (fixing) the carrier power at its maximum rated value (100%), at the peak of sync and adjusting the gain of the system so that its value at the reference white level is 7.25%. The *rated* transmitted power is the value at the peak of sync. The *average* transmitted power is less and will depend on the picture content, being least for an all-white picture and greatest for all black.

Although the sync and blanking waveforms are very effective in fulfilling their intended functions, they are extravagant in their use of transmitter power. On a voltage scale with the peak of sync at 1.4 and the blanking pulses at 1.0, the rms voltage of the blanking intervals over a complete frame is about 1.25. The rms voltage during the active picture intervals depends on the

Table 15.3 FCC power/antenna height limits

Channels	Maximum ERP	Maximum Height
2–6	100 kW	1000 ft.
7–13	316 kW	2000 ft.
14–68	5000 kW	2000 ft.

The power limitations apply to the horizontally polarized component
of the radiation. Stations employing circular polarization are
permitted to radiate an added equal amount of vertically polarized
energy.
See Section 15.15.2 for definition of antenna height.
See Figures 15.4 and 15.5 for power limitations if antenna height
exceeds the maximum.

picture content, but it typically would be around 0.2. In this example, the
average power level during blanking intervals would be proportional to 1.25^2
or 1.56, while the average picture power would be proportional to 0.2^2 or only
0.04. Although the blanking intervals occupy only 25% of the time, they utilize
97.5% of the transmitter power averaged over an entire frame. With a
transmitting system having an effective radiated power of 100 kW, only 2.5
kW is used for transmitting the picture information.

Co-channel interference is also a problem with the high-powered sync and
blanking pulses. They are the major source of interference because the

Figure 15.4 FCC power/antenna height maximums (zone 1)

Figure 15.5 FCC power/antenna height maximums (zones 2 and 3)

lower-level picture signal from the desired station must override the higher-power synce and blanking pulses from adjacent interfering stations.

Both the specing between co-channel stations and the average transmitted power could be reduced substantially by using digital broadcasting as proposed for the United States HDTV system and described later in this chapter.

15.4.2 Vestigial sideband format

The spectrum of the visual carrier and its sidebands is shown in Figure 15.1. Both the upper and lower sidebands are transmitted for video signal frequency components up to 0.75 MHz. The lower sideband is attenuated for frequency components above 0.75 MHz, and only the upper sideband is transmitted for signal components above 1.25 MHz. This is known as a *vestigial sideband* format, and its purpose is to reduce the amount of spectrum space required for the channel.

The effect of transmitting only a single sideband is shown on the vector diagrams in Figure 15.3. Double sideband operation is shown in Figure 15.3(a). The sidebands are represented by two vectors rotating in opposite directions around the carrier vector. When both sidebands are present, the amplitude of the vector sum varies by ±twice the sideband amplitude and its phase is constant. If one sideband is removed, the amplitude variation of the resultant is reduced by one-half and phase modulation known as *quadrature phase distortion* is introduced. [Quadrature phase distortion is not the same

Figure 15.6 Analog transmitter functional diagram

as *incidental phase distortion*, the variation of the phase of the carrier with level. Incidental phase distortion can be corrected (see Section 15.10.3.3).]

The loss in amplitude can be compensated by the amplitude response of the IF amplifier in the receiver. Typically it is tuned so that the response at the carrier is 6 dB below the maximum. Quadrature distortion cannot be easily corrected, but this is not necessary because of the relatively low amplitudes of the high-frequency components in a video waveform. For most picture contents, the maximum instantaneous high-frequency sideband amplitude occurs for the color burst signal.

15.5 The Aural Carrier

The aural carrier is frequency modulated with a peak deviation of ±25 kHz as compared with ±75 kHz for FM radio broadcasting. The lower deviation is specified to make as much spectrum space as possible available to the visual signal, and higher deviation of the aural carrier is unnecessary because the visual signal establishes the limiting signal-to-noise ratio parameter for the system.

The power of the aural transmitter is normally 20 percent of the peak visual power. This is usually more than adequate to produce an acceptable aural S/N in any location with a satisfactory visual S/N.

Table 15.4 Carrier frequency tolerances

	Visual	Aural
FCC requirements	±1,000 Hz	(Visual + 5.5 MHz) ±1,000 Hz
Industry objective	±200 Hz	(Visual + 4.5 MHz) ±200 Hz

Table 15.5 Representative TV transmitter power ratings

VHF low band	VHF high band	UHF
5 kW	15 kW	30 kW
15 kW	25 kW	60 kW
25 kW	45 kW	120 kW

15.6 Elements of a TV Transmitting System

A TV transmitting system includes a transmitter that generates the modulated visual and aural carriers, a filter that removes a portion of the lower sideband, a diplexer that combines the visual and aural carriers, and an antenna that radiates the signal.

In older systems, the modulated visual carrier, after amplification, passed through a sharply tuned *vestigial sideband* filter that removed a portion of the lower sideband and produced the vestigial sideband visual carrier. This wasted a lot of power. In current designs the lower sidebands are removed from the carrier at an intermediate frequency where the power level is low by a *surface acoustic wave* (SAW) filter (see Section 15.8.5).

The aural and visual carriers are combined in a multiplexer, usually a *notch diplexer* that has high attenuation of the lower subcarrier sideband, for transmission to the antenna in a single transmission line. Some antennas require two feeds separated in phase by 90°. The phase shifter for this follows the multiplexer. In most cases the line is two coaxial copper tubes with the ratio of their diameters chosen to give it a *characteristic impedance* of 51.5 or 75 ohms.

The antenna is invariably directional in the vertical plane to concentrate the radiated energy toward the horizon and along the earth rather than wasting it in space.

Some stations, especially in the UHF band, employ antennas that are directional in the horizontal plane as well. The FCC restrictions on this mode are severe, and antenna directivity cannot be used to permit closer station spacing. It can be used, however, to increase the station's coverage in a desired direction, and this provision of the FCC rules is sometimes used by UHF stations.

The *gain* of the antenna is the ratio of the maximum radiated energy density to the radiated energy density from a reference antenna with equal power input. The reference antenna used for broadcast systems is the *half-wave dipole*. A half-wave dipole radiating 1 kW in free space produces a field intensity of 137.6 mV/m at a distance of one mile.

The half-wave dipole reference differs from the *isotropic radiator*, the reference for microwave and satellite systems that radiates equally in all directions. An isotropic radiator is a more fundamental reference and it

Table 15.6 Amplitude-frequency response tolerances

Frequency Range (MHz)	Tolerance
0.5 – 4.1	±1.0 dB
4.1 – 4.18 (VHF)	+ 0.0, -2.0 dB
4.1 – 4.18 (UHF)	+ 0.0, -4.0 dB

The upper sidebands between 4.75 and 7.75 MHz and the
lower sidebands between 1.25 and 4.25 MHz should be
attenuated at least 20 dB below the visual carrier, using the
200 kHz sideband as reference.

simplifies calculations, but it cannot be physically constructed. The gain of a
half-wave dipole is 2.15 dB with respect to an isotropic radiator.

The FCC rules specify limitations on the *effective radiated power* or ERP,
the visual carrier power input to a half-wave dipole that would result in the
same radiated energy density. It is given by the equation:

$$\text{ERP} = P_T G_A E_{eff} \tag{15.1}$$

where: ERP is the effective radiated power

P_T is the output power of the visual transmitter

G_A is the gain of the antenna with respect to a half-wave
dipole

E_{eff} is the efficiency of the transmission lines, diplexer, and
other system elements between the output of the transmitter and the input to the antenna

Table 15.7 Standard envelope delay vs. frequency

Frequency (MHz)	Delay
0.05 – 3.0	0 ns
3.0 – 4.18	increases linearly with frequency with 0 at 3.0 MHz and −170 ns at 3.58 MHz.

The tolerance goals for these values are:

Frequency (MHz)	FCC, ns	Goal, ns
0.05 – 2.1	±100	±50
2.1 – 3.0	± 70	±35
3.0 – 3.58	± 50	±25
3.58 – 4.18	±100	±60

15.7 FCC ERP/Antenna Height Limitations

The FCC limits the ERP and antenna height of television stations to provide the basis for an orderly plan for assigning channels to cities in the United States. These limits are higher in high-band VHF, Channels 7–13, than at low-band, and higher at UHF than at VHF. This reflects the less favorable propagation characteristics of the higher frequencies. In consideration of these limitations, the FCC established criteria for mileage separation between co-channel and adjacent channel stations to minimize interference while permitting a reasonable number of assignments. The radiated power limitations for antenna heights up to indicated maximums are shown in Table 15.3. Additional power limitations apply for heights above these maximums as shown in Figures 15.4 and 15.5. The zones on which the limitations are based are defined in Part 73.699 of the FCC rules. Zone I includes the northeast and midwest. Zone III includes the Gulf Coast states while zone II includes the remainder of the country.

15.8 Television Transmitters

The designs of television transmitters differ widely depending on their frequency band and power rating. Other differences result from proprietary choices of manufacturers for the basic system arrangement and components. All transmitters, however, must include the following components:

1. Aural preemphasis to preemphasize the high-frequency components in accordance with FCC requirements.

2. Processing amplifiers to correct deficiencies such as improper sync level in the input visual signal and to provide DC restoration.

3. Visual and aural carrier generators with the frequency tolerances specified by the FCC.

4. Visual and aural modulators.

5. Filters to remove the lower sidebands of the modulated visual carriers and to remove spurious frequency components from both carriers.

6. Linearity correction devices in the visual chain to compensate for incidental phase modulation and differential gain. Ideally, these corrections should be made in both the video and RF domains.

7. Intermediate and final power amplifiers. Figure 15.6 is a functional diagram of a representative analog television transmitter that shows the locations of these functions.

Figure 15.7 Transmission line voltages and currents

15.8.1 Exciter modulator

The exciter modulator is the heart of the transmitter and includes functions 1 to 6 in Section 15.8. Its outputs are the modulated visual and aural carriers at low power levels.

15.8.1.1 Aural preemphasis The aural preemphasis must be in accordance with FCC requirements. The time constant of the preemphasis circuit is 75 microseconds.

15.8.1.2 Stabilizing amplifier The stabilizing amplifier regenerates sync pulses and stabilizes their amplitude. The signal is DC restored by clamping the signal at black level. White peaks are clipped is applied to avoid overloading the modulator.

15.8.1.3 Linearity and frequency response correction   Linearity and frequency response correction is made in both the video and RF domains.

15.8.1.4 Aural and visual carrier generation   The aural and visual carriers are generated at IF frequencies. Their frequency tolerances are small (see Table 15.4), in part because of their use of offset carriers (see Section 3.17.2), which require the difference in carrier frequencies between adjacent co-channel stations to be stabilized at ±10 kHz. Stations are assigned frequencies on this basis and the technique would not be effective if the difference in carrier frequencies greatly exceeded this tolerance.

It is equally important that the frequency difference between aural and visual carriers be maintained within close tolerances. This is accomplished by maintaining their frequency difference at 4.5 MHz so that the carrier frequencies drift together.

15.8.1.5 Visual modulator   The engineering profession was originally divided on the issue of low-level vs. high-level modulation, i.e, whether the carrier should be modulated at a low or high power level. With the passage of time, technical advances have favored low-level modulation, and as of this writing (1996) most new transmitters employ this modulation mode.

The advantages of low-level modulation are numerous. They include the elimination of high-power modulators and vestigial sideband filters, and the ability to correct linearity and frequency response distortions with compensation in both the video and radio frequency domains. The disadvantage is the requirement for linear power amplifiers.

15.8.1.6 SAW filter and sideband removal   The device that finally tipped the balance for low-level modulation was the *surface acoustic wave* (SAW)

Table 15.8 Transmission line diameters

Type	Z_c (ohms)	Outer diameter (in.)
560	75	7/8
561	75	1-5/8
562A	75	3-1/8
563	75	6-1/8
573	50	6-1/8

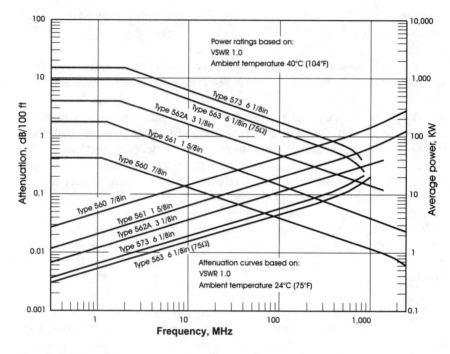

Figure 15.8 Coaxial transmission line power ratings and attenuation

filter. SAW devices are tuned filters that employ piezoelectric semiconductors. SAW filters can produce amplitude and envelope delay responses that are unachievable with conventional RLC filters. They have steep skirts at the band edges, but they do not introduce the envelope delay distortion that is inherent in RLC filters. Transmitter designers have taken full advantage of this ability to provide the vestigial-sideband amplitude-frequency response required by the FCC for the visual carrier and to make frequency and linearity corrections in the RF domain. The insertion loss of the SAW filter is high, typically 25 dB, and low-noise amplifiers must be used with it to achieve a satisfactory signal-to-noise ratio.

15.8.1.7 Up-converter The up-converter is a conventional heterodyne circuit that mixes the outputs of the visual and aural modulators with that of a local oscillator operating at the sum of the IF and final carrier frequencies. The up-converter outputs at the final carrier frequencies are then sent to the visual and aural intermediate power amplifiers.

15.8.2 Power amplifier drivers

Power amplifier drivers are linear amplifiers that provide the input to the final power amplifier, the PA. They must have sufficient output power to drive the PA to its full rated power, and their frequency response and linearity must be consistent with the performance specifications of the transmitter.

The power requirements for the driver vary from less than 25 watts for klystron UHF power amplifiers to nearly 2 kw for some tetrode VHF power amplifiers. In common with the trend for all electronic devices, electron tubes are being replaced by solid-state devices as soon as the latter can meet the power level and other performance specifications at an economic price. The technology of solid-state devices has now progressed sufficiently to meet these conditions, and most new commercial transmitters use solid-state drivers.

15.8.3 Power amplifiers

In common with all television technologies, transmitter power amplifiers (PAs) have benefitted from continuous technical improvement. Early VHF transmitters used tetrodes that were grid-modulated when high-level modulation was employed. Different models of early UHF transmitters employed tetrodes and *klystrons*, a tube that employs *velocity modulation* of an electron beam to amplify the carrier. The features of klystrons include stability, high output power, reasonable efficiency (about 40 percent), and high gain so that very little driving power is required. This characteristic eliminated the need for a long chain of preamplifiers with their linearity problems and was one of the factors that hastened the trend to low-level modulation. Unlike tetrodes, whose performance deteriorates with increasing frequency, klystrons work better at higher frequencies. This led to their early choice as the power amplifier for UHF transmitters.

Figure 15.9 TE$_{10}$ waveguide transmission mode

Table 15.9 Waveguide power ratings and attenuation

Type	Width Inside (in.)	Freq. Range MHz	Power Rating MW	Attenuation dB/100 ft.
WR 1800	18.0 – 9.00	410 – 625	93.4 – 131.9	0.056 – 0.038
WR 1500	15.0 – 7.00	490 – 750	67.6 – 93.3	0.069 – 0.050
WR 1150	11.5 – 5.75	640 – 960	35.0 – 53.8	0.125 – 0.075

The *klystrode*, a member of a class known as inductive output tubes or IOTs has been introduced recently that employs both a control grid and velocity modulation. This has extended the use of the klystron principle to VHF transmitters and provides higher efficiency than the klystron at UHF.

In the meantime, solid-state technology continues to advance, and in time solid-state devices will probably supersede electron tubes of all types. They are already being used in PA drivers and in low-power PAs, and their power and frequency limits are continuously being expanded.

Solid-state devices as used in transmitters are not particularly efficient, and a considerable amount of heat must be dissipated, which is difficult because of the small size of the devices. The *heat pipe* is one solution. A liquid in a wick transfers heat from the device to cooling fins that can be fan cooled if necessary.

15.9 Representative Transmitter Power Ratings

Although the FCC does not specify power ratings for transmitters, certain power ranges have been established in commercial transmitters that are strongly influenced by the maximum ERP permitted by the FCC and practical antenna gain values.

Table 15.5 shows typical power ratings for commercial TV transmitters in the United States. Note that they become progressively higher in the higher frequency bands. This is not because of equipment limitations but because the propagation characteristics of higher frequencies are not as favorable — a fact that is recognized in the power limitations specified by the FCC.

15.10 Transmitter Performance Standards

It is important to make a distinction between *equipment* performance and *system* performance in interpreting performance standards. Degradations and distortions are cumulative in analog systems, and for many criteria, e.g., signal-to-noise ratio, the performance of each element in the system must be better than that of the total system. The standards for transmitter perform- ance in this section are for the transmitter only.[2] The standards for the

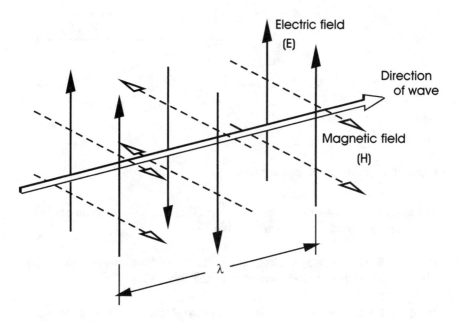

Figure 15.10 Electromagnetic wave

radiated signal, which includes the effect of other components of the system, would be somewhat lower. The transmitter standards should be compared with the EIA standards for point-to-point transmission systems (such as microwave) described in Chapters 17 and 18.

The following paragraphs summarize the most significant performance standards for television transmitters as contained in two sources: the FCC rules, Section 73, and informal industry equipment goals[3] that have been established based on the actual performance of high quality transmitters. The industry goals are more complete and generally more stringent than the FCC rules.

15.10.1 Frequency related standards

15.10.1.1 Carrier frequency tolerance   See Table 15.4 for FCC requirements and industry goals for frequency tolerances.

15.10.1.2 Amplitude-frequency response   The amplitude-frequency response of the transmitter as specified by the FCC is shown in Figure 15.1.

Figure 15.11 Tropospheric propagation

The goals for the maximum deviation from this response (as measured with a sideband analyzer) are shown in Table 15.6.

15.10.1.3 Envelope delay vs. frequency The standard *envelope delay* (see Chapter 4) and its tolerances are shown in Table 15.7.

15.10.1.4 Frequency response vs. brightness Measurements of frequency response are made with a modulating sine wave having a peak-to-peak amplitude of 25 percent of reference black to reference white and with axes on the brightness scale of 25, 45, and 65 percent. The reference is the 45 percent axis.
   Tolerance: FCC—not specified; goal—±1 dB

15.10.2 Signal-to-noise ratio, visual signal

The visual signal-to-noise ratio is defined as:
   S[blanking to reference white]/N[weighted rms] dB
   (See Section 4.15.3 for definition of weighted noise.)
   Tolerance: FCC—not specified; goal—55 dB

15.10.3 Linearity standards

Linearity standards specify the variations in gain or phase delay with brightness or other signal parameter.

15.10.3.1 Differential gain and low-frequency nonlinearity Differential gain is the measure of variations in the transmitter's gain at the color subcarrier frequency (see Section 5.10) for different luminance levels of the video signal. It is measured by superimposing a 3.58 MHz sine wave, having a peak-to-peak amplitude of 20 percent of blanking-to-reference white, on a stairstep luminance signal and measuring the differences in the sine wave amplitude at the transmitter output on each of the steps.

Low-frequency nonlinearity is the measure of the variation in the transmitter's gain with brightness for lower video frequencies. It is also measured with a stairstep waveform, and the nonlinearity is indicated by variations in the height of the steps at the transmitter's output.

Tolerance: FCC—not specified; goal—5 percent

15.10.3.2 Differential phase Differential phase is the complement of differential gain and is the variation in the color subcarrier phase delay with brightness (see Section 5.10.4). The same waveform is used for measurement.

Tolerance: FCC—not specified; goal—±3°

15.10.3.3 Incidental phase modulation Incidental phase modulation differs from differential phase distortion in that differential phase refers to the phase shift of the color *subcarrier envelope* whereas incidental phase modulation refers to the phase of the *carrier*. It is usually larger than the differential phase.

Tolerance: FCC—not specified; goal—± 7°

15.11 Transmission Lines

The output energy of the transmitter is conveyed to the antenna by means of a *transmission line*. It is usually two coaxial copper tubes that are insulated from each other, although *wave guides* are occasionally used for high-power UHF stations (see Section 15.12).

15.11.1 Coaxial line transmission modes

When the radio-frequency output of the transmitter is applied to the input of a coaxial line with a gaseous dielectric and a diameter that is small compared with the wavelength, it generates a voltage and current wave that travels at approximately the velocity of radiation in free space. (The deviation results from the fact that the transmission line is not a perfect conductor. In most practical cases it is small enough to be ignored.) If the dielectric is solid, the velocity and the wavelength are reduced by the square root of the dielectric constant.

In this primary mode, the electric field is perpendicular to the axis of the line. If the wavelength is equal to or less than the circumference of the line, a secondary mode can occur in which the electric field is circular and coaxial with the line. This is known as *multimoding*. Multimoding is undesirable because it increases the losses and reduces the power handling capability of the line. In practice, this limits the usefulness of coaxial lines to frequencies with wavelengths greater than the circumference of the line, a factor that must be considered at the upper end of the UHF band.

The voltage and the current waves are in phase and their ratio, E to I, is Z_C, the *characteristic impedance* of the line. If the impedance of the load at the end of the line equals the characteristic impedance, all of the energy in the wave is absorbed. If the load impedance does not equal the characteristic impedance, some of the energy is not absorbed, and voltage and current waves are reflected back to the transmitter. The magnitude and phase of the voltage and current at each point on the line is the vector sum of the direct and reflected waves, and a stationary *standing wave* is formed with maxima and minima separated by one-half wavelength.

15.11.2 Voltage and current relationships

Figure 15.7 shows the relationship between voltage and current vectors at the load and 1/4 wavelength toward the source for four load impedances: characteristic impedance, short circuit, open circuit, and double the characteristic impedance.

The ratio of the voltage maximum to the voltage minimum in the standing wave is called the *voltage standing wave ratio* or VSWR. It is a fundamental measurement of transmission line performance.

The VSWR is calculated by Eq. (15.2):

$$\text{VSWR} = \frac{1 + |C_R|}{1 - |C_R|} \tag{15.2}$$

where $|C_R|$ is the magnitude of the *reflection coefficient*.

C_R is calculated by Eq. (15.3):

$$C_R = \frac{Z_{\text{LOAD}} - Z_C}{Z_{\text{LOAD}} + Z_C} \tag{15.3}$$

where Z_{LOAD} and Z_C are the load and characteristic impedances.

Figure 15.12 Vertical antenna directivity and gain

Z_C has a small reactive component as the result of the ohmic resistance of the transmission line conductors. This is usually so small that it can be ignored.

If Z_{LOAD} equals Z_C, the standing wave pattern disappears, and the voltage and current are constant except for a gradual diminution toward the load owing to losses in the line.

If Z_{LOAD} is a pure resistance and greater than Z_C, a voltage maximum will appear at the load and at 180° intervals toward the load on the line. Similarly, if Z_{LOAD} is a resistance less than Z_C, a voltage minimum will appear at the load and at 180° intervals along the line. If Z_{LOAD} has a reactive component, the maxima and minima will be shifted so that the standing wave at the load will be neither a maximum nor a minimum in the standing wave pattern.

The input impedance undergoes an inversion every quarter wavelength. For example, an open-circuit quarter-wavelength line will have zero input impedance, and a short circuited quarter-wavelength line will appear to be an open circuit.

The relationships between load impedance, VSWR, and input impedance in the general case can be determined by the use of a *Smith chart*.

15.11.3 Characteristic impedance calculation

The characteristic impedance of transmission lines is $(L/C)^{1/2}$, where L and C are the inductance and capacitance of the line per unit length. It is determined by the ratio of the diameters of the conductors and the dielectric constant of the insulating material between them:

$$Z_C = \frac{138 \log_{10}\left(\dfrac{b}{a}\right)}{\sqrt{\mu}} \tag{15.4}$$

where b and a are the diameters of the outer and inner conductors, and

               μ is the dielectric constant of the insulator between them.

The ratio of outer and inner conductor diameters for lines with air dielectric, $(\mu = 1)$, is 3.52 for a Z_C of 75 ohms and 2.30 for a Z_C of 50 ohms.

15.11.4 Standard transmission line diameters

The de facto standard dimensions that have been developed for rigid coaxial transmission lines are tabulated in Table 15.8. Rigid transmission lines normally come in 20-ft. lengths that are joined by flanges welded to their ends. A wide variety of elbows, T-connections, and other fittings are available.

Figure 15.13 Superturnstile antenna

15.11.5 Power ratings and attenuation

The power ratings and attenuation of transmission lines are normally established for a VSWR of 1.0. If the VSWR exceeds 1.0, the higher currents and voltages at their maxima will cause greater attenuation and a lower power rating. Figure 15.8 shows the power rating and attenuation as a function of frequency for the standard transmission line sizes in the VHF and UHF television frequency ranges, all with a VSWR of 1.0. The rapid increase in attenuation for 6 1/8" line at the upper end of the UHF band is caused by multimoding.

Note that the power ratings are based on *average* transmitter power. The peak-power rating used for television transmissions would be higher, but on the basis of an all-black picture and after allowing for the aural carrier, the difference is small.

15.11.6 Mechanical features

A number of mechanical features of coaxial transmission lines should be noted.

The inner and outer conductors are subjected to different temperatures so that differential expansion occurs. On a long run the difference can be large, and provision must be made to compensate for it. One solution is to join the

Figure 15.14 Panel antenna

inner conductors with slotted plugs that fit inside the two ends. This allows the conductors to expand or contract without breaking the electrical connection.

Provision must also be made for expansion of the outer conductor, and the line is usually hung from the tower with spring hangers rather than being rigidly attached to it.

Moist air inside the line can cause condensation and leakage across the inner conductor spacers. To prevent this, most lines are pressurized with dry air. This requires the joints to be sealed so that they are air tight.

15.12 Waveguides

It is unfortunate that the trends of the power capacity and attenuation vs. frequency curves of coaxial transmission lines are in the wrong direction for UHF transmissions. More power is required for UHF channels, but the power rating of coaxial lines is lower at the higher frequencies and their attenuation is greater. The use of larger diameter lines at the upper end of the UHF band to reduce losses and increase the power rating is limited by multimoding. As a result, waveguides, which are better suited for higher frequencies, are sometimes used for the upper UHF channels.

Rectangular waveguides operating in the TE_{10} mode (see Figure 15.9) are the most commonly used. In this mode, the width of the waveguide must be more than 1/2 wavelength at the transmission frequency. The frequency at

Figure 15.15 Slot antenna

which the wavelength equals twice the width of the waveguide is its *cut-off frequency*, and the waveguide will not transmit lower frequencies.

The losses in waveguides are much lower than transmission lines at these frequencies. The dimensions, frequency range, power rating, and attenuation of aluminum waveguides operating in the TE_{10} mode and in the standard sizes that cover the UHF broadcast band are shown in Table 15.9. Note that the attenuation *decreases* and the power rating *increases* as the frequency increases.

Like coaxial transmission lines, waveguides are subject to moisture condensation as they cool. This can be avoided by pressurization, but the amount of pressure is limited by the mechanical weakness of the rectangular structure. Higher pressures can be used with *double truncated waveguides*. The corners of a rectangular waveguide are rounded, and it is fitted inside a circular tube. The circular structure not only permits greater pressurization but also reduces the wind loading.

Figure 15.16 Vertical pattern with beam tilt

15.13 Properties of VHF and UHF Radiation

This section provides a brief review of the properties of VHF and UHF electromagnetic radiation as an introduction to a description of television broadcast antennas and the coverage areas of broadcast stations.

15.13.1 Electric and magnetic fields

As its name implies, electromagnetic radiation is a moving wave of electric and magnetic fields that are at right angles to each other and to the direction of travel (see Figure 15.10). The path of the wave is a straight line unless altered by refraction, diffraction, or reflection.

15.13.2 Frequency and wavelength

The frequency and wavelength of electromagnetic radiation in free space are related by Eq. (15.5):

$$\lambda = \frac{c}{f} \tag{15.5}$$

where: λ is the wavelength in meters

c is the wave velocity in meters/sec.

f is the frequency in hertz

Figure 15.17 Vertical pattern with null fill

15.13.3 Velocity and index of refraction

The velocity of electromagnetic radiation in free space, c, is 3×10^8 meters/sec. Its velocity through gases, liquids, or solids is c/n, where n is the *index of refraction* of the transmission medium.

The index of refraction of the atmosphere varies in accordance with altitude, barometric pressure, and humidity. Typically it is 1.003 at sea level and decreases exponentially toward exactly 1.000 with altitude. The deviation from unity is small, but is sufficient to have a major effect, normally an increase, on the coverage of television broadcast stations because of *refraction*.

15.13.4 Refraction

Refraction occurs when the radiation path passes obliquely across a boundary between two media having different refractive indices. The path is bent toward the denser medium that has a higher index of refraction.

The path of a broadcast television signal is directed tangentially along the earth's surface at the transmitter and passes through progressively less dense layers of the *troposphere*, the lower portion of the atmosphere. Because of refraction, it is bent downward and passes beyond the geometric horizon; this is known as *tropospheric transmission* or "tropo" (see Figure 15.11).

Under normal atmospheric conditions the effect of refraction can be approximated by assuming that the earth's radius is 4/3 of its actual value, thus increasing the distance to the calculated radio horizon. Atmospheric

Table 15.10 Grades of television service

Service	F(50, 50) dBμ Channels		
	2 – 6	7 – 13	14 – 68
Principal community 90% of locations 90% of time	74	77	80
Grade A 70% of locations 90% of time	68	71	74
Grade B 50% of locations 90% of time	47	56	64

abnormalities sometimes exist in which the atmospheric density *gradient* deviates from its normal pattern. The most common is an *inversion* in which the temperature of the air increases with height, thus causing the density to decrease more rapidly than normal and the distance to the radio horizon to increase correspondingly.

15.13.5 Diffraction and reflection

Electromagnetic radiation travels in a straight line in free space, but if a part of the wave front is blocked by an intervening object the path will bend, and some radiant energy will appear behind the geometric shadow of the object. This is one of the effects of a phenomenon known as *diffraction*.

Radiant energy is reflected, and, as the result of the combined effects of reflection, refraction, and diffraction, television signals can be received at most points in a station's service area even though not all of them have a line-of-sight path to the transmitting antenna.

15.13.6 Multipath transmission

The signal at the receiving antenna is the vector sum of a direct signal and one or more reflected signals. If the difference in path lengths is small, the signal strength will be increased or decreased depending on the relative phase of the incoming signals. If the difference in path lengths is large, e.g., as the result of a reflected signal from a large distant object, the time difference will be great enough to cause a *ghost*. This effect is called *multipath*, and it can be serious in mountainous or congested urban areas.

15.13.7 Polarization

The *polarization* of a wave describes the orientation of its field vectors, and by convention it is expressed as the orientation of the electric field. Two

Figure 15.18 FCC F(50,50) field intensity curves, high-band VHF

Figure 15.19 FCC F(50,50) field intensity curves, low-band VHF and UHF

Table 15.11 Distance (miles) to coverage contours

Channels	Height	Service Grade		
		Prin. Com.	A	B
2 – 6	500 ft.	19	28	43
	1,000 ft.	28	37	57
7 – 13	500 ft.	25	36	52
	1,000 ft.	33	41	60
14 – 68	500 ft.	31	41	60
	1,000 ft.	41	52	70

	ERP
2 – 6	100 kW, 20 dBkW
7 – 13	316 kW, 37 dBkW
14 – 68	5,000 kW, 37 dBkW

polarization modes are permitted for television broadcasting signals, horizontal and elliptical. With elliptical polarization, the electrical vector rotates as the wave passes a fixed point. Circular polarization is a special case of elliptical in which the vector has an equal value in all orientations.

FCC rules permit the same ERP for the horizontal component of circularly polarized radiation (or the component on the major axis for elliptical polarization) as for horizontal polarization, so that the total radiated power can be twice as great.

Circular polarization has a number of advantages. Since the total radiated power is twice as great, the received signal is usually greater with "rabbit ears" or other indoor antennas that are not optimized for horizontal polarization. Improvements of up to 3 dB in received signal strength can be achieved by using circularly polarized receiving antennas that derive energy from both the horizontal and vertical components.

Circularly polarized receiving antennas also reduce the magnitude of ghosts since the direction of rotation is inverted upon reflection, and the reflected wave is not received.

As a result of these advantages, there has been an increasing trend toward the use of circular polarization.

15.13.8 Radiated field strength and energy

The intensity of electromagnetic radiation is usually specified by the strength of its electric field. The usual units for the magnitude of the fields encountered in broadcasting are millivolts/meter or microvolts/meter.

Figure 15.20 HDTV (ATV) broadcast channel

The field strength of an electromagnetic wave in free space is inversely proportional to the distance from the antenna. For radiation from a ½ wave dipole, it is:

$$E = \frac{137\,P^{1/2}}{D} \qquad (15.6)$$

where: E is the field strength in millivolts/meter

   P is the radiated power in kilowatts

   D is the distance in miles

The strength of a wave can also be measured by the power passing through a unit area, usually a square meter, at right angles to the wave's path. Field strength and energy in space or in the atmosphere are related by Eq. (15.7):

$$\text{Power} / \text{m}^2 = \frac{E^2}{377} \qquad (15.7)$$

where: power is measured in microwatts

   E is the rms value of the field strength measured in millivolts/meter

The constant, 377 (120π), has the dimension of ohms, and it is sometimes described as the impedance of free space.

Table 15.12 Aspect ratios

Designation	Frame Size (mm)	Aspect Ratio
16 mm	10.26 × 7.49	1.37
Super-16	12.52 × 7.42	1.69
35 mm	21.95 × 16.00	1.37
Super-35	24.89 × 18.67	1.33
Cinemascope		2.35
70 mm		2.20

15.14 Antennas

The role of the antenna is to couple the transmitter to space by converting its electrical output energy to electromagnetic radiation. The primary antenna specifications are vertical directivity and gain, horizontal directivity, polarization, input impedance and bandwidth, and power handling capacity.

15.14.1 Vertical directivity and gain

Television transmitting antennas increase the signal strength on the earth's surface by minimizing the power that is radiated into space and concentrating it at angles toward or just below the horizon. This property, the antenna's *directivity*, is indicated by a *radiation pattern*, a plot of the radiation intensity at different elevations.

Figure 15.12 shows the vertical radiation patterns for three representative antennas, a low-band, low-gain VHF antenna (see Section 15.6 for definition of "gain"), a high-band, medium-gain VHF antenna, and a high-gain UHF antenna.

The points on the pattern where the field strength is 70 percent of the maximum are known as the *half-power points*, and the angular separation of these points is the *beam width* (see Figure 15.12).

The gain of the antenna is increased by narrowing the beamwidth to concentrate the radiated power in the desired directions, and the product of the gain and beamwidth tends to be constant.

The directivity of a television broadcast antenna can be calculated by defining its *aperture*, an imaginary cylinder in space surrounding the radiating elements. The cylindrical aperture is *illuminated* by the radiating elements, and for analytic purposes each point on the aperture can be considered to be a source of radiant energy.

The antenna's directivity is determined by the dimensions of the aperture and by the amplitude and phase of the radiant energy at each point on its surface. In the simplest case, all the points on the aperture are illuminated by a field of the same phase and of equal amplitude. The beamwidth, the

Table 15.13 HDTV (ATV) receiver scanning standards

Aspect ratio	16:9
Progressive scanning alternative	
Active scanning lines	720
Pixels per scanning line	1280
Field/frame rate	60/second
Interlaced scanning alternative	
Active scanning lines	1080
Pixels per scanning line	1920
Field rate	60/second
Frame rate	30/second

angular separation between the half-power points at which the radiation intensity is 70 percent of that at the beam maximum, is inversely proportional to the ratio of the aperture height to the wavelength. If the illumination of the aperture is uniform, the beamwidth in degrees is:

$$\text{Beamwidth} = 48\left(\frac{\lambda}{h}\right) \tag{15.8}$$

where: λ is the wavelength

           h is the height of the aperture measured in the same units as λ

If the illumination of the aperture is not uniform, Eq. (15.8) does not apply precisely, but in general the longer the aperture relative to the wavelength, i.e., the taller the antenna, the narrower the beam.

15.14.2 Horizontal directivity

The FCC rules make a distinction between antennas that are intentionally designed to be directional and those that are directional by virtue of inevitable imperfections in the design. There are no firm rules with respect to the circularity of the horizontal pattern of "nondirectional" antennas, but most of them maintain it within ±3 dB or less.

As the result of its unfortunate experience with AM directional antennas, which were overused to increase the number of stations on each frequency, the FCC ruled that directional antennas *cannot* be used in television to reduce the spacing between stations. Also, the maximum ERP from directional antennas cannot exceed the maximum allowed for nondirectional antennas. Their major purpose of horizontal directivity is to avoid wasting energy over sparsely populated areas and to provide the maximum allowed ERP to densely populated areas with lower total power.

Table 15.14 NTSC/HDTV comparative service

	NTSC		HDTV	
	VHF	UHF	VHF	UHF
Transmitter ERP	10 kW*	500 kW*	0.63 kW**	31.6 kW**
Sites measured	169	199	169	199
Satisfactory reception	39.6%	76.3%	81.7%	91.5%

* Peak power
** Average power

15.14.3 Bandwidth, VSWR, and echoes

Most antenna components are frequency sensitive, and achieving the proper input impedance over the bandwidth of a single channel was one of the early challenges of television technology. In subsequent years, techniques have been developed that make it possible to achieve satisfactory response over more than one channel.

The conventional indicator of antenna bandwidth is the range of frequencies at which its input impedance approximates the characteristic impedance of the transmission line. This is measured by the VSWR of the transmission line feeding the antenna, and it is desirable that this be less than 1.05 for frequencies from 0.75 MHz below to 5.45 MHz above the visual carrier. The VSWR at the color subcarrier 3.58 MHz above the carrier is particularly important.

A low VSWR is not an absolute guarantee of a satisfactory match between antenna and transmission line, and reflections from the antenna may be great enough to produce visible echoes or ghosts without causing the VSWR to exceed otherwise acceptable limits. The visibility of the echoes depends on their amplitude and their separation from the primary signal. For a transmission line length of 1,000 ft., echoes having an amplitude of 1.5 percent or more will be visible. For a 500 ft. length, visibility begins for echoes having an amplitude of 3.0 percent or greater.[4]

15.14.4 Radiating systems

Antennas can be classified by the type of radiators that illuminate the aperture, and these can be divided into two categories, dipoles and slots.

15.14.4.1 Dipole radiators This section describes representative dipole radiators.

1. Superturnstile antenna. The *superturnstile* antenna was one of the first to be introduced in the early years of television broadcasting, and it is still in wide use for VHF stations. The dipole is a pair of radiators, sometimes called *bat wings* because of their shape, mounted on oppo-

site sides of a supporting pole. The *radiator* consists of two dipoles mounted at right angles, and the *antenna* consists of two to twelve layers of dipoles (see Figure 15.13). The spacing between layers is approximately one wavelength, and the gain is approximately equal to the number of layers. The dipoles are plane surfaces rather than cylinders to increase their bandwidth. The dipole pairs are fed in quadrature and generate a reasonably circular pattern in the horizontal plane.

2. Panel antenna. The superturnstile antenna requires a cylindrical mounting pole of limited diameter so that it is not well suited for the lower element of a stacked array where two or more antennas are mounted above one another. The panel antenna (see Figure 15.14) solves this problem by mounting the dipole in front of a reflector and mounting a dipole/reflector assembly on each face of a tower that can be either 4- or 3-sided. Panel antennas are used for both high- and low-band VHF stations.

3. Circularly polarized antennas. In one form of circularly polarized antenna, the single horizontally polarized dipole in the panel antenna is replaced by crossed dipoles that are fed in quadrature, thus producing a circularly polarized wave.

15.14.4.2 Slot radiators Slot radiators are most frequently used for UHF stations but are also used for VHF. They consist of a hollow cylindrical column in which vertical slots are cut. A conductor in the center of the column, which is connected to the transmitter output, is capacitively coupled to the slots by radial pins mounted on their edges (see Figure 15.15).

An electric field is produced across the slots with its amplitude and phase determined by the length and position of the pins. The electric field across the slots and the current around the cylinder generate the radiated energy.

There are two types of slot antennas, *standing wave*, commonly used for UHF and *traveling wave*, commonly used for VHF. The inner conductor of the standing wave antenna is connected to the outer cylinder at the end opposite the feed point, and a standing wave pattern is produced. The inner conductor of the traveling wave antenna is terminated by a capacitive connection to the end slots that absorbs any remaining energy and provides a characteristic impedance load. The amplitude of the voltage in the standing wave antenna is a standing wave pattern with diminishing maxima amplitudes. The amplitude of the voltage in the traveling wave antenna diminishes continuously from feed point to termination.

15.14.4.3 Helical radiators Helical antennas are occasionally used. The radiating elements are two helixes, one above the other, wrapped around but separated from a supporting pole. The two are wrapped in opposite directions with the result that the vertical components of the radiation from the two halves cancel each other while the horizontal components are reinforced.

15.14.5 High-gain UHF antennas

The gain of VHF antennas is lower than UHF because, first, the longer wavelengths would lead to impractical dimensions and, second, because the lower power limit for VHF stations reduces the necessity for high gain. A gain of 12 is common for high-band VHF antennas as compared with a gain of 50 or more for UHF antennas.

UHF antenna gains and their accompanying narrow radiation patterns create a number of problems.

High-gain antennas must be adjusted with a high degree of precision because of the leverage involved in the geometry of the radiation path.

Practical antenna radiators can only approximate the theoretical illumination patterns and the antenna patterns must be verified before the antenna is erected.

If the pattern is adjusted so that its main lobe is horizontal, the pattern maximum will be above the surface of the earth and miss it because of the antenna height, even at great distances. This problem can be solved or alleviated by *beam tilt*.

Nearby areas will receive a weak signal since they will be in the null of the pattern. This problem can be solved by *null fill*.

15.14.5.1 Beam tilt   The beam is tilted below the horizontal so that it is aimed toward the horizon or slightly below it. The depression angles and distances to the horizon for different antenna heights are:

Height (ft.)	Angle (deg.)	Distance (mi.)
500	0.34	31.6
1,000	0.49	44.7
2,000	0.69	63.2

The beam is often pointed below the horizon, and beam tilts of 1.0° or more are common. Figure 15.16 shows a high-gain pattern with 2° of beam tilt. This pattern is based on uniform illumination of the aperture, but the phase varies linearly from +75° at the top of the aperture to –75° at the bottom.

15.14.5.2 Null fill   The second problem of high-gain antennas, close-in coverage, is solved by null fill, which deliberately increases the signal in the nulls of the pattern to increase ERP at the angles for close-in points. By careful adjustment, the pattern can be shaped so that the product of distance and $(ERP)^{1/2}$, the indicator of received field strength, remains fairly constant out to a considerable distance. Figure 15.17 shows the radiation pattern of an antenna with null fill. Here the phase is constant, but the relative amplitude rises exponentially from 0.15 at the bottom to 1.0 at the top.

15.14.6 Power ratings

The power handling limitations of television antennas are established by corona and voltage breakdown and by heating due to ohmic resistance. Both must be taken into account in the design and construction of the antenna, and the manufacturer's published specifications should be carefully followed.

15.15 TV Station Service Areas

15.15.1 FCC definitions

A TV station's service area includes the locations, usually contiguous, in which the signal level is adequate to produce a satisfactory signal-to-noise ratio in the receiver, and that are not subject to an unsatisfactory level of co-channel or adjacent channel interference. The FCC has established standards for both parameters, and they are generally accepted by the industry.

The specification of the signal level required for a service area must recognize that signal levels vary widely and randomly within relatively short distances because of differences in absorption, reflection, diffraction, and refraction at adjacent points on the highly irregular surface of the earth. At greater distances they also vary with time because the refractive index of the atmosphere changes with the weather.

As the result of these variations, it is necessary to define service areas on a statistical basis, and signal levels are usually defined by the term, F(50,50). A level of F(50,50) means that the field strength, F, is exceeded at 50 percent of the potential receiver locations for 50 percent of the time within a limited area. A receiving antenna height of 30 feet is assumed. F is usually defined in dBμ, the number of decibels above a field strength of 1 μV/m.

The FCC defines three grades of television service: Principal Community, Grade A, and Grade B. These grades are defined in Table 15.10. As shown in the table, a different probability of service is defined for each of these grades, but all are specified in terms of the corresponding F(50,50) values.

The specified values are based on the signal levels that were required to overcome both thermal and man-made noise with the monochrome receivers available in the early 1950s.[5] Modern receivers produce less thermal noise, but viewers have become accustomed to lower noise levels and man-made noise is not affected by receiver performance. Accordingly, as a subjective standard, the signal levels specified in the 1950s are still reasonable.

Table 15.10 indicates important differences between the propagation characteristics of the three bands. For example, higher values of F(50,50) are required at the higher frequency bands because the smaller receiving antennas require greater field strength to produce the same voltage at their terminals. Note that the difference between the F(50,50) value for the Principal Community and Grade B services is 27 dB for channels 2–6 but only 16 dB at UHF. In part this is because the level of man-made noise is higher in the low-band VHF channels than in UHF, particularly in urban areas.

15.15.2 FCC field intensity curves

The FCC developed and published families of curves on the basis of measurement data that show the values of F(50,50) for a range of distances and transmitting antenna heights. One set of curves is for high-band VHF, channels 7 to 13 and the other, somewhat paradoxically, for low-band VHF and UHF.

The FCC also published F(50,10) curves that show the values of field strength that are occasionally exceeded. These are mainly of value in calculating interference between stations since a signal level that exists only 10 percent of the time has little value in providing service but can be very annoying as interference. These curves were used in constructing the FCC Table of Assignments.

It must be emphasized that these curves show statistical results only and do not make allowance for specific terrain features. They have little value, therefore, in forecasting the field strength at a particular point at a particular time. With an understanding of this limitation, however, they fill a useful role in estimating the gross features of television coverage.

Figures 15.18 and 15.19 show the FCC F(50,50) curves for high-band VHF and for low-band VHF and UHF. The abscissa is the antenna height, which is defined as its height above the "average terrain," the average elevation in the 2 to 10 mile distance from the antenna.

The antenna height has a major effect on the received field strength, in part because it increases the distance to the horizon but also because it results in a more favorable phase relationship between the direct waves and those reflected from the earth. As an example, according to the curves for UHF coverage, increasing the antenna height from 500 to 1,000 feet increases the field strength by nearly 8 dB or 8 times.

The predicted distance to the Principal Community, Grade A, and Grade B contours for antenna heights of 500 and 1,000 feet and the maximum permitted ERP are shown in Table 15.11. Achieving this coverage for UHF stations requires a costly transmitting system. Also, UHF transmissions are more adversely affected in rough terrain or built-up areas because of their shorter wavelength. For the same reason, variations in signal strength within short distances are greater and more rapid. (These problems may be greatly reduced with digital modulation, however; see below.)

15.16 Digital Broadcasting

15.16.1 Current status

The Grand Alliance, the industry committee that assumed the responsibility of developing and recommending HDTV transmission standards, has proposed a digital format for the channels to be assigned for HDTV (or ATV). This proposal is subject to additional field testing and review before it is finally adopted. Nevertheless, it is important to consider it at this time because it is probable that it includes the policies and standards that will finally be adopted.

15.16.2 HDTV (ATV) channels

The Grand Alliance proposal is based on the assumption that a second set of channels will be assigned in the same areas of the VHF and UHF spectra that are now allocated for NTSC broadcasting.

As originally conceived, these channels would be assigned to existing broadcasters and used exclusively for HDTV transmissions during a 15-year transition period. During this interval broadcasters would provide programming on both their existing NTSC channels and the new HDTV channels. At the end of the 15-year period, broadcasting would become all HDTV and broadcasters would surrender their NTSC channels.

Some segments of the industry are strongly advocating that the new channels be made available for non-HDTV applications in addition to HDTV, e.g., data transmission or multiple standard-definition broadcasts. As a symbol of this position, it has been proposed that the new channels be described as ATV for "advanced television" rather than HDTV. The extent of the permitted uses of these channels is largely a policy rather than a technical issue, and as of this writing it has not been resolved.

15.16.3 Digital format advantages

The use of a digital rather than an analog format for broadcast signals as proposed by the Grand Alliance has a number of major advantages that are summarized here and described more fully elsewhere in the book.

15.16.3.1 Bandwidth conservation Perhaps the most obvious advantage of digital formats is their major conservation of bandwidth by the use of data compression algorithms. For example, the display formats proposed by the Grand Alliance for HDTV would require a *video* bandwidth of nearly 24 MHz for conventional transmission. With digital transmission, the *modulated* *carrier*, including video and high fidelity stereo sound signals, can be accommodated in a 6 MHz channel (see Section 3.11).

15.16.3.2 Interference immunity Digital signals are both more *benign* and more *robust* than analog, i.e., they both cause less interference *to* other signals and are less susceptible to interference *from* other signals. Sufficient tests of these signals have been made to give reasonable assurance that the added HDTV (ATV) channels can be fitted into the spectrum gaps in the NTSC assignment table that were created by adjacent channel and UHF taboo restrictions (see Section 15.2.3).

15.16.3.3 Signal-to-noise performance Digital transmission systems have the characteristic (see Section 3.8) that they add no noise to the signal provided the bit-error rate, BER, remains low enough to be within the range of the error-correction circuits. Above this threshold, a BER of about 3×10^{-6} for the encoding formats proposed for broadcasting, the correction circuit loses control and the picture is lost. This is sometimes known as the *cliff effect*. The result is that signals transmitted digitally are either excellent or they disappear. This is not entirely an advantage because fringe areas receive noisy reception with analog signals but no reception with digital.

15.16.3.4 Signal distortion Analog transmission inevitably causes at least some nonlinear and frequency distortion of the signal. Digital transmission should not introduce any.

15.16.3.5 Display format flexibility In the analogue systems that are presently in use, the transmission and display formats are "coupled," and the display format is rigidly controlled by the transmitted signal. The Grand Alliance proposal includes a major change in this policy, and the formats will be decoupled. The display format will be chosen by the receiver, and it can differ considerably from the transmission format. This flexibility is one of the characteristics of digital signals.[*]

15.16.3.6 Compatibility with interactive multimedia systems A number of interactive and multimedia systems are now under consideration (see Chapter 14). All of them employ a digital format for their signals, and it is desirable,

---

[*] The concept of variable display formats is not entirely new. In an effort to solve the major problem of incompatibility of the CBS sequential color system, the FCC proposed that receivers be designed with "bracket" standards, i.e., that they could operate with either the current monochrome standards or with the CBS color standards. This proposal was almost unanimously opposed by the industry, and it was not adopted. In the present digital era, it is not expected that such opposition will be repeated.

if not essential, that the transmission format employed for broadcasting be compatible.

15.17 HDTV (ATV) Transmission Standards

15.17.1 Channel bandwidth

A bandwidth of 6 MHz, the same as for analog NTSC channels, is specified for the HDTV (ATV) channels. Among other advantages of this choice is the simplification of the transition from NTSC to HDTV.

15.17.2 Bit rate

A bit rate of approximately 20 Mb/s is specified for HDTV transmission. This is high enough (with bandwidth compression) to provide the picture definition required for HDTV systems. It is low enough to be accommodated in a 6 MHz channel provided a suitable modulation method is used.

15.17.3 Modulation type

8-VSB modulation (see Section 3.7) has been chosen as the modulation type to be used for the HDTV (ATV) channels. The factors included in the choice were the signal speed (bs/Hz), the power per bit (E_B/N_O)[**] required to achieve a satisfactory bit error rate, and the benignity and robustness of the desired signal in the presence of other signals.

The choice was narrowed to 8-VSB and 32-QAM. They were approximately equal in signal speed (3.1 bits/Hz) and signal-to-noise performance, but 8-VSB was superior in its benignity and robustness, both for co-channel and adjacent channel signals.[6]

15.17.4 Channel occupancy

The amplitude/frequency characteristic of a 6-MHz HDTV (ATV) channel is shown in Figure 15.20. It should be compared with the amplitude/frequency characteristic of an NTSC channel as shown in Figure 15.1. The main difference is the absence of the aural and visual carriers. The aural carrier is unnecessary because the audio signal is transmitted by a digital bit stream. The 8-VSB modulation employed for the video signal transmitted with a suppressed carrier. A small in-phase pilot having an amplitude 11.3 dB below

---

[**] E_B = energy per bit; N_O = noise spectral density.

the level of the digital bit stream provides the reference for recovering the carrier at the receiver in the presence of noise and interference.

15.18 Display Formats for HDTV (ATV) Systems

The decoupling of the transmission and display formats gives the system designer wide latitude in the choice of display formats. In practice, because of economic factors and the desire to limit the number of receiver models, the GA has limited the transmission format choices to two variations of two numbers of active scanning lines (720 and 1,080—see Section 6.4). A choice of progressive and interlaced scanning is offered for 1,080-line systems while 720-line system choices all use progressive scanning.

15.18.1 Aspect Ratio

As noted in Chapter 1, the standard aspect ratio for NTSC broadcasts is 4:3 or 1.33. The Grand alliance has specified a wide screen format, 16:9 or 1.78, regardless of the scanning standards.

A number of aspect ratios are employed for films that are produced for direct projection. They are shown in Table 15.12. All of the wide screen film formats on this list have an aspect ratio that is much greater than the NTSC TV standard and also, although to a lesser extent, than the 1.78 HDTV ratio. These incompatibilities can be handled in one of three ways:

1. A TV image is produced with the same aspect ratio as the film by leaving blank areas on the top or bottom of the TV frame. This is called letter-box transmission. This conversion can be accomplished photographically by producing a film print for TV reproduction.

2. The full film frame height is reproduced and the sides of the image are cropped to produce the correct aspect ratio. This can also be done photographically.

3. The horizontal location of an aperture on the film image having the TV aspect ratio is moved dynamically so that the portion of the image with the greatest interest is reproduced. This can be done while making a film-to-tape transfer with a process called *pan and scan*. This method produces the best results, but it is laborious and it requires excellent editorial judgment.

None of these methods is ideal, and it is expected that films for HDTV will be produced with a 16:9 aspect ratio once standards have been established.

15.18.2 Scanning standards

The decoupling of the HDTV transmission and display formats in the Grand Alliance proposal is reflected in the flexibility of its scanning standards. For example, any transmission format permits the receiver to choose either progressive or interlaced scanning for the display. Two receiver scanning choices that represent the range available are tabulated in Table 15.13. The *transmission* standards in this example could be either progressive or interlaced, and the conversion, if necessary, is made at the receiver.

15.19 HDTV (ATV) Channel Assignments

In January 1995, the Association for Maximum Service Television (AMSTV) published a proposal for the assignment of 1,691 HDTV (ATV) channels to existing broadcasters and license holders, although final criteria for these assignments had not been established. They were chosen with the aid of a computer program that had three goals:

1. to provide an ATV channel for each current NTSC channel
2. to provide an ATV service area (see Section 15.20) that was at least comparable with that of the NTSC station
3. to minimize interference to existing NTSC service

The proposed assignments are overwhelmingly, although not quite totally, in the UHF portion of the spectrum.

15.20 HDTV (ATV) Station Coverage

The most complete data currently available for calculating the coverage of HDTV (ATV) stations are included in a set of measurements made in Charlotte, North Carolina, in September 1994 under the sponsorship of the Grand Alliance.[7] It compared the coverage of 8-VSB digital broadcast signals and NTSC signals in urban and rural areas in the vicinity of Charlotte. Measurements were made both at VHF (Channel 6) and UHF (Channel 53).

The results of the measurements are summarized in Table 15.14. The HDTV *average* power was adjusted to be 12 dB below the NTSC *peak* power, which is the ratio of the average to the peak power of an NTSC signal with typical picture content. This ratio was estimated to provide "equivalent" service with the two modulation methods.

The criterion for "satisfactory" reception for HDTV (ATV) transmissions in Table 15.14 was a bit error rate of less than 3×10^{-6}, the rate at which it is expected that the error-correction circuits would become ineffective. For NTSC transmission, the criterion was subjective, and was chosen to be the point at which expert observers judged noise and interference to be "slightly annoying."

The measurements were not sufficiently extensive to be totally conclusive, for example they did not cover extremely densely populated urban areas such as Manhattan, but they give a strong indication that the coverage with digital transmission will be as good or better as with analog. (They do not reveal, however, that digital signals will either be excellent or absent while analog signals suffer "graceful degradation"; some degree of analog reception could be achieved at locations having a S/N ratio where digital signals would disappear.)

The comparatively poor performance of the VHF NTSC signals was attributed to a number of local factors including interference from a cable system and an educational FM station.

15.21 HDTV (ATV) Channel Utilization Alternatives

The 19.3 Mb/s signal speed in the HDTV channel, together with digital bandwidth compression, opens up a wide variety of possibilities in addition to HDTV for the utilization of this channel. It is because of these possibilities that broadcasters are opposed to limiting their use to HDTV. They have not been fully explored at the present time, but a few of the options are:

1. With additional compression, a standard-definition channel in addition to HDTV

2. Two or more standard-definition channels

3. A multiplicity of digital data channels

The decision as to the uses of these extra channels involves issues of economics and public policy that are beyond the scope of this book.

15.22 HDTV (ATV) Transmitters

Since the HDTV channels are in the same portions of the VHF and UHF spectrum as the NTSC channels, there are some similarities in the designs of NTSC and HDTV transmitters, but there are also major differences as described in the following sections.

15.22.1 Visual and aural channels

NTSC transmitters have separate aural and visual power amplifiers. In a digital HDTV transmitter, the aural and visual bit streams are combined at the Transport layer (see Section 6.6) and only a single power amplifier is required.

15.22.2 Signal processing and carrier modulation

The processing of analog signals for transmission and the modulation of carriers in NTSC transmitters is accomplished in the exciter-modulator (see Section 15.8.1). The performance of equivalent functions in digital HDTV transmitters is described in Section 6.7.3 and is considerably more complex.

15.22.3 Transmitter power ratings and levels

Digital signals are characterized by a high ratio of peak-to-average power. The power rating of digital transmitters, therefore, must specify both peak and average power.

15.22.4 Transmitter performance standards

The need for good linearity in amplitude-modulated analog transmitters is obvious, since distortion of the signal waveform causes direct distortion of the gray scale. The importance of linearity in digital signals is not so obvious, but for the Grand Alliance digital modulation system, it is equally important.

With 8-VSB signals, the amplitude as well as the presence or absence of pulses is significant in transmitting information and, like noise, amplitude nonlinearity can cause bit errors. Nonlinearity can also create cross-modulation products that are seen by the system as noise. The overall effect of nonlinearities, therefore, is to create noise in the signal that is added to the ordinary transmission noise. This causes the threshold bit error rate to be reached at a lower level of transmission noise, which in turn results in a reduction of the range of the transmitting system.

References

1. *Broadcasting and Cable*, January 16, 1995.

2. Donald G. Fink and Donald Christiansen (eds.), *Electronics Engineers' Handbook*, 3d edition, McGraw-Hill, New York, 1989, Chapter 21, para. 41.

3. Electronics Industry Association, *EIA/TIA Standard, Electrical Performance Standards for Television Transmission Systems, EIA/TIA-250-C* (1990).

4. K. Blair Benson (ed.), *Television Engineering Handbook*, McGraw-Hill, New York, 1986, Chapter 8 (for a more complete description of antenna echoes).

5. Federal Communications Commission, "Sixth Report and Order," Docket No. 8736, 8975, 8976, 9175, April 11, 1952.

6. Grand Alliance, "Transmission Subsystem Comparison Testing, *(January 13–February 11, 1994),*" FCC Advisory Committee on Advanced Television Service.

7. Jules Cohen and Victor Tawil, "Testing of the Grand Alliance HDTV Subsystem," *J. SMPTE*, May, 1995.

16

CATV Systems

16.1 A Brief History of CATV

Cable television or *CATV* began in the very late 1940s as *community TV*, a system for bringing television service to small communities that did not receive adequate off-the-air signals in individual homes, either as the result of their distance from broadcasting stations or because of intervening hills or other terrain features. Local TV dealers or other entrepreneurs erected high-gain antennas on mountain tops or towers to provide usable signals to cable systems that delivered it to local homes.

Within a few years the variety of program services available to community television systems was increased by importing signals from distant stations by microwave, and service was delivered to communities far outside the normal service areas of the originating stations.

In the late 1970s, communication satellites made it possible for cable systems to receive programs from all over the world. The variety of available programming became great enough to be attractive to viewers in large cities, even though they had several off-the-air sources available. At this time community TV was no longer descriptive of the service, and cable TV or CATV became the accepted nomenclature. CATV is now a large business, and more than 50 million homes in the United States are receiving programs by cable.

Looking to the future, the introduction of digital technology will improve the performance of CATV systems, increase their capacity, and extend their functions to include interactive multimedia services.

16.2 Cable System Elements

Figure 16.1 shows the elements of an analog cable TV system. The incoming signals—off-air, microwave, and satellite and locally originated—are received

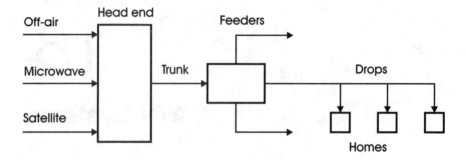

Figure 16.1 Cable system elements

at the *head end*. Except for the off-air signals, they are in the baseband mode and they modulate carriers in accordance with broadcast standards. The carrier frequencies of the off-air signals are usually shifted before multiplexing with the other signals.

The cable distribution system is a combination of *trunks, feeders,* and *drops*. Trunks are cables that feed major portions of the service areas. They are connected to feeders that serve individual streets and the drops that make the connections to individual homes. At the present time, with the exception of drops that may employ flat twin lead, these system components employ coaxial cable or fiber optics.

Some systems have *super trunks* to interconnect more distant parts of the system or even different systems. Super trunks may employ coaxial cable, fiber-optic cable (see Chapter 18), or a type of microwave known as AML for *amplitude modulated link*.

16.3 Cable Television Channels

The first community antenna systems employed the five low-band VHF channels, 2 to 6, because of the low attenuation of cable at these frequencies (see Section 16.4.2.2). As the demand for channels grew, systems expanded into the high-band VHF, channels 7 to 13. The spacing between line amplifiers had to be reduced, but unmodified standard receivers could be used.

As the demand for channels grew still further, the industry had the choice of adding channels in the UHF spectrum or of utilizing channels between low-band and high-band VHF and between high-band and UHF. The latter course was chosen because of extremely high cable attenuation in the UHF band.

A standard CATV configuration utilizes *frequency division multiplex* with fifty-five adjacent channels in the NTSC format, the highest located just below the UHF band. More channels can be added by expansion into the UHF band (see Table 16.1).

In addition, four 6-MHz channels are available in the 5 to 30 MHz region that are sometimes used for *upstream* transmissions to provide two-way capability. The region from 30 to 55 MHz is reserved as a guard band to separate the upstream and downstream channels.

Table 16.1 CATV channels

Channel Designation	Frequency (MHz) Visual Carrier	Channel Designation	Frequency (MHz) Visual Carrier
Low-band VHF		Superband	
2	55.25	J	217.25
3	61.25	K	223.25
4	67.25	L	229.25
5	77.25	M	235.25
6	83.25	N	241.25
Mid-band VHF		O	247.25
A-2	109.25	P	253.25
A-1	115.25	Q	259.25
A	121.25	R	265.25
B	127.25	S	271.25
C	133.25	T	277.25
D	139.25	U	283.25
E	145.25	V	289.25
F	151.25	W	295.25
G	157.25	Hyperband	
H	163.25	AA	301.25
I	169.25	BB	307.25
High-band VHF		CC	313.25
7	175.25	DD	319.25
8	181.25	EE	325.25
9	187.25	FF	331.25
10	193.25	GG	337.25
11	199.25	HH	343.25
12	205.25	II	349.25
13	211.25	JJ	355.25
		KK	361.25
		LL	367.25
		MM	373.25
		NN	379.25
		OO	385.25
		PP	391.25
		QQ	397.25
		RR	403.25

Adjacent channels are not assigned to the same geographic location in off-the-air broadcast practice because receivers may not be able to discriminate between a weak desired signal and a strong undesired one on an adjacent channel. In CATV systems, however, adjacent-channel signals can be maintained at approximately equal levels and adjacent-channel reception is possible. It requires very careful control of signal levels to avoid adjacent-channel interference and of system linearity to avoid cross-modulation between channels. This is an inherent weakness of analog systems that should be solved by the introduction of digital transmission.

Most of the channels in the CATV configuration cannot be received directly by a standard TV set designed for broadcast reception. *Set-top converters* (see Section 16.11) heterodyne all incoming channels to a single output channel, usually channel 3 or 4. Alternatively, *cable-ready* receivers that can tune to the cable channels are also available.

16.4 Coaxial Cable

Coaxial cable is most commonly used for the distribution of CATV signals.

16.4.1 Physical construction

The physical construction of a coaxial cable of the type commonly used in CATV systems is shown in Figure 16.2. The inner conductor is copper-clad aluminum. Copper, with its low resistivity, is used on the outside of the inner conductor where the current density is highest. Aluminum is used for the outer conductor and core of the inner conductor because of its light weight.

The insulating material between conductors is polyethylene foam. Gas is sometimes injected to prevent water from migrating through the insulation.

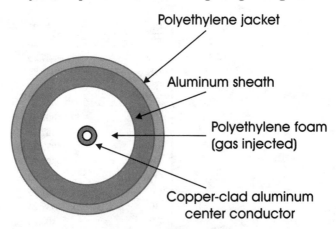

Figure 16.2 Coaxial cable construction (*from Eugene R. Bartlett,* Cable Television Technology and Operation, *McGraw-Hill, New York, 1990*)

A flexible steel tubing is sometimes placed over the polyethylene jacket for mechanical protection. Standard cable diameters are 0.50", 0.75", 0.875", and 1.00". Cables, particularly in the larger diameters, are only semi-flexible, and the bending radius is an important specification.

Cable expands and contracts with temperature change, approximately 13 parts per million per degree Fahrenheit, and provision must be made to allow for this in the installation of the cable. The problem is complicated by the fact that the steel cable (called a *messenger*) that supports an overhead cable expands at a different rate, and allowance must be made for this also.

16.4.2 Electrical characteristics

16.4.2.1 Characteristic impedance The characteristic impedance of flexible transmission line has the same electrical significance as for the solid-tube lines used to connect transmitters to antennas (see Section 15.6). The standard characteristic impedance for CATV transmission line is 75 ohms.

16.4.2.2 Attenuation The signal is attenuated as it passes down the cable owing to ohmic losses. The amount of loss depends on the diameter of the cable, the frequency, the standing wave ratio, and the temperature. The attenuation with a VSWR of 1.0 and a temperature of 68°F is shown in Table 16.2. Note that the attenuation (in dB) is approximately proportional to the

Table 16.2 Cable attenuation (dB/100 ft.)

Frequency MHz	Cable Diameter (inches)			
	0.500	0.750	0.875	1.000
5	0.16	0.11	0.09	0.09
30	0.40	0.26	0.24	0.23
50	0.52	0.35	0.32	0.30
110	0.75	0.52	0.47	0.44
150	0.90	0.62	0.55	0.52
180	1.00	0.68	0.60	0.59
250	1.20	0.81	0.72	0.68
300	1.31	0.90	0.79	0.75
400	1.53	1.05	0.91	0.87
450	1.63	1.12	0.98	0.92
500	1.70	1.17	1.04	0.97
550	1.80	1.23	1.09	1.02
600	1.87	1.29	1.15	1.06

square root of the frequency. The attenuation increases with temperature at the approximate rate of 1 percent per 10° F change.

16.4.2.3 Return loss The return loss is a measure of the deviation of the actual value of the cable's characteristic impedance from its nominal value. It is equal to the ratio of the incident power, P_I, to the reflected power, P_R, in decibels:

$$L_R = 10 \log \frac{P_I}{P_R} \qquad (16.1)$$

The closer the actual characteristic impedance of the line matches its nominal value, the smaller the P_R and the larger the return loss. For a perfect match, $P_R = 0$ and $L_R = $ infinity. No line is perfect, and practical values of L_R vary from 28 to 32 dB. If the return loss is too low, reflections occur along the line causing variations in voltage or *signal ripple*.

16.4.2.4 Loop resistance The primary power for the line amplifiers that are installed to compensate for cable attenuation is usually supplied by low-voltage ac or dc transmitted along the cable with the RF signal. Since the amplifiers operate at relatively low voltage, typically 45 V, and high current, the *loop resistance*, the combined resistance of the inner and outer conductors to dc or power-frequency ac, is an important cable specification.

The loop resistance of coaxial cable with copper-clad aluminum inner conductors is slightly greater than all-copper conductors as shown in Table 16.3.

16.5 Line and Bridging Amplifiers and Equalizers

Line amplifiers are inserted in the cable at intervals to compensate for attenuation, and their specifications are a key element in determining the performance of the system. Each amplifier includes an equalizer that compensates for the increase in cable attenuation with frequency.

Bridging amplifiers have a high-impedance input and are used to tap the signal from a trunk without disturbing its performance.

Table 16.3 Loop resistance, coaxial cable, ohms/1000 ft., 68°F

Outer diameter (inches)	Solid copper center conductor	Copper-clad aluminum center conductor
0.412	1.75	1.93
0.500	1.23	1.68
0.750	0.56	0.760

16.5.1 Technical requirements

The technical requirements for amplifier/equalizer combinations are demanding because of the cumulative degradation of a large number of amplifiers in series:

1. They must operate over a wide frequency range, and their gain must be high enough at the top of the frequency range to enable the amplifiers to be separated by a reasonable distance.

2. The equalizers must compensate precisely for the variations in cable attenuation with frequency.

3. The linearity of the amplifiers must be good to minimize cross talk and the generation of beat frequencies between channels.

4. Automatic gain and frequency response, (*slope correction*) must be provided to compensate for variations in these parameters with temperature or equipment aging (see Section 16.5.3).

5. The signal-to-noise ratio of individual amplifiers must be high enough so that the cascaded noise level of all the amplifiers in a trunk meets industry standards (see Section 16.8).

16.5.2 Gain-frequency response

The gain of the amplifiers in typical amplifier/equalizer combinations is essentially constant at 20 dB from 50 MHz to 450 MHz. Assuming that the cable loss between amplifiers increases from 6.9 dB at 50 MHz to 20 dB at 450 MHz, the equalizer is designed to reduce the gain of the combination to 6.9 dB at 50 MHz so that the frequency response of the transmission line/amplifier-equalizer combination is approximately constant over the entire frequency band.

16.5.3 Automatic gain and slope control

If a system has many amplifiers in series, a small change in the gain or frequency response of each amplifier resulting from a common cause such as a temperature change will produce a major change in the gain and frequency response of the system because of the compounding effect. To maintain these parameters within close tolerances, inverse feedback is introduced by the addition of pilot tones near the bottom and top of the frequency band. Using these tones as references, the gain and frequency response of each amplifier is automatically maintained at a stable level.

16.5.4 Signal-to-noise ratio

The system noise amplitude is the root sum square of the noise amplitudes of the system components. The signal-to-noise and carrier-to-noise ratios are usually expressed in decibels, and the signal-to-noise ratio for n identical cascaded amplifiers, each with a signal-to-noise ratio of S_o/N_o is:

$$\frac{S_n}{N_n} = \frac{S_o}{N_o} - 10\log n \tag{16.2}$$

16.5.5 Distortion and cross-modulation

Nonlinearities in system amplifiers can cause distortion, both of the amplitude-modulated carrier envelope and of the carrier itself. The most serious effects of envelope distortion are differential gain and differential phase distortion (see Sections 5.10.3 and 5.10.4). Distortion of the carrier causes "beats" and other cross-modulation effects between the closely spaced channels. The beat frequencies retain some of the modulation characteristics of the original signals. With 50 or more channels, the possibilities for the production of spurious signals by cross-modulation are almost endless. A common type is a *triple beat* caused by the interaction of the signals on three channels.

A number of rules, some empirical and others based on theory, have been developed for estimating the cumulative distortion of a number of amplifiers in series. If $d_1, d_2, \cdots$ are the distortions of successive elements in a system, the total distortion, D, may be the linear sum:

$$D = d_1 + d_2 + d_3 + \cdots \tag{16.3}$$

or follow a 3/2 power rule:

$$D = \left(d_1^{\frac{3}{2}} + d_2^{\frac{3}{2}} + \cdots \right)^{\frac{2}{3}} \tag{16.4}$$

or the root-sum-square rule:

$$\sqrt{d_1^2 + d_2^2 + \cdots} \tag{16.5}$$

Differential gain and phase distortion increase in accordance with the 3/2 rule.

16.6 Trunk Circuits

The frequency response of a trunk circuit is the product of the responses of its individual components.

Line amplifiers divide trunk circuits into segments, each having the length at which the attenuation at the highest transmission frequency is equal to the gain of the amplifiers. Equalizers are added that reduce the gain at lower frequencies so that the gain of each cable-segment amplifier/equalizer combination is 0 dB. As an example, assume the system bandwidth is 50 MHz to 450 MHz, the gain of each amplifier is 20 dB, and the cable attenuation rises from 3.1 dB/1,000 ft. at 50 MHz to 10 dB/1,000 ft. at 450 MHz. The spacing between amplifiers would be (1,000 ft.) × (20 dB/10 dB) or 2,000 feet. An equalizer would be added to each amplifier with an attenuation varying from 6.9 dB at 50 MHz to 0 dB at 450 MHz. If the slopes of the equalizers are properly designed, the combined gain of the cable, amplifier, and equalizer would be 0 dB over the entire band.

16.7 Reverse Path Circuits

It is sometimes desired to bring signals from remote program sources back to the head end in a reverse direction. This can be accomplished on the frequencies below channel 2 (6.0 to 30.0 MHz) where there is space for four 6-MHz channels. These channels have the advantage of lower attenuation but the disadvantage of higher ambient noise levels.

Special care must be exercised in the design of two-way systems. Every drop to a receiver is a noise source, and every return path must be blocked except the one carrying the desired signal. Otherwise the S/N of the system drops to a level that makes it unusable.

16.8 Video Performance Standards

As the cable TV industry has matured, various industry and governmental organizations have issued performance standards for CATV systems. Table 16.4 summarizes the basic performance standards that have been issued by the FCC, the NCTA, and the EIA for the transmission of NTSC signals over cable. It illustrates the fact that there is not complete industry agreement on performance standards or on parameters that should be specified.

The standards for carrier-to-cross-modulation products and co-channel interference in part 76.605 of the FCC rules are particularly important specifications for CATV systems because there are no guard bands between channels.

Like many industry standards, Table 16.4 is a compromise between what is desired and what is realistically possible in the current state of the art. Meeting them requires meticulous adjustment and maintenance of the system with very little "headroom" for misadjustment. The introduction of fiber-optic cables and/or digital transmission will make it possible to bring about significant improvements in CATV system performance (see Section 16.14) and also to make it more insensitive to minor misadjustments.

16.9 Head Ends

The head end of a CATV system collects the signals of all its program sources, processes them for transmission, generates and modulates channel carriers for signals that are in baseband form, shifts the channel frequency where necessary, and combines the resulting channels for transmission on the system.

16.10 Head-End Equipment

16.10.1 Broadcast channel antennas and receivers

The enormous expansion of CATV service has been made possible by satellite-transmitted signals, but the distribution of off-air broadcast signals continues to be essential to its popularity. A CATV system head end, therefore, requires high quality receiver-antenna combinations for off-air pickup.

The antenna must be located at a sufficient height to receive strong signals from the desired broadcast stations. It may be a commercial version of an all-channel home antenna, although the all-channel feature is often not satisfactory—one problem is that the off-air stations may be located in different directions—and most CATV systems use single-channel, high-gain antennas or antennas that receive only a few channels.

The most common antenna formats are the familiar *YAGI*, which is frequently used in home antennas, and the *log periodic* in which each element is driven, thus giving better control over the antenna's directivity and bandwidth.

The principal components of the receiver are the preamplifier and signal processor. If the field strength is weak at the antenna location, the S/N of the system can be improved by placing the preamplifier at the antenna location.

Table 16.4 CATV system performance standards

	FCC 76.605	NCTA	EIA RS250B
Frequency response	±2 dB	—	±0.7 dB
Differential gain	—	10%	4%
Differential phase	—	5°	1.5°
Cross modulation (C/XMO)	46 dB	49 dB	—
Composite triple beat	—	52 dB	—
Second-order beat	—	60 dB	—
Co-channel interference	36 dB	—	—
C/N (unweighted)	36 dB	44 dB	—
C/N (weighted)	—	55 dB	56 dB

The signal at the output of the preamplifier is converted to the IF frequency for further amplification and signal processing. The IF signal is then heterodyned to the desired RF channel where it is mixed with other signals for transmission and distribution over the system.

16.10.2 Video, microwave, and satellite signals

The signals from local program sources, microwave systems, and satellite systems are in the video and audio baseband formats, and after processing they are used to modulate channel carriers. The modulated signals should meet FCC standards for broadcast signals.

16.11 Set-Top Converters

Set-top converters provide an interface between the cable system and a standard television receiver. Their original function was to shift the frequencies of the incoming cable channels on a tunable basis to a single output channel. This is done by a heterodyning process and it means that cable channel selection is done by controlling the converter while the TV receiver remains on the specified single channel. Subsequently, the functions of set-top converters were expanded as described below.

16.11.1 Converter functions

16.11.1.1 Increased system capacity Because of the converter's heterodyne frequency shifting described above, cable systems are not limited to broadcast channels for signal distribution. This allows 55 channels in the 450 MHz cable bandwidth as listed in Table 16.1.

16.11.1.2 Off-air signal interference reduction Cable signals are subject to interference from off-air signals picked up by imperfectly shielded receivers. Since cable signals travel more slowly than off-air signals, there can be a time difference of a few microseconds between the cable signal and the off-air signal at the receiver. This can cause ghosts or, in less severe cases, a smearing of horizontal detail in the picture. With a converter in use, the receiver is tuned to a different channel than the original broadcast channel and off-air interference is greatly reduced or eliminated.

Further protection from off-air signal interference (as well as interference from other cable signals) is achieved by using a double heterodyning process in the converter (see Figure 16.3).

16.11.1.3 Design objective conflicts There is a major conflict in the sensitivity design objectives between off-air and cable operation of receivers. Off-air reception needs high sensitivity since signals may be weak, but the

FCC channel assignment policies protect an off-air receiver from adjacent-channel interference. By contrast, cable reception does not require high sensitivity because cable signals are strong, but with 55 channels located side by side, a cable receiver must have excellent sensitivity. The converter's design is optimized for cable reception, thus eliminating this requirement for the receiver.

16.11.1.4 Pay-TV services Subscriber and pay-per-view services require a means for controlling access to signals. Set-top converters with controlled descramblers provide this capability.

16.11.1.5 Digital signal distribution It can be expected that cable systems will eventually convert, at least in part, to digital distribution because of its many advantages (see Section 16.16). The conversion will be costly because of the public's investment in analog receivers. Set-top converters equipped with digital decompression and D-A circuits will ease this problem considerably since only a new converter will be required at each receiver. (Of course, this approach will limit the performance to that of the analog receiver, so there will still be incentive for all-digital receivers that can offer higher picture and sound quality.)

16.11.1.6 Interactive and other special services A number of interactive services for cable TV are now in the planning phase (see Chapter 14 and Section 16.16). All of them require an interface at the receiver, which can become a function of the set-top converter.

16.11.1.7 Set-top converter summary Since the introduction of set-top converters in the early history of CATV, their importance has grown enormously. Once considered to be a temporary expedient to increase system capacity and reduce interference from off-air signals, they are now an essential component of cable systems because of their many additional useful purposes. This is indicated by the fact that there are 25 to 30 million converters in use (not including VCRs, which also have many converter functions—see Chapter 10).

16.11.2 Converters vs. cable-ready receivers

All of the functions of set-top converters can be incorporated in receivers, and many recent receiver designs include at least the most essential converter features. Receivers that have the ability to receiver all the standard CATV channels are known as *cable-ready*. However, many of the additional features listed above are specific to particular cable systems, so it has so far been difficult to include them in standard TV receivers. Efforts are being made in the development of digital systems to include features such as access control

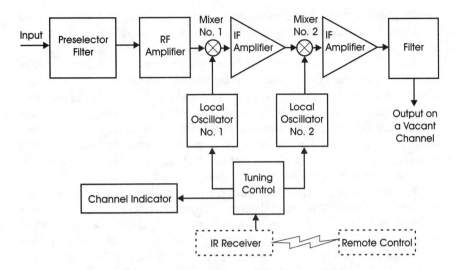

Figure 16.3 Set-top converter block diagram

in a standardized way, so that they may be included in future digital receivers and eliminate the need for the set-top converter.

16.11.3 Converter operation

Figure 16.3 shows a block diagram for a set-top converter of moderate complexity. The input from the cable system is first passed through a preselector filter that removes signals outside the CATV passband. The signal is then amplified and passes through two mixers that perform double heterodyning to provide image channel rejection and increased selectivity. The output of the second mixer passes to the receiver through a second filter that is tuned to the output channel. The infrared (IR) remote control shown in the figure is an optional feature.

Three optional features available on some converters are also not shown in the figure. The first is a connection for an upstream signal at the converter input (see Section 16.3). The second is a baseband video output obtained by detection of the signal from the second mixer, and the third is a descrambler for pay-TV services.

16.12 Pay-TV Systems

Most cable systems offer a choice of pay-TV services, usually known as *basic, premium subscription,* and *pay-per-view.* Basic service is offered for a fixed monthly base rate and comprises a group of channels, ordinarily including off-the-air broadcast signals, for which there is no extra charge. Premium subscription service is optional and includes channels that carry special programs for an additional monthly charge. As its title indicates, pay-per-view service is offered for individual programs for which customers are billed.

All pay-TV services must provide a means for limiting the availability of service to subscribers who have paid for it. This can be done rather simply for premium service, which is billed on a monthly basis, by *trapping*—installing a filter at the subscriber's drop that blocks premium channels for nonsubscribers or by *jamming*, adding a signal between the sound and picture carrier that is removed by a trap for desired signals. A serious disadvantage of these techniques is that a technician must pay a costly visit to the subscriber's premises to add or disconnect service.

More sophisticated systems eliminate the requirement for visiting subscribers' homes by *scrambling* the signal, i.e., making it unusable unless it is descrambled. The set-top converters are provided with addressable decoders that descramble the signal upon command from the system head end. This system can be used either for premium channels or for pay-for-view.

A variety of scrambling techniques have been developed that differ in security and cost. An example is *sync attenuation*, in which the amplitude of the horizontal sync pulses is reduced for transmission sufficiently to make the signal unusable until it is restored by the descrambler.

If digital transmission is adopted as the standard for CATV, scrambling systems of even greater sophistication can be introduced with an increase in flexibility and security.

16.13 Wireless Cable

"Wireless cable" is an oxymoron that is used to describe cable-like television distribution systems that are transmitted by radio rather than by cable.[1]

16.13.1 Wireless cable services

Wireless cable services include four point-to-multipoint media for program distribution—ITFS (Instructional Television Fixed Service), MDS (Multipoint Distribution Service), MMDS (Multipoint Multichannel Distribution Service), and OFS (Operational Fixed Service)—and one multichannel point-to-point service, CARS (Cable Television Relay Service). All of these services employ single-sideband amplitude modulation with 6 MHz bandwidths. They differ in their spectrum allocations and usage authorizations.

16.13.1.1 ITFS The ITFS was first authorized in 1963 for the distribution of instructional and educational programs to schools and colleges. Twenty channels are allocated for ITFS, but they are not all used full-time for instructional purposes. FCC rules permit them to be leased for commercial use by their licensees with the proviso that they be used part-time for the distribution of a minimum amount of instructional and educational programs (20 hours per week per channel two years after they begin service).

The requirement for part-time use of these channels for education means that a commercial lessor cannot have full-time use of a channel. This problem is solved by channel sharing. A channel map is prepared that indicates the time periods during which each channel is in use for educational programming. Full-time schedules can be developed for commercial services by combining available time periods from two or more channels. The program being transmitted is switched to an unused channel and the subscribers' receivers are automatically retuned to it. Several channels may be used in the course of a single program. With this technique, as many as three full-time commercial services can be distributed by four ITFS channels.

16.13.1.2 MDS MDS was first authorized in 1969 with two channels, MDS-1 that was available in the entire United States and MDS-2 that was available in the fifty largest metropolitan areas. It was used primarily for the distribution of pay-TV programming. It was so successful that there were strong pressures for the authorization of additional channels for commercial programs. The FCC made three responses to these pressures:

1. The part-time use of ITFS channels for commercial programs was authorized (see above);

2. Eight ITFS channels were reallocated to the MMDS service (see below); and

3. Three channels from the Operational Fixed Service portion of the spectrum were authorized for commercial program distribution.

MDS is allocated 10 MHz of spectrum space from 2,150 to 2,160 MHz.

16.13.1.3 MMDS Eight channels are allocated for full-time commercial use. They were made available for commercial use in 1983 by the reallocation of ITFS channels. They are very much in demand, and obtaining a new assignment is difficult if not impossible in many areas.

MMDS is allocated 48 MHz (eight amplitude-modulated TV channels) from 2,596 to 2,644 MHz. It is used for the distribution of television programs in sparsely populated areas where cable is uneconomic.

16.13.1.4 OFS Three channels were allocated in 1983 for commercial point-to-multipoint distribution of television programs in the Operational Fixed Service portion of the spectrum. Like the MMDS channels, they are very much in demand.

16.13.1.5 CARS CARS is an inexpensive microwave service for transmitting a large number of CATV program signals from one site to another, e.g., from head end to head end. The amplitude modulated CATV channels from

Figure 16.4 Configuration of a wireless cable system

112–300 MHz are heterodyned upward to 12,758.5–12,946.5 MHz and transmitted as microwave channels. This is called an *amplitude modulated link* or AML. For relatively short distances, an AML is usually more economic than cable or conventional microwave service.

16.13.2 Spectrum allocations

The spectrum allocations of wireless cable distribution services are shown in Figure 16.4. The channels for these services all have a bandwidth of 6 MHz, and their usage format—vestigial sideband for video and FM for audio—is identical with that in CATV systems except that the position of the video and aural subcarriers is inverted in the two MDS channels in which the aural subcarrier is below the visual.

16.13.3 Wireless cable distribution system configuration

The configuration of wireless cable distribution systems is shown in Figure 16.4. It is identical with most CATV systems up to the head end. At the head end, the signal is delivered to the carrier transmitter, typically a highly linear broad-band device that amplifies 8 channels simultaneously, rather than to a cable network. The linearity is required to avoid the production of intermodulation products by the simultaneous amplification of multiple channels.

The transmitter power level is usually 10 to 100 watts, typically generated by a GaAsFET (gallium arsenide, field effect transistor) amplifier.

The connection from the transmitter to transmitting antenna may be by wave guide or cable.

Since the propagation of signals at the MMDS/IFTS frequencies is essentially line-of-sight, the transmitting antenna must be high enough to clear major terrain and man-made obstacles.

Either horizontal or vertical polarization may be used.

Like broadcast antennas, it is usually highly directional in the vertical plane (which requires that it have a large vertical dimension), and its vertical beamwidth may be so narrow that beam tilt is required to cover close-in subscribers as with high-gain UHF broadcast antennas (see Chapter 15). The antenna's directivity in the horizontal plane depends on the angular sector coverage; it may be from 90° to 360°. The antenna gain is determined by the combined vertical and horizontal directivity and is typically 15 dBi.

The receiving antenna located at the subscriber's premises should be high enough to clear local obstructions. It usually consists of a dipole-reflector combination, but unlike the transmitting antenna it is usually directional in the horizontal plane since it receives signals from a fixed direction. Typical gains are 18 to 24 dBi.

The receiver antenna feeds a block downconverter, which should be mounted near the antenna to minimize losses in the cable to the receiver. It translates the carrier frequencies to the standard CATV channel configuration (see Section 16.3). At this point the signal is functionally identical to a subscriber feed from a CATV system. Pay-TV signals pass through a decoder at this point.

16.13.4 System coverage

In the absence of terrestrial obstacles, the theoretical distances (miles) to the 45, 50, and 55 dB S/N contours with *fresnel zone* (see below) path clearance and typical system parameters are shown in Table 16.5. Note that these distances are theoretical maxima, and they assume a path that is far enough above the line-of-sight to provide fresnel zone clearance. This is given by Eq. (16.6):

$$F = 72.1 \sqrt{\frac{d(D-d)}{fD}} \qquad (16.6)$$

where: $F =$ fresnel zone in feet

$D =$ path length in miles

$f =$ frequency in gigahertz

$d =$ distance from point of minimum path clearance to the nearest path end

As an example, the height of the fresnel zone on a five-mile path with the minimum path clearance at the path mid-point and a frequency of 2.5 GHz is 51 feet. With a 20-mile path it would be 102 feet.

Table 16.5 Distance (miles) to S/N contours with fresnel zone path clearance

	Transmitter power			
	100 watts		50 watts	
Receive antenna gain (dBi) ⟶	24	18	24	18
S/N contour (dB)				
45	60	30	36	19
50	33	17	20	10
55	18	9	12	6
Transmit antenna gain		15 dBi		
Antenna/downconverter noise figure		7.5 dB		
Operating frequency		2.6 GHz		

If, as is often the case, the path clearance is less than the fresnel zone, the signal will be substantially attenuated.[2] If there is a major obstruction in a path, it sometimes can be cleared by a beam bender. This is a low power repeater that picks up the signal at the top of the obstruction and retransmits it into the shadowed area.

16.14 Digital Transmission and Fiber-Optic Cable

The principal technologies employed by current CATV systems—analog signal formats, FDM modulation, and coaxial cable distribution—have served the industry well for more than 40 years. They have limitations, however, that are being overcome by newer technologies that will replace or supplement them in the years ahead. The most significant of these are digital transmission and fiber-optic cable.

As with broadcast systems, the use of digital transmission on cable is almost a necessity for HDTV and it will be equally important for interactive pay-per-view systems (see Section 16.15.4). The current work by the Grand Alliance (see Chapter 6) has recognized this and includes standards for cable digital transmission.

16.14.1 Digital transmission

Formats and standards for the encoding, compression, and modulation of digital signals on CATV systems have not been established for either broadcast or HDTV services. The Grand Alliance, however, in parallel with its planning for broadcast HDTV, has proposed standards for CATV HDTV that are a variant of the terrestrial broadcast standards. Like the broadcast standards, they specify MPEG-2 compression and VSB modulation, but they are modified somewhat to take advantage of the higher signal levels on cable systems.

The system parameters of the broadcast and CATV HDTV formats are compared in Table 6.5. Both require channel bandwidths of 6 MHz. The most important differences are:

	Broadcast	CATV
	HDTV	HDTV
Modulation mode	8-VSB	16-VSB
Data rate	19.29 Mb/s	38.58 Mb/s
C/N threshold	14.9 dB	28.3 dB

The result is that the digital data rate of a 6-MHz cable channel is twice that of a broadcast channel. The additional data rate can be used in a number of ways—one example is to deliver two complete programs in each 6-MHz cable channel. This would effectively double the capacity of the cable system without needing any increase in its bandwidth.

To date (1996), standards for digital transmission of standard-definition signals compatible with NTSC scanning have not been proposed. With the compression and modulation methods of HDTV, however, it may be possible to transmit as many as 12 standard-definition video signals in a 6-MHz cable channel.

16.14.2 Fiber optics

Fiber optics offers additional potential for increasing system capacity and performance improvement. These enhancements will be essential for the distribution of interactive multimedia programs or for the merging of television program and telephone service on the same transmission facilities. Fiber optics provides substantially more transmission bandwidth and greater distances between line amplifiers, but it is more expensive. Because of this, most systems employing fiber optics will probably be a hybrid, using both fiber and cable transmission.

16.14.3 Hybrid cable systems

Hybrid cable systems' territories are divided into serving areas, each containing perhaps 500 subscribers within a limited geographical area. Signals are distributed within the serving area by coaxial cable from *nodes* that are close enough to the subscribers to be within the range of cable transmission. The nodes are connected to the head end by fiber optics. The conversion from the optical to the electrical domains is then necessary only at the nodes.

As will be described later, this division of the system's territory into small serving areas is a basic technique for providing individual pay-per-view service to all subscribers.

16.15 Future Cable Systems

The remainder of this chapter is devoted to a forecast of some of the most important features of future CATV systems. It includes a description of the services they may offer, their utilization of the cable bandwidth, and their functional design.

16.15.1 Service offerings

It seems likely that future cable systems will offer the following services:

- Standard-definition programs that may be offered as basic or premium service
- High-definition programs also may be offered as basic or premium service
- Pay-per-view programming in which subscribers pay fees for receiving individual programs
- Interactivity—the ability of subscribers to communicate directly with the head end. This service might be used, for example, for ordering programs on the pay-per-view channel or for purchasing items from a catalog
- Limited data and telephony service

16.15.2 Bandwidth utilization

There are numerous possibilities for the future utilization of the bandwidth available in cable systems. One of them, based on a 5 to 750 MHz passband, is shown in Figure 16.5.

In this example, the band from 5 to 40 MHz is used for upstream signals from subscribers to head end (see Section 16.7). The 40 to 50 MHz band is unused for transmission and it serves as a buffer between the downstream and upstrean channels.

The standard analog NTSC cable channels, 2 to RR, are transmitted in the 50 to 450 MHz band. As in the broadcast plan, digital HDTV channels will be interspersed with the NTSC channels in this band.

Figure 16.5 Future cable system bandwidth allocation

Figure 16.6 Diagram of a future CATV system

The 200-MHz band from 450 to 650 MHz is used for digital pay-per-view signals. As an option, the band from 650 to 750 MHz could be used for telephony and data transmission.

16.15.3 System functional diagram

Figure 16.6 is a functional diagram of a CATV system that includes the services described above. The output of the video server includes basic and premium analog NTSC and digital HDTV signals, together with the switched pay-per-view signals. These are transmitted by fiber-optic cable to the nodes from which they are distributed by coaxial cable to individual subscribers.

16.15.4 Pay-per-view services

A computerized control center receives orders for specified programs from individual subscribers. When the center receives an order, it directs the server, which may have 500 or more titles in storage, to feed the specified title to the fiber-optic trunk to the node for the service area where the subscriber is located.

To completely avoid the equivalent of a busy signal on a telephone line, the fiber trunk must have the capacity to carry a separate program for each subscriber connected to its node. As with telephony, however, it is possible to limit the capacity of the trunk to a lower value based on statistical analysis of subscriber usage. A standard is established that specifies the maximum acceptable percent of time that the entire trunk or a particular service area network may be fully loaded. The necessary number of circuits can then be calculated by standard statistical methods. A typical requirement for a 500-subscriber service area is 110 circuits, the assumption being that 500 subscribers will very rarely wish to see more than 110 different programs at any given time.

References

1. Baylin, Frank and Berkoff, Steve, *Wireless Cable and SMATV*, Baylin Publications, Boulder, CO, 1992.

2. Inglis, Andrew F. (ed.), *Electronic Communications Handbook,* Chapter 17, McGraw-Hill, New York, 1988.

Satellite Video Communications

17.1 Satellites and Television

The use of satellites for television broadcasting and CATV program transmission and distribution began in the mid-1970s, and its subsequent growth has been phenomenal. Satellites are now indispensable to both media and their utilization continues to grow.

The value of satellites for transmission and distribution results from four unique properties:

1. The cost of a *satellite circuit* (an end-to-end transmission path via satellite) is independent of its length. It costs no more to send a signal across the country than across the street. This property generally makes satellites cost effective for long-haul, point-to-point service.

2. Except for a short segment at the beginning and end of each circuit, the path of the satellite signal is far above the surface of the earth and is unaffected by the nature or accessibility of the intervening terrain. This property makes satellites especially useful for paths over oceans, jungles, mountains, and arctic regions.

3. Once the satellite is in place, a new circuit can be established very quickly. This makes satellites valuable for *satellite news gathering*, SNG, where circuits must be set up quickly and with little preplanning. In this role they are of equal value for local and international news events.

4. Finally, satellites are especially adapted for point-to-multipoint service. When the satellite is in place, service can be supplied to an indefinite number of receiving points for the relatively low cost of *receive-only*

earth stations (see Section 17.6). The most prominent use of this capability to date has been the distribution of programs to CATV systems. It is also used by the television broadcast networks for distribution of programs to their affiliates.

Point-to-multipoint services have also been introduced for broadcasting directly from satellites to homes. The generic term for these services is *direct broadcast service* (DBS). The most rapidly growing type of DBS employs digital transmission. With digital transmission and high-power satellites, it is possible to receive a satisfactory signal with a very small, c. 18 inch, receiving antenna. The digital transmission mode employs *video compression* technologies (see Chapters 3 and 6) that greatly increase the channel capacity of the satellite. As of this writing, (early 1996) more than 150 video program services are being distributed by the digital satellite service and millions of receivers are in use.

Part-time channels in the broadcast service are leased on an *occasional use* basis, primarily for news (SNG) and sporting events.

17.2 Satellite Communication Systems Elements

The elements of a satellite communication system are shown in Figure 17.1. An earth station *uplink* transmits a signal to the satellite in a very narrow beam. Here it is received and amplified, and its frequency shifted to the *downlink* frequency by a *transponder*. The signal is retransmitted to earth by the downlink, covering an area known as the satellite's *footprint*.

The boundary of the footprint is the contour of the *effective isotropic power*, EIRP, which is high enough to provide satisfactory reception. The EIRP is the ratio in decibels between the actual signal strength and the signal from an isotropic antenna (a physically unrealizable antenna that would radiate equally in all directions) radiating one watt.

17.3 The Geosynchronous Orbit

17.3.1 Orbit location

The *geosynchronous orbit* is a circle in the *equatorial plane*,[1] 22,300 miles above the earth's surface (see Figure 17.2). It has the property that the centrifugal force acting upward on objects located on the orbit and rotating with the earth equals the gravitational force acting downward. Under these

---

1 Sometimes the orbit becomes inclined slightly with respect to the equatorial plane as a result of fuel saving for the N-S station keeping of the satellite (see Section 17.5.1).

Figure 17.1 Satellite communication system

conditions, satellites in the orbit will appear to be stationary when viewed from the earth.

17.3.2 Orbital slots

Satellites are assigned fixed positions on the geosynchronous orbit known as *orbital slots* that are identified by their longitude. By international agreement, each country is allocated an arc of the geosynchronous orbit for its satellites, and the slots within this arc are assigned to individual licensees by the country's regulatory authority, the FCC for the United States.

The United States is allocated the arcs 62°–103° and 120°–146° W. Long. (west longitude) for C-band satellites and 62°–105° and 120°–136° W. Long. for K_u-band satellites. (See Section 17.4.2 for definitions of the C and K_u bands.) The slots allocated to Canada are located in gaps in the U.S. arcs.

The spacing between orbital slots is critical, and it should be as small as possible to maximize the number of satellites that can be accommodated in

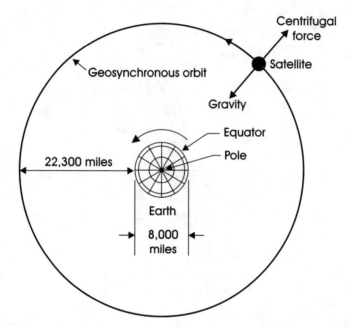

Figure 17.2 Geosynchronous orbit

the geosynchronous orbit. The minimum spacing is determined by the directivity of uplink and downlink antennas, since all satellites in a given band operate on the same frequencies and must depend on antenna directivity for isolation. Initially, C-band satellites were spaced 4°, and K_u band 3° on the orbital arc. (The K_u-band spacing was smaller because a narrow antenna beam can be produced more easily at its shorter wavelength.) Later, the spacing in both bands was reduced to 2°.

The spacing between DBS satellites in the K_u band is wider (9°) because smaller receiving antennas with wider beam angles are used in this service.

17.3.3 Elevation and azimuth angles

The *elevation* or *look angle* of a satellite is the angle above the horizontal made by the line of sight from a point on the earth to the satellite. The *azimuth angle* is the longitudinal position of the line of sight measured in degrees of west of Greenwich from 0° to 360°.

To calculate the elevation and azimuth angles, from a point on the earth's surface, first calculate the path length L and the angle ß:

$$\beta = \cos^{-1}[\cos(\delta long) \cos(lat)] \tag{17.1}$$

$$L = (18.2 - 5.4 \cos \beta) \times 10^{-4} \text{ km} \tag{17.2}$$

where: δ*long* is the difference between the earth station and satellite
                longitudes and

        lat is the latitude of the earth station.

The elevation and azimuth angles are then:

$$elevation = \cos^{-1}\left(\frac{4.22 \times 10^4}{L}\right)\sin\beta \tag{17.3}$$

$$azimuth = 180° + \tan^{-1}\left(\frac{\tan(\delta long)}{\sin(lat)}\right) \tag{17.4}$$

Assuming there is a choice, it is desirable that an orbital slot be chosen that results in as high an elevation angle as possible. With low elevation angles there may be difficulty in clearing obstacles, the path through the troposphere is longer, which may result in more rain attenuation (especially in the K$_u$ band), multipath distortions, and electrical noise generated by the earth's heat.

It is considered good design practice to utilize a minimum elevation when possible of 5° for C band and 10° to 20° for the K$_u$ band.

17.3.4 Prime orbital arcs

The prime orbital arc for a location is the range of orbital slots that have an elevation angle in excess of 5°. The prime C-band arcs for the continental United States (CONUS), Alaska, and Hawaii are:

	Degrees west longitude
CONUS	55-138
CONUS plus Hawaii	80-138
CONUS plus Anchorage	88-138

The prime arcs for K$_u$ band are shorter because of the higher minimum angle. The arcs are also shorter at more northerly latitudes (satellites are below the horizon at the north pole).

17.3.5 Solar eclipses

Solar eclipses occur when the earth comes between the satellite and the sun. They are not of direct interest to satellite users, but they have a major effect on satellite design. Satellites depend on *solar cells* for electrical power, and they must be equipped with on-board batteries to supply power during the eclipses.

Eclipses occur near midnight at the satellite's longitude for a period of 21 days centered on the spring and autumn equinoxes. Their duration is a maximum of 70 minutes.

17.3.6 Sun outages

Sun outages occur when the sun and satellite are in line as seen from the receiving location. The high temperature of the sun causes it to transmit a high-energy noise signal to earth station receiving antennas whenever it passes behind the satellite. The increase in noise is so great that a signal outage usually results.

The length and duration of the outages depends on the latitude of the earth station and the size of the receiving antenna. At an average latitude of 40° in the United States and with a 10-meter antenna, outages occur for a period of 6 days with a maximum duration of 8 minutes. Outages are longer with smaller receiving antennas.

Outage times for specific locations can be calculated or can be obtained from the satellite operator.

Sun outages cannot be avoided, and if continuous service is to be maintained the traffic must be switched to another satellite.

17.4 Satellite Classifications

Satellites are classified by their *usage* and by their *technical* characteristics.

17.4.1 Usage Classifications

The FCC defines two usage classifications for the communication satellites used for television: Fixed Satellite Service (FSS), a broad classification that includes most commercial communication satellites, and Broadcast Satellite Service, a specialized FSS classification for satellites designed for broadcast directly to homes.

The broadcast satellite service may employ either analog or digital transmission. If analog transmission is used, it is usually described as DBS, direct broadcast service. If digital transmission is used, it is described here as digital DBS.

17.4.2 Technical classifications

The most important technical classifications of satellites are their spectrum bands and service areas.

17.4.2.1 Spectrum bands   Three spectrum bands are used for commercial television satellite communications: C-band FSS, K_u-band FSS, and broadcast (B/C) satellite service. The uplink and downlink frequencies allocated to these bands are:

Service	Uplink (GHz)	Downlink (GHz)
C-band FSS	5.925–6.425	3.700–4.200
K_u-band FSS	14.0–14.5	11.7–12.2
B/C satellite service (DBS)	17.3–17.8	12.2–12.7

Direct-to-home broadcasting is also permitted on the FSS bands—in fact the C-band FSS spectrum was the first to be used for this purpose—but the wider spacing (see above) and higher power authorized for satellites in the broadcast service band permit the use of much smaller receiving antennas.

The characteristics of satellite service in these three bands are quite different with respect to propagation effects, equipment performance, and FCC regulation.

C and K_u bands have both advantages and disadvantages. C-band satellites share spectrum space with microwave systems, and the location and directional patterns of C-band earth stations must be *coordinated* or "cleared" with nearby microwave stations. It is sometimes impossible to clear a site in crowded metropolitan areas, and it is often not practical to use C-band uplinks for satellite newsgathering where the uplink location cannot be planned in advance. In addition, the EIRP of C-band downlinks is limited to prevent interference to microwave systems. Finally, C-band uplink earth stations require larger antennas because of their longer wavelength.

Offsetting these problems, the C band has significant advantages, the most important being virtual immunity from signal outages resulting from rain attenuation. C-band components are generally lower in cost, in part because of larger mechanical tolerances, although this difference is diminishing as the production volume of K_u-band components is increasing.

17.4.2.2 Service areas Satellite service areas vary widely in size, ranging from a metropolitan area to most of a hemisphere. The size of the service area depends on the size of the downlink antenna on the satellite—the larger the antenna the narrower the beam and the smaller the service area. One classification defines four service area sizes: global, regional, national, and spot.

Global beams cover the entire area that is visible from the satellite. They are used for international communications.

Regional beams cover a group of countries, for example, western Europe.

National beams cover all or a significant portion of a single country. Most domestic U.S. satellites are in this category.

Spot beams cover a limited area and are used principally for point-to-point voice and data communications. They are seldom used in television service.

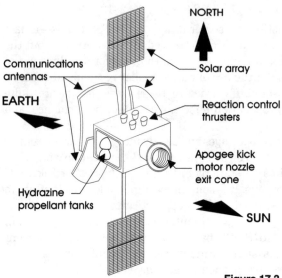

Figure 17.3 Satellite construction

17.5 Satellite Construction

The construction of a communication satellite is shown in Figure 17.3. It includes two primary subsystems: the *bus* and the *payload*. The bus includes the satellite housing and the subsystems that provide power and maintain the satellite in its orbital slot.

The payload includes the communications subsystems—the antennas, the receivers, and the *transponders* that shift the frequency and amplify the uplink signals for retransmission.

17.5.1 Satellite bus

The satellite housing can be rectangular as shown in Figure 17.3 or cylindrical, depending on the choice of the stabilizing mechanism that maintains the *attitude* of the satellite. The satellite in Figure 17.3 employs internal gyroscopes and *three-axis* stabilization. An alternative is *spin stabilization* in which the housing is a slowly rotating cylinder and the orientation of the antenna is held fixed through a rotating joint.

The power system consists of arrays of solar cells with battery back-up for use during the solar eclipses. The solar cells can be mounted on flat panels as in Figure 17.3 or on the cylindrical surface of a spin stabilized satellite.

Although the forces on a geosynchronous satellite are nominally in balance, there are minor perturbations, e.g., the effect of irregularities on the earth's surface, that would cause it to leave its slot if uncompensated. Maintaining the satellite in its slot is known as *station keeping*, and it is achieved by ejecting small amounts of gas through the *reaction thrusters*. The amount of gas that can be stored in the satellite is one of the elements that limits its life.

Substantial savings in fuel and a corresponding increase in the useful life of the satellite can be achieved by omitting N-S station keeping, thus allowing the satellite to depart from the equatorial plane. The result is a build-up in the inclination of the satellite's orbit with respect to the equatorial plane of about 0.85° per year. This causes cyclic annual variations in the azimuth and elevation of the satellite as seen from the earth that may be great enough to require periodic repointing of the earth station antennas.

17.5.2 Satellite payload

A functional diagram of the satellite's payload, its reason for being, is shown in Figure 17.4. It includes the transmitting and receiving antennas, the receivers, and the transponders. Separate channels are provided for vertically and horizontally polarized signals or, alternatively circularly polarized antennas with opposite rotations.

17.5.3 Frequency reuse

Frequency reuse nearly doubles the satellite's communications capacity by employing both horizontal and vertical polarization (or right- and left-handed circular polarization for digital DBS systems). The de facto standard for the utilization of the downlink spectrum of a C-band satellite is shown in Figure 17.5. It includes 24 40-MHz channels—12 in each of the two polarities. Adequate isolation between channels sharing the same frequency is achieved by a combination of cross-polarization and offsetting the channel centers.

Additional isolation between channels on adjacent satellites is also achieved through cross-polarization.

K_u-band satellites do not have a de facto or official channel bandwidth standard. For FSS service, it ranges from 36 to 72 MHz, with 54 MHz as a common choice. As with C band, cross-polarization is used to double the satellite capacity, but the lack of standardization of channel bandwidth makes it impractical to increase the isolation between adjacent satellites by cross-polarization.

Digital DBS satellites have 16 transponders (or 32 with cross-polarization), each with a bandwidth of 24 MHz as described in Section 17.9.

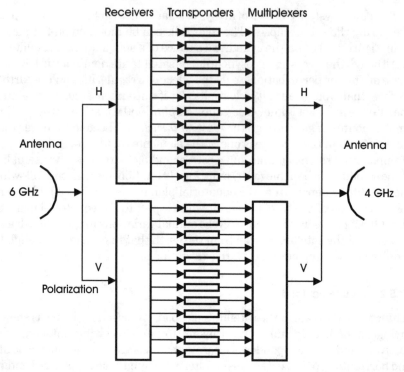

Figure 17.4 Satellite payload

17.5.4 Satellite antennas

The size and shape of the satellite's footprint is determined by the directivity of its transmitting antenna.

17.5.5 Satellite receivers

The satellite normally receives a strong signal from the earth station, and the demands on its receiver's sensitivity are not extreme. There are usually two pairs of receivers in each satellite, two (one spare) for each group of channels with like polarization.

17.5.6 Satellite transponders

The transponder receives an IF signal for a single channel (sometimes two channels in K_u-band systems) from the satellite receiver, shifts its frequency to the downlink channel, and amplifies it for retransmission to the earth. Transponders originally employed traveling wave tubes, but, as with much of the electronics industry, a transition has been made to solid-state power amplifiers, SSPAs.

Figure 17.5 C-band spectrum utilization

Most transponders exhibit the property of *saturation*, a power level above which an increase in input causes no further increase in output. The output of the receiver should drive the transponder to saturation.

Transponder power is determined by FCC regulations, economics, and technical feasibility.

In the C band, the FCC (and similar international organizations) specification on downlink power density limits the transponder power to about 8 watts. The power density limit is stated in terms of power per kHz of bandwidth per unit area. For FM signals with energy dispersal (see Section 17.7), it translates to EIRPs ranging from 38 dBw at elevations below 25° to 48 dBw at higher elevation angles.

The energy density limit is much higher in the K_u band, and economics and technical feasibility usually establish the actual transponder power. Forty watts is a typical choice for FSS service. Even higher power is required for DBS and transponder powers up to 250 watts have been proposed (see Section 17.9).

17.6 Earth Stations

Earth stations come in a wide range of sizes, costs, and performance. Digital TVROs (*television receive-only*), small earth stations that sell for less than $1,000 and are designed for home reception, are at one end of the range (although they employ technically sophisticated technology). At the other end are multimillion-dollar installations owned by satellite operators and program suppliers.

Figure 17.6 Earth station antenna directivity

17.6.1 Earth station antennas

The antenna is a basic component of both uplink and downlink earth stations. Most earth station antennas employ the familiar format of back-yard TVROs, a parabolic or quasi-parabolic reflector *illuminated* by a feed horn. Their performance, however, differs widely depending on their size and design. The directional properties of earth station antennas are shown in Figure 17.6. They are rated by their gain with respect to an isotropic radiator (or alternatively with respect to a half-wave dipole), their half-power beamwidth (HPBW), their null beamwidth (NBW), and their side-lobe amplitude (see Figure 15.12 for definitions).

The shape of the pattern depends on the size of the antenna and the uniformity of the illumination of the antenna aperture by the feed horn. The width of the beam is inversely proportional to the antenna diameter and is a minimum when the illumination is uniform. Uniform illumination results in rather large side lobes, however, and these can be reduced at the expense of greater beamwidth by tapering the illumination from the center to the edge of the reflector.

The half-power beamwidth is given by Eq. (17.5):

$$\text{Half - power beamwidth} = C_{\text{HP}} \frac{\lambda}{D} \tag{17.5}$$

where: C_{HP} is a factor that varies from 50° for uniform reflector illumination to 70° for typical practical designs

λ is the wavelength
D is the antenna diameter measured in the same units as λ
The null beamwidth is given by Eq. (17.6):

$$\text{Null beamwidth} = C_{\text{null}} \frac{\lambda}{D} \qquad (17.6)$$

where: C_{null} varies from 114° to 170° depending on the uniformity of
the reflector illumination

The gain of the antenna is inversely proportional to the square of the beamwidth (or the product of the vertical and horizontal beamwidths if different). It is also affected by the antenna efficiency, the fraction of the energy that is radiated after subtracting ohmic losses and energy intercepted by the feed horn. Table 17.1 lists specifications of representative uplink and downlink antennas assuming 70% efficiency and non-uniform illumination.

17.6.2 Uplink functional diagram

A functional diagram of a typical earth station uplink for frequency-modulated carriers is shown in Figure 17.7. The functions of the components in an earth station with digital transmission are considerably different and the FM exciter is replaced by a series of components having functions that generate and process the bit stream for transmitting the audio and video signals (see Section 17.9).

Table 17.1 Antenna specifications

	Antenna diameter (meters)	HPBW (degrees)	Null BW (degrees)	Gain (dBi)
C-band downlink	2	2.6	6.4	36.6
(4 GHz)	5	1.05	2.5	44.5
	7	0.75	1.8	47.4
	10	0.52	1.3	50.6
C-band uplink	5	0.70	1.7	48.0
(6 GHz)	7	0.50	1.2	50.9
	10	0.35	0.85	54.0
Ku-band downlink	1	1.75	4.2	40.0
(12 GHz)	1	0.88	2.1	46.0
	5	0.35	0.95	54.0
	7	0.25	0.60	56.9
Ku-band uplink	2	0.75	1.8	47.4
(14 GHz)	5	0.30	0.72	55.3
	7	0.21	0.51	58.4

Figure 17.7 Earth station uplink

The video and audio signals may be *scrambled* to prevent unauthorized reception.[2] They then pass to the exciter, the heart of the earth station. It generates the IF carrier, typically at 70 MHz for the visual carrier, adds *preemphasis* and a *sideband energy dispersal* waveform (see Section 17.7), modulates the IF carrier and *up-converts* it to the final carrier frequency and drives the high-power amplifier (HPA).

The power of the HPA in combination with the uplink antenna gain should be sufficient to drive the satellite transponder to saturation. This depends on the characteristics of the satellite, but a typical figure is 83 dBw. If the uplink antenna gain were 54 dB (see Table 17.1), the required power delivered to the antenna would be 29 dBw and a rated power of 1 kW would probably be chosen.

Traveling wave tubes have been frequently used for HPAs, although there is now a trend to solid-state devices.

Uplink antennas must meet the FCC specification shown in Table 17.2, which limits the off-axis gain of uplink antennas to avoid interference to adjacent satellites.

17.6.3 Downlink functional diagram

A functional diagram of a satellite downlink earth station for FM modulation is shown in Figure 17.8. Digital downlinks are more complex and are described in Section 17.9.

---

2   For a summary of security techniques, see *Electronic Communications Handbook*, A.F. Inglis (ed.), Chapter 17, McGraw-Hill, New York, 1988.

Table 17.2 Uplink antenna maximum off-axis gain

Off-Beam Angle (degrees)	Maximum Gain (dBi)
1	29.0
2	21.5
3	17.0
4	14.0
5	11.6
6	9.6
7	8.0

17.6.4 G/T, the downlink figure of merit

Other factors being equal, the signal-to-noise performance of a downlink earth station is proportional to the ratio, G/T, where G is the gain of the antenna, see Table 11.1, and T is the *equivalent noise temperature* of the system in degrees Kelvin [see Eq. (17.7) below].

Most of the noise in a satellite system is random and has the same characteristics as thermal noise, the electrical signals generated by hot objects. The equivalent noise temperature, T, is a convenient measure of the noise contribution of a component or system, even though it is nonthermal in origin.

T is given by the equation:

$$T = 293° (F_n - 1) \tag{17.7}$$

where: F_n is the *noise figure* expressed as an arithmetic (not logarithmic) ratio and is equal to the quotient of the signal-to-noise ratio at the output of a system to the S/N at the input normalized to an input temperature of 293°

Figure 17.8 Earth station downlink

Table 17.3 Antenna noise temperature

Elevation angle	C band	K$_u$ band
5°	58°K	80°K
10°	50°K	58°K
30°	30°K	38°K

Because of its universality, noise temperature can be used to specify the noise contribution of all the components of a system. The system noise temperature is then equal to the sum of the noise temperatures of its components.

The major sources of noise in practical systems are the antenna and the input stage of the receiver. The antenna noise is electrical energy radiated from the earth because of its warmth and picked up by the antenna. The input stage of the receiver is a *low noise amplifier* (LNA), chosen for its low equivalent noise temperature.

The amount of electrical noise power received from the earth by the antenna and its equivalent noise temperature depends on its directivity and its elevation angle. Noise also originates in heavy rainfall, which is a source of noise in the K$_u$ band.

Typical antenna noise temperatures are shown in Table 17.3. Important progress has been made in recent years in the development of reasonably priced LNAs. Among these developments are the field effect transistor (FET) and the parametric amplifier. For the most demanding applications, the parametric amplifier can be artificially cooled. Equivalent noise temperatures for typical LNA types are shown in Table 17.4. The LNA is mounted at the antenna and provides the initial amplification. Alternatively, as in Figure 17.8, a *low-noise block converter,* LNB, amplifies and translates all the incoming signals to a lower frequency range, often 950–1,450 MHz, for transmission to the receiver by coaxial cable.

The remainder of the receiver functions are conventional.

17.7 Transmission Formats

The transmission format used by satellite systems may be either analog or

Table 17.4 LNA noise temperature

LNA Type	C band	K$_u$ band
Field effect transistor	85°K	150°K
Parametric amplifier	60°K	120°K
Cooled parametric amplifier	29°K	43°K

f_R = reference carrier frequency
f_V = maximum video frequency

Figure 17.9 Video sideband spectrum

digital. Video signals are inherently analog, and the format initially used for satellite transmission of these signals was frequency modulation. This format still predominates in the FSS service, but with recent developments in digital technology the use of digital transmission is enjoying increasing usage, particularly in the digital DBS services (see Section 17.9).

The advantages of frequency modulation as compared with amplitude modulation are relative noise immunity, insensitivity to system nonlinearities, and the ability to spread the transmitted energy over the entire channel by *sideband energy dispersal* (see below). Its major disadvantage as compared with amplitude modulation is its greater bandwidth requirement. Its performance as compared with digital transmission is described in Section 17.9.

The baseband waveform and the sideband spectrum for a carrier that is frequency modulated with a video signal are shown in Figure 17.9. The sync tip is clamped to a reference frequency so that the dc component of the signal is retained.

Dispersal of the sideband energy from the FM signal across the channel bandwidth is required in the C band to meet the FCC regulatory limits on EIRP per kHz. It is achieved by adding a low-frequency, 30-Hz (25-Hz for PAL), sawtooth waveform to the video signal to cyclically move the range of the frequency deviation of the carrier. This spreads the sideband energy over a larger portion of the spectrum and prevents the heavy concentration around the carrier that would severely limit the amount of downlink power because of the FCC limitation on energy density. The added sawtooth is removed at the receiver by the dc restoration circuit.

17.8 FM Performance Characteristics

The defining specifications of FM transmission systems are bandwidth and signal-to-noise ratios.

17.8.1 Bandwidth

The bandwidth of a frequency-modulated signal is greater than the peak-to-peak frequency deviation and is determined by the spectrum of its sidebands. In theory they extend to infinity in both directions, but almost all of the sideband energy is confined to a frequency band extending from the extremes of the instantaneous frequency deviation waveform by an amount approximately equal to the video baseband bandwidth. The approximate total IF (or RF) bandwidth B is:

$$B_{IF} = \delta f_V + 2B_V \tag{17.8}$$

where: δf_V is the peak-to-peak frequency deviation

B_V is the video baseband bandwidth

The same frequency units should be used for all terms.

A complete 40-MHz C-band transponder is usually used for each analog video signal, although two could be compressed into this spectrum space at the expense of S/N. It is common practice to transmit two analog video signals in a single 54 MHz K_u-band transponder.

17.8.2 Signal-to-noise ratio

The procedure for determining the signal-to-noise ratio of a satellite circuit that employs frequency modulation begins by performing a *link analysis* to determine the carrier-to-noise ratio, C/N, of the unmodulated carrier as described in Section 17.8.2.1 below. The signal-to-noise ratio is then calculated by adding the *FM improvement, preemphasis,* and *noise weighting* factors as shown in Eq. (17.9):

$$\frac{S}{N} = \frac{C}{N} + FM_{FM} + F_{DEEMP} + F_W \tag{17.9}$$

17.8.2.1 Carrier-to-noise ratio   The carrier-to-noise ratio is calculated by means of Eq. (17.10):

$$\frac{C}{N} = (EIRP - 20\log f_{GHz} - 10\log B_{Hz} + 45) + \frac{G}{T} \tag{17.10}$$

where:

EIRP is the effective radiated power expressed in decibels relative to 1 watt from an isotropic radiator, i.e., one radiating equally in all directions

f_{GHz} is the carrier frequency in gigahertz

B_{Hz} is the bandwidth expressed in hertz

The C/N ratio and hence the fade margin (see Section 17.8.3) decrease with increasing frequency and bandwidth. As a result, C/N rather than S/N is sometimes the limiting factor in the performance of K_u-band systems. It can be increased by reducing the receiver bandwidth.

17.8.2.2 FM improvement factor   The presence of the FM improvement factor, F_{FM}, is a major advantage of the FM transmission mode. It is the ratio (in dB) of the S/N of the demodulated receiver output to the *carrier-to-noise* ratio of the unmodulated carrier—usually measured at the receiver IF. It is calculated by Eq. (17.11):

$$F_{FM} = 20\log(\frac{\delta f_V}{2B_V}) + 10\log(\frac{B_{IF}}{B_V}) + 7 \tag{17.11}$$

where: F_{FM} is   the FM improvement factor (dB)

B_{IF} is   the bandwidth of the radio-frequency channel

f_V is   the video frequency

δf_V is   the video peak-to-peak frequency deviation

B_V is   the video bandwidth

The ratio $\delta f_V/f_V$ is the *modulation index*

F_{FM} can be quite large, sometimes more than 20 dB.

17.8.2.3 FM noise weighting factor   The noise weighting factor, F_W, for video signals, which converts unweighted to weighted noise, is described in Section 4.15.3. The noise spectrum for frequency-modulated video signals is triangular, i.e., the amplitude of the noise components increases with frequency. Since the acuity of the eye is less for fine-grained high-frequency components, the noise weighting factor is greater for FM noise than for flat noise. The weighted signal-to-noise ratio for an NTSC signal transmitted by a frequency-modulated carrier, calculated by the EIA standard noise weighting curve, is about 12 dB higher than the unweighted value.

17.8.2.4 Frequency preemphasis/deemphasis   The signal-to-noise ratio of an FM transmission system can be improved further by the use of frequency preemphasis/deemphasis. The amplitude of the higher frequency components of the video baseband is increased or emphasized for transmission and

Table 17.5 CCIR Rec. 405 preemphasis standard

Baseband Frequency	Relative Response (dB)
10 kHz	-10
20	-10
50	-9.5
100	-8.8
200	-6.8
500	-2.0
1 MHz	+1.4
2	+2.8
5	+3.6

deemphasized at the receiver. The deemphasis (F_{DEEMPH}) offsets the increase in amplitude of the high-frequency signal components and reduces the amplitude of the high-frequency noise components.

Standardization is required to obtain the full benefits of preemphasis/deemphasis and the CCIR has issued Recommendation 405 (see Table 17.5) for this purpose.

17.8.3 Fade margin

The fade margin is the difference (in dB) between the receiver input power under nonfading conditions and the receiver's *threshold*. The threshold is a property of FM receivers and is the point at which the signal-to-noise ratio, S/N, of the output signal begins to drop much more rapidly than the C/N. The fade margin is given by Eq. (17.12):

$$\text{Fade margin} = P_{\text{R}} - T_{\text{S}} \qquad (17.12)$$

where: P_{R} is the power in dBm at the terminals of the receiver under nonfading conditions. It is directly proportional to the carrier-to-noise ratio, C/N, and a high C/N is required for a good fading margin

T_{S} is the *threshold sensitivity* in dBm of the receiver, the point at which the S/N at the output of the receiver begins to drop much more rapidly than the C/N

Since the signal quality decays rapidly at power levels below the threshold, it is important that the fade margin be sufficient to maintain service, even during fading from refractive effects or heavy rainfall.

The length of the path of satellite signals through the troposphere is relatively short, and tropospheric fading is not a severe problem except for K_{u}-band signals in regions of torrential rainfall. For C-band and many K_{u}-band satellite circuits, a 3 dB fade margin is adequate.

Many receivers have *threshold extenders* that reduce the threshold power, typically from 12 to 8 dBm. These work by reducing the receiver bandwidth as the signal level approaches the normal threshold value.

17.8.4 FM satellite system performance

This section illustrates the principles that establish the S/N and fade margin of FM satellite links by applying them to representative examples:

1. A C-band trunk circuit of high quality and reliability such as would be used for interconnecting two major switching centers.

2. A C-band downlink to a CATV head end.

3. A K_u-band ENG circuit. One-half of a 54 MHz transponder is assumed.

4. A C-band homesat (earth station used for home reception) receiving transmissions intended for CATV systems.

5. A K_u-band homesat operating with a higher EIRP than is permitted on the C band—sometimes called medium-power DBS. As with the ENG circuit, one-half of a 54 MHz transponder was assumed.

6. A high-power analog K_u-band DBS system (see Section 17.9) operating in the broadcast satellite service band.

The circuit parameters that have been assumed for these examples are shown in Table 17.6 together with the calculated weighted signal-to-noise ratios and fade margins.

Thirteen dB was assumed for $F_{DEEMP} + F_W$.

A receiver threshold of 8 dB was assumed except for the CATV and homesat downlinks where 4 dB of threshold extension was assumed.

17.8.4.1 Trunk circuits Trunk circuits are designed to add negligible noise to the signal since they are in series with other system elements, e.g., CATV downlinks, which for economic reasons leave very little margin for noise in the trunks. This is recognized in EIA/TIA-250-C performance standards, which specify a minimum *unweighted* S/N of 56 dB for point-to-point satellite transmission circuits. The example in Table 17.6 with a *weighted* S/N of 68 dB meets this standard.

Since the EIRP is limited by FCC rules, a high S/N must be achieved with a receiving system having a high G/T—in this example with an 11-meter antenna and a cooled parametric amplifier for the LNA. This not only increases the S/N but permits a full bandwidth IF amplifier to be employed without adversely affecting the C/N and the fade margin.

The fade margin is 19 dB even without a threshold extender.

17.8.4.2 CATV downlinks Downlinks are a significant part of the cost of CATV systems, but their G/T is usually chosen so that their S/N exceeds current industry standards for system performance. A weighted S/N of 54

Table 17.6 Representative circuit parameters and performance—FM analog transmission

	Trunk	CATV	ENG	Homesat	Homesat	DBS
Link Parameters						
Band	C	C	Ku	C	Ku	DBS
Downlink EIRP	38	38	43	38	43	54
Antenna						
Diameter (m)	11	4.5	5	2	1	0.5
Gain (dB)	55	42	54	36	40	34
Noise Temp. (°K)						
Antenna	40	40	50	40	50	50
LNA	50	85	150	125	150	150
G/T (dB/°K)	35	21	31	14	17	11
B_{IF} (MHz)	32	32	24	24	24	27
Bv (MHz)	4.2	4.2	4.2	4.2	4.2	4.2
δfv (MHz)	21	21	15	21	21	20
Rcv Threshold (dBm)	12	8	12	8	8	8
Link Performance						
F_{FM}	24	24	20	25	23	23
$F_{DEEMP} + F_W$	13	13	13	13	13	13
C/N (dB)	31	17	24	11	10	14
S/N (dB—weighted)	68	54	56	46	45	48
Fade Margin (dB)	19	9	12	3	2	6

dB, the performance of the CATV downlink in the example in Table 17.6, exceeds these standards. With the prospect of more stringent requirements for HDTV in the future, it is likely that the S/N standard for CATV downlinks will be increased.

The fade margin of 9 dB is satisfactory, even without the use of a threshold extender.

17.8.4.3 ENG circuits An examination of the CATV and ENG system specifications in Table 17.6 illustrates some of the differences between C-band and K_u-band performance parameters.

The gain of the K_u-band antenna is 5 dB higher because of the shorter wavelength, but this is offset by a greater space loss and a higher antenna noise temperature. The higher C/N of the K_u-band system results from the greater EIRP of its footprint, and its smaller IF bandwidth, the latter chosen to increase the fade margin to compensate for rain attenuation.

The fade margin of 12 dB is adequate except in areas of extremely heavy rainfall.

17.8.4.4 C-band homesats C-band satellite systems are not primarily designed (or even intended) to provide service directly to homes, and Table 17.6

clearly shows the compromises that must be made to operate a C-band homesat with a small antenna.

The IF bandwidth is reduced to increase the fade margin but at the expense of frequency response. Even with this increase, the use of a threshold extender would be desirable.

17.8.4.5 Ku-band analog FM homesats   The FSS Ku band has been proposed for analog direct-to-home transmission, but it has not been widely used for this purpose. The example shown in Table 17.6 assumes a 1-meter antenna and two video channels in a 54 MHz transponder. The 45 dB weighted S/N is marginal as is the 2 dB fade margin.

17.9 Digital DBS Services

The technical and economic practicality of direct-to-home broadcasting from satellites was increased enormously by the introduction in 1994 of a digital transmission format for DBS services. Its growth has been extremely rapid (digital DBS receivers have been constantly in short supply), and the rate shows no sign of diminishing.

The following reasons for using digital transmission for DBS service have been advanced in filings with the FCC:

1.   More efficient use of the spectrum by transmitting more video channels per transponder as the result of bandwidth compression (see Section 17.9.9).

2.   A major improvement in the achievable signal-to-noise ratio (see Section 17.9.11).

3.   Smaller receiving antennas.

4.   More flexibility in the assignment of channels.

5.   More easily integrated with audio and other services.

The high technical quality of digital DBS broadcasts is encouraging public interest in high definition television, and it will probably be a major factor in the future development of HDTV systems.

The following sections describe the basic technologies of digital satellite broadcasting.

17.9.1 Spectrum allocation

Digital DBS shares the spectrum (see Section 17.4.2.1), 12.2–12.7 GHz, that has been allocated for broadcast satellite services.

17.9.2 Orbital slots

Eight orbital slots have been allocated to the United States for DBS:

Deg. W.Long.

61.5

101

110

119

148

157

166

175

Only three of these slots—101°, 110°, and 119°—provide coverage for CONUS. 61.5° covers only the east while 148°, 157°, 166°, and 175° cover the west. Transponders on the 61.5° satellite can be paired with transponders on the four western satellites, however, to provide complete CONUS coverage from these satellites.

17.9.3 Transmission mode

It was initially expected that DBS satellites would employ analog FM transmission, the standard format for the FSS services. A number of different combinations of transponder EIRP and downlink receiver G/T were proposed. Table 17.6 shows an example in which the EIRP is 52 dB (as compared with 43 dB for a typical K_u-band FSS satellite). With the assumptions used in the example, the fade margin would be 4 dB and the weighted S/N would be 48 dB. This is acceptable although not excellent performance, and it would be satisfactory in the absence of competition from signal sources with better performance. The plan to use FM for this service, however, was overtaken by the demonstrated advantages of digital transmission, which have caused it to be the preferred mode.

17.9.4 Satellite configuration

32 channels are allocated to each orbital slot as shown in Table 17.7. Alternate channels have right- and left-hand circular polarization. The transponder bandwidth is 24 MHz and the spacing between transponders of the same polarity is 29.16 MHz, thus providing a 5.16 MHz guard band. The spacing between transponders of opposite polarity is 14.87 MHz, which means that their spectra overlap as shown in Figure 17.10. The use of alternate left-hand and right-hand polarization and offsetting the transponder centers provide the necessary isolation between adjacent transponders in the overlap area (equivalent to the use of vertical and horizontal polarization in FSS satellites).

Table 17.7 DBS channel allocations

Transponder Number	Polarization	Frequency (MHz)
1	RH	12,224.00
2	LH	12,238.80
3	RH	12,253.16
4	LH	12,267.74
5	RH	12,283.32
6	LH	12,296.90
7	RH	12,311.48
8	LH	12,326.06
9	RH	12,340.64
10	LH	12,355.22
11	RH	12,369.80
12	LH	12,384.38
13	RH	12,398.96
14	LH	12,413.54
15	RH	12,428.12
16	LH	12,442.70
17	RH	12,457.28
18	LH	12,471.86
19	RH	12,486.44
20	LH	12,501.02
21	RH	12,515.60
22	LH	12,530.18
23	RH	12,544.76
24	LH	12,559.34
25	RH	12,573.92
26	LH	12,588.50
27	RH	12,603.08
28	LH	12,617.66
29	RH	12,632.24
30	LH	12,646.82
31	RH	12,661.40
32	LH	12,675.98

17.9.5 Satellite locations

The two initial digital DBS satellites launched in 1994 are nominally collocated at 101° W. Long., but their precise locations are at 100.8° and 101.2°—approximately 700 miles apart in space—to avoid a collision. Each satellite has 16 120-watt transponders, one-half of those assigned to the orbital slot, to reduce the satellite's power requirements to a reasonable level. From the standpoint of frequency allocation, the two are treated as though they were a single satellite.

Figure 17.10 Digital DBS spectrum utilization

17.9.6 System capacity

With 8 orbital slots and 32 transponders per slot, the allocation described in Section 7.10.4 provides spectrum space for 256 transponders. Only three of these slots have full CONUS coverage, but the equivalent of an additional full coverage slot can be obtained by pairing the eastern slot with one of the western slots. Full coverage can then be achieved with the equivalent of 128 transponders, 96 single transponders and 32 transponder pairs. With video compression, these could transmit more than 1,000 NTSC signals simultaneously, the total depending on the amount of compression employed (see Section 17.9.8). The paired slots, of course, require the use of dual TVROs.

17.9.7 Transponder power

Each of the initial digital DBS satellites has a total power capacity of approximately 2kW. This can be configured to power sixteen 120-W or eight 240-W transponders, which compares with 40 watts for the transponders in typical K_u FSS satellites. With the 120-watt option, the maximum downlink EIRP in the United States with complete CONUS coverage is 54 dBW.

17.9.8 System architecture

The architecture of the uplink of a digital satellite broadcast system is shown in Figure 17.11. The video signal is first converted to a digital format by sampling, quantizing, and compression (see Chapter 3). Compression is

Figure 17.11 Digital DBS uplink architecture

followed by encryption—scrambling or otherwise making the signal unusable by all but authorized viewers. The signal is then divided into packets to provide the mechanism for error correction and multiplexing. The resulting video packets are interleaved with similar-sized digital audio packets.

The result is a complete audio-video bit stream for a single program source. It is then multiplexed with the bit streams for other program sources by further packet interleaving and the complete bit stream modulates the transmitter.

The architecture of the receiver is the converse of the transmitter.

The receiver is tuned to the desired program, a process that includes selecting the antenna polarization, the transponder, and the channel. The audio and video packets are separated, errors are corrected in the video bit stream, the encryption is defeated for the desired signal, the signal is decompressed and finally reconverted to the analog format.

17.9.9 System parameters

Typical parameters of a satellite digital transmission format for NTSC signals are shown in Table 17.8.

The 4 f_{SC} sampling rate with 8 bits per sample, giving a net data rate of 114.56 Mbs (for signal information only), are common choices for these parameters.

The gross data rate, which includes bits for error correction and "overhead" for the packet headers, (see below), can be computed from the packet structure. A packet includes 147 bytes—127 for the audio and video payload, 17 for error correction, and 3 for the header (see Figure 17.12). The header includes 12 bits to identify the program, 4 bits to identify the encryption, 4 bits to identify the use of the packet, and a 4-bit continuity counter.

The MPEG signal compression algorithm is the same as recommended for broadcast use. The system designer has a wide variety of choices for the

Table 17.8 Digital signal transmission parameters—NTSC signals

Parameter	Value		
A–D sampling rate (4 $f$sc)	14.32 MHz		
Bits/sample (2^8 brightness levels)	8		
Single-channel video net data rate (uncompressed, no audio or error correction)	114.56 Mb/s		
Single-channel gross data rate (includes audio and overhead)	132.60 Mb/s		
Packet length (see Section 3.7.1 and Figure 17.12)	147 bytes		
Compression algorithm (see Section 3.12)	MPEG-1		
Transponder data capacity (QPSK modulation, 24 MHz bandwidth) (see Table 3.4)	45.6 Mb/s		
Compression ratio (amount is optional. Maximum is determined by desired picture quality and image motion)	2:1 to 24:1		
Representative compression ratios:	9:1	12:1	24:1
Compressed bit rate (Mb/s)	14.73	11.05	5.53
Channels per transponder	3	4	8
Encryption for signal security (addressable)	"Smart Card"		
Error correction algorithm (see Section 3.8.3)	Reed-Solomon		
Multiplex format	Packetizing		
Modulation mode (see Section 3.7.2)	QPSK		

compression ratio. There is no sharply defined limit, and as it is increased there is a gradual deterioration of image quality and the ability to handle rapid motion. Compression ratios up to 24:1 usually do not significantly affect image quality, and even greater compression is practical.

The encryption method must control access not only to complete program services but also to individual programs, as in pay-per-view. One technique is to use a "smart card" on which the authorized program services or program

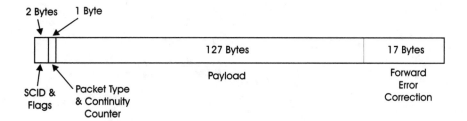

2 Bytes   1 Byte

| | | 127 Bytes | 17 Bytes |

Payload

SCID & Flags

Packet Type & Continuity Counter

Forward Error Correction

Figure 17.12 Digital DBS packet

times are recorded remotely by the program supplier. To view an authorized program, the smart cart also stores the necessary information to control the receiver's digital processing to undo whatever digital scrambling may have been used at the uplink.

17.9.10 Digital DBS for HDTV

The high quality of digital DBS signals makes it an attractive potential medium for HDTV. In fact, there is one view that digital DBS will be a catalyst that will lead to the successful introduction and growth of HDTV service. No transmission standards have been adopted or even officially proposed for digital DBS transmission of HDTV signals, but it seems almost inevitable that they will be in time. Since the Grand Alliance proposal for broadcast HDTV specifies a digital format (see Chapters 6 and 15), the adaptation of this format to satellite transmission would be relatively straightforward.

17.9.11 Digital DBS link performance

Noise in digital transmission circuits is the result of bit errors. Error correction can be so effective below a *threshold BER* that the noise contribution of the circuits is negligible. When the BER threshold is exceeded, error correction quickly becomes ineffective and the S/N of the signal deteriorates rapidly in a manner analogous to below-threshold FM signals. As with FM signals, the fade margin is the difference between the strength of the received signal under nonfading conditions and its level at the point where error correction fails. Most of the digital DBS receivers now in use provide a large fade margin even with small (18") receiving antennas.

18

Fiber-Optic Transmission Systems

18.1 Overview

Fiber optics is one of the many new technologies that are revolutionizing the television industry. Because its transmission medium is infrared radiation having an extremely high frequency in electromagnetic terms, fiber-optic transmission systems have enormous bandwidths—20 GHz is typical. They are especially adapted for digital signals although analog transmission is possible as well (see Section 18.5.2).

The cost of fiber optics now limits its applications to point-to-point transmission and CATV trunks (see Chapter 15), but as its technology advances and its cost reduces, it can be expected that it will be used for distributing voice, message, video, and audio service to individual homes.

Fiber-optic systems are simple in principle as shown in Figure 18.1. An optical energy source—a laser or LED—is modulated with the video signal. It illuminates the end of an optical fiber, and the energy is transmitted by

Figure 18.1 Fiber-optic system

Figure 18.2 Optical fiber construction. (*From Ira Jacobs, "Fiber-Optic Transmission Systems," Chapter 8 in A. F. Inglis (ed.),* Electronic Communications Handbook, *McGraw-Hill, New York, Figure 8.1)*

internal reflections within the fiber to its opposite end. Here it is detected and demodulated and the original baseband signal is recovered.

Unlike electromagnetic transmission systems, in fiber-optic systems it is necessary to demodulate and remodulate the signal at repeater points. This is not a serious problem for digital transmission, but analog transmission is only used in single-hop systems because of this.

18.2 Fiber-Optic Construction and Operation

18.2.1 Construction

A fiber (see Figure 18.2) consists of an exceedingly fine glass filament that is surrounded by a concentric *cladding*, also made of glass but having a slightly lower index of refraction. The index-of-refraction profile for single-mode fiber is known as the *step index*. The index for the multimode fiber in Figure 18.1 is carefully shaped because it determines the bandwidth and frequency response of the fiber (see Section 18.4.2). It is known as *graded-index* multimode fiber.

The diameter of the fibers depends on the transmission mode (see Section 18.3). The cores of *single-mode* fibers typically have diameters of 7 to 10 μm while *multimode* cores range from 50 to 85 μm. One standard for multimode

Figure 18.3 Fiber-optic cable. (*From Donald G. Fink (ed.),* Electronic Engineers' Handbook, *McGraw-Hill, New York, 1989, Figure 22.25a*)

fibers is a 50 μm core and a 125 μm cladding. These diameters are comparable with those of human hair.

A fiber-optic cable consists of one or more fibers bundled together and surrounded by a metallic or plastic sheath for mechanical protection. The bundle may include a steel wire for mechanical strength (see Figure 18.3). In one configuration, 12 groups, each containing 12 fibers, are bundled together. Needless to say the bandwidth and hence the communication capacity of such a cable is enormous.

18.2.2 Operation

Infrared radiation is injected into the end of the fiber and is totally reflected when it strikes the cladding with its lower index of refraction. It is contained within the core and continues down the fiber until it is completely attenuated by losses in the glass.

18.3 Single and Multimode Fibers

The operation of single-mode and multimode fibers is illustrated in Figure 18.2. The single-mode fiber has a smaller diameter, and only one mode, a

Figure 18.4 Fiber-optic attenuation. (*From Ira Jacobs, "Fiber-Optic Transmission Systems," Chapter 8 in A.F. Inglis (ed.),* **Electronic Communications Handbook,** *McGraw-Hill, New York, Figure 8.3*)

single path down the fiber, is possible. The *shortest* wavelength that can be transmitted in a single mode is:

$$\lambda_{min} = 2.6r(n_1^2 - n_2^2) \qquad (18.1)$$

where: r is the radius of the core and

n_1 and n_2 are the indices of refraction of the cladding and core

Larger diameter fibers permit multiple reflections and multimode operation is possible. Each mode has advantages and disadvantages.

Single-mode fiber requires far greater mechanical precision in splicing. A lateral offset of only 1.5 μm in the ends of the fiber cores results in 0.5 dB attenuation. An offset of 7 μm would be required to produce the same loss in multimode fiber.

On the other hand, at wavelengths in excess of 1.1 μm, the attenuation of single-mode fiber is substantially less (see Figure 18.4). Also, single-mode fiber is not subject to *modal dispersion* (see Section 18.4), which would reduce its bandwidth. The practical result is that single-mode fibers are commonly used for wider bandwidth (higher speed) applications and for longer paths.

18.4 Attenuation and Dispersion

18.4.1 Attenuation

Both single-mode and multimode fibers suffer from attenuation as shown in Figure 18.4 for a representative glass type. For this glass, the attenuation of single mode fibers reaches a low of about 0.2 dB per km at wavelengths in the 1.5–1.6 μm infrared range. Long distance signal transmission by fiber-optic cables was made possible by the development of energy sources in this region of the spectrum together with the availability of glass with lower attenuation and the development of techniques for splicing the tiny single-mode fibers.

18.4.2 Modal dispersion

The dispersion encountered in fiber-optic circuits results from differences in the arrival time of different rays or different frequency components. It can be caused by differences in the path length of different modes, *modal dispersion*, or by differences in transmission speed for different wavelengths, *chromatic dispersion*.

The effect of dispersion is to cause pulses to spread, and this limits the maximum pulse rate. The effect can also be described as a bandwidth limitation. The relationship between bandwidth and pulse spreading is given by Eq. (18.2):

$$B = \frac{0.44}{i} \tag{18.2}$$

where: B = bandwidth in MHz

          i = the width in microseconds between the points on the edges of the pulse where it is one-half its maximum value

The bandwidth of multimodal fiber, when determined by modal dispersion, is a function of the wavelength and the index-of-refraction profile. The profile can be adjusted so that the index of refraction is greatest and the transmission velocity is the least on the shortest paths. This partially equalizes the transmission time, reduces the pulse spread, and increases the bandwidth for a given circuit length. Figure 18.5 shows three examples of bandwidth×length vs. wavelength for different index-of-refraction profiles.

18.4.3 Chromatic dispersion

Chromatic dispersion affects both single-mode and multimode fibers, and it is the principal bandwidth limitation for single-mode fibers. It occurs because fiber-optic light sources are not monochromatic, and the transmission velocity varies with the wavelength.

Figure 18.5 Fiber-optic bandwidth. (*From Ira Jacobs, "Fiber-Optic Transmission Systems," Chapter 8 in A.F. Inglis (ed.),* **Electronic Communications Handbook,** *McGraw-Hill, New York, Figure 8.3)*

Two effects are present that affect the variation of transmission velocity with wavelength. The first is *material dispersion* due to the variation in the refractive index of the fiber with wavelength. The second is *waveguide dispersion,* which is caused by a graded index fiber in which the index of refraction varies across its diameter. Material and waveguide dispersion are in opposite directions, and the waveguide dispersion can be controlled so that the variation in delay is equal to zero at some wavelength. The wavelength at which this occurs can be adjusted within limits by appropriate choice of the refraction index profile.

Figure 18.6 compares the dispersion in picoseconds per km for each nanometer of source spectral width for two fibers, one with a uniform index of refraction across the core and the other with an index profile that has been adjusted to move the zero dispersion wavelength from 1.3 μm to 1.6 μm.

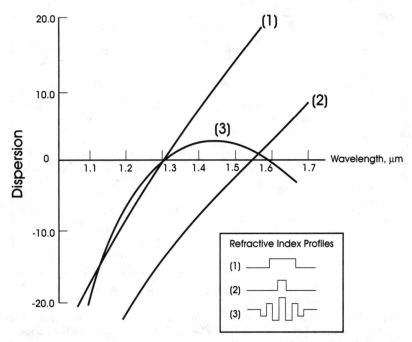

Figure 18.6 Spectral dispersion. (*From Ira Jacobs, "Fiber-Optic Transmission Systems," Chapter 8 in A.F. Inglis (ed.),* **Electronic Communications Handbook,** *McGraw-Hill, New York, Figure 8.4)*

18.5 Signal Formats and Modulation Modes

18.5.1 Digital format

The digital format is uniquely suited to the characteristics of fiber optics. Fiber-optic systems have a surplus of bandwidth but are limited with respect to power and receiver sensitivity. *Pulse code modulation*, PCM, trades bandwidth for signal-to-noise ratio, and is almost universally used except for short single-hop paths. Present optical sources are not practical for frequency-shift or phase-shift keying.

Digital video signals with any of the standard bit rates described in Chapter 3 can be transmitted by fiber optics. They generally employ a simple binary code (see Chapter 3) with *on-off keying*, OOK. Other modulation modes such as *quaternary pulse-position modulation*, QPPM, can be used to effect a different trade-off between S/N and bandwidth. Modes involving phase-shift or frequency-shift modulation are not practical for use with an optical carrier.

Figure 18.7 Range of analog systems. (*From Ira Jacobs, "Fiber-Optic Transmission Systems," Chapter 17 in A.F. Inglis (ed.),* Electronic Communications Handbook, *McGraw-Hill, New York, Figure 17.19)*

18.5.2 Analog format

The analog format is more economical because A-to-D and D-to-A conversion is not required, and it is practical for some single-hop systems. Amplitude modulation can be used with intensity modulation of the optical source. Alternatively a subcarrier that is frequency modulated by the signal can be created by intensity modulation of the light source. This eliminates problems of linearity and provides an increase in signal-to-noise ratio by the FM improvement factor (see Chapter 17).

Frequency modulation has the additional advantage that several video signals can be multiplexed on the same fiber by the use of FDM-FM, frequency division multiplex.

The range of a single-hop system is sometimes known as its *reach*, which is established by the signal-to-noise ratio. This calculation for multichannel FDM-FM systems is complex and beyond the scope of this book. The reach of these systems is indicated in Figure 18.7, which shows the S/N based on empirical results.

Table 18.1 Lasers and LEDs

Parameter	Laser	LED
Light output	6 dBm	6 dBm
Coupling loss	3 dB	20 dB
Spectral width		
At 0.8 mm	2 nm	40 nm
At 1.3 mm	4 nm	100 nm
Maximum modulation rate (approx)	5 Gb/s	500 Mb/s
Feedback power control	Yes	No
Temperature sensitivity	Strong	Weak
Failure mechanisms	Many	Wears out

18.6 Path Performance, Digital Format

The measure of performance for a digital transmission system is the bit error rate rather than the signal-to-noise ratio as in analog systems. This is determined on a statistical basis and is expressed as a *bit error probability*.

Bit errors can be caused not only by system noise but also by *signal-dependent noise* including such noise sources as intersymbol interference and improper setting of decision thresholds. In total these are known as *eye degradation*. The calculation of bit error probability in fiber-optic systems is beyond the scope of this book, and the reader should consult the references.

The ultimate criterion of the performance of fiber digital transmission systems is the required repeater spacing for multi-hop systems. This is not an exact calculation because it must be based on statistical results and it will depend on the performance standards. Approximate maximum spacings for single and multi-mode fibers with laser and LED optical sources (see Section 18.7) are shown in Figure 18.8. This graph clearly shows the increase in range that has resulted from the development of laser sources in the 1.3 to 1.6 μm range and the introduction of single-mode fibers.

18.7 Transmitters for Fiber-Optic Systems

Two types of optical transmitters are used in fiber-optic systems, lasers and light emitting diodes (LEDs). Both are semiconductors, but there is a fundamental difference in that the radiation from lasers is *coherent*, that is it retains its wavelength and phase over a period of time.

The use of lasers was limited at first because they were not available in the 1.5 to 1.67 μm wavelength range, the area of lowest attenuation (see Figure 18.4). The development of lasers that operated at these wavelengths was a significant breakthrough in fiber-optic technology.

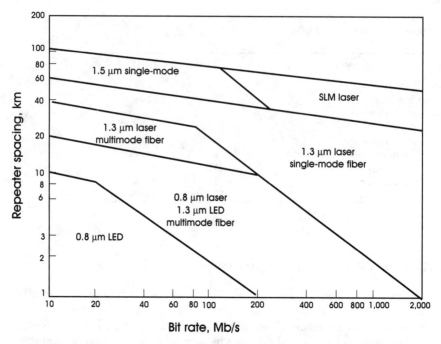

Figure 18.8 Repeater station spacing. (*From Ira Jacobs, "Fiber-Optic Transmission Systems," Chapter 8 in A.F. Inglis (ed.),* Electronic Communications Handbook, *McGraw-Hill, New York, Figure 8.17)*

Some important comparisons of lasers and LEDs are summarized in Table 18.1. In both of these examples, the same output power is emitted but the coupling of the power into the fiber is far more efficient for lasers.

The spectral width of laser emission is much narrower, an important advantage because it minimizes the loss of bandwidth due to chromatic dispersion.

The maximum rate at which lasers can be modulated is greater by a ratio of about 10:1. However, the maximum rate for LEDs—approximately 500 Mbs—is high enough for most purposes.

The output power of lasers can be stabilized by feedback, an operation that is not possible with LEDs. On the other hand, LEDs are more rugged, less sensitive to temperature variations, and probably have a longer life. As a result of these differences, LEDs are used in short single-hop path lengths while lasers are used for long paths and multihop systems.

Although the spectral bandwidth of laser emission is comparatively narrow, it is not monochromatic, and its fine-grained spectrum shows a number of discreet spectral lines. Figure 18.9 shows a representative example of the spectrum of a laser emitter at different power levels.

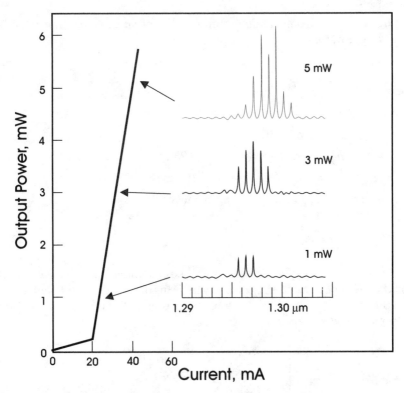

Figure 18.9 Laser emitter spectrum. (*From Ira Jacobs, "Fiber-Optic Transmission Systems," Chapter 8 in A.F. Inglis (ed.),* Electronic Communications Handbook, *McGraw-Hill, New York, Figure 8.7)*

With a spectral distribution of this width, the bandwidth is reduced as the result of chromatic dispersion (see Section 18.4.3) and this can be a significant problem on long-distance high-speed circuits. For these circuits, the output spectrum can be limited to a single spectral line, often 1.55 μm, by the use of an external cavity or a resonant structure within the laser. The result is a *single longitudinal mode* or *SLM* laser.

18.8 Receivers for Fiber-Optic Systems

A fiber-optic receiver consists of a *photodetector* that converts an optical to an electrical signal, followed by stages of electronic amplification. The critical element in the receiver is the photodetector since it determines the receiver's sensitivity.

The sensitivity of a fiber-optic receiver is defined as the minimum received power, P_R, required to achieve an error probability, P_E, at a bit rate B. P_R

can be calculated from the minimum number of *photons*, n_p, that must be received per pulse:

$$PR = \frac{n_p Bhc}{\lambda} \qquad (18.3)$$

where: λ = wavelength

B = bandwidth

h = Planck's constant = 6.6252×10^{-34} Joule-sec

c = velocity of light = 3×10^8 m/s

n_p is calculated from the error probability:

$$PE = \frac{1}{2} e^{-n_p}$$

The most common photodetectors are the *avalanche* photodetector, APD, and the *PIN junction*. Their sensitivity in terms of the number of photons/bit and the received power level for a bit error probability of 10^{-9} for PIN and APD photodiodes is shown in Figure 18.10.

Figure 18.10 Receiver sensitivity. Photons per bit required at the receiver to achieve 10^{-9} bit error rate. Translation to received power level (in dBm) shown by dashed lines. (*From Ira Jacobs, "Fiber-Optic Transmission Systems," Chap. 8 in A.F. Inglis (ed.),* Electronic Communications Handbook, *McGraw-Hill, New York, Figure 8.8*)

The demands on the performance of a fiber-optic receiver are severe. As an example, consider an APD receiver operating at a bit rate of 100 Mbs. From Figure 18.10 it can be seen that it would require 10^2 photons per bit or 10^{10} photons per second to achieve a bit error probability of 10^{-9}. If each photon were converted to an electron by the photodetector (an optimistic assumption), the current would be 1.6×10^{-9} amperes or 1.6 nanoamperes. This compares with an output signal of more than 100 nanoamperes for a typical photoconductive image tube in broadcast cameras. A current of this level requires a high degree of amplification in the electronic system following the photomultiplier.

18.9 Passive Components

A number of passive devices, including connectors, optical attenuators, couplers, and wavelength-division-multiplexing (WDM) filters, some necessary and some optional, are available to complement the active devices and complete the system.

18.9.1 Connectors

The design of connectors that could place and maintain the ends of very tiny fibers in almost perfect alignment at joints has been a challenging engineering assignment. The success of single-mode fiber was made possible by the development of tools and techniques for splicing extremely thin fibers with precision. Permanent connections are bonded while demountable connectors utilize cylindrical bushings or conical plugs. Techniques have been perfected that have reduced the losses at splices to 0.05 dB or less.

18.9.2 Optical attenuators

Because of the wide amplitude range of received signals, a means of adjusting the optical input power is required for the operation of practical systems. This is accomplished by attenuators, both fixed and adjustable neutral density filters.

18.9.3 Couplers

Couplers divide the power in a single fiber between two or more paths. This can be accomplished with a Y junction or, if only a small amount of power is desired in one of the branches, by putting a sharp bend in the fiber so that reflection is less than total and the power that escapes is captured by a second fiber.

18.9.4 Wavelength-division-multiplexing filters

The purpose of wavelength-division-multiplex filters is to separate the radiation from the optical source into radiation bands, each of which forms a signal channel that is modulated separately. The channels are multiplexed and transmitted by a single fiber. This is a very effective technique for utilizing the large bandwidth of fibers to increase their information handling capacity.

The filters are thin-film dielectric materials that transmit one wavelength and reflect the other—as with the dichroic surfaces used in color television cameras.

The Future of Video Systems

19.1 Introduction

The goal of video engineers is to create systems that speak to the human senses of hearing and vision to entertain, educate, or convey information. The accomplishments of the first 50 years, described in the preceding chapters, are ample proof of their success. But what lies beyond that?

Digital HDTV is just around the corner—its technical capabilities have been proven—but success depends on many other factors that are not yet concluded. Will consumers buy HDTV? Will programs be produced for it? Will it replace present TV? How soon will it happen? These questions and many others probe the combination of technology and marketing factors that determine a new product's success or failure. This chapter explores those questions and other issues that may affect the path of video engineering into the next century.

A warning to the reader: This chapter contains forecasts of the future, which must inherently be shaped by the writers' opinions since there is nothing factual about the future until it becomes present or past. Therefore, the reader is free to disagree as much as he or she pleases. However, this discussion should prove stimulating and may trigger some surprising new thoughts on the part of the reader.

19.2 Technologies in Video Engineering

One of the interesting aspects of video engineering is the breadth of technology that it involves (see Table 19.1). No single engineer can be master of all these skills, but a senior video engineer must be familiar with them all at a

Table 19.1 Video engineering technologies

Technology	Applicability
Audio	Creation, recording, receiving
Colorimetry	Cameras, displays
Communications	Transmission, reception, recording
Digital	Everywhere
Ergonomics	Cameras, recorders, control panels, user interfaces, receivers
Materials	Cameras, recording, transmission
Mechanical	Cameras, tape transports, enclosures, antennas
Optics	Cameras, recorders, displays, transmission
Solid-state electronics	Custom LSI, everywhere
Systems	Everywhere

level sufficient to understand their interactions and when more in-depth expertise may be required. This list of technologies is also useful when considering trends for the purpose of forecasting what may happen in the near future. The following sections discuss such considerations.

19.2.1 Trends in solid-state devices

The progression of solid-state technology described in Figure 7.1 is expected to continue well into the next century. More circuits per chip, larger chips, lower cost per device, all result from this. General-purpose devices such as microprocessors and RAM chips will continue to become more powerful and lower in cost. There also will be progress in improving and simplifying the techniques for making custom chips so that more discrete functions can be integrated on custom chips. This will continue the trend of reducing size and cost of the electronic portions of all products. The result will be that products can employ more sophisticated electronic functions while, at the same time, they become less expensive. The current introduction of digital HDTV depends on this trend for its market success.

The trend of solid-state technology underlies much of the progress in video systems and their use. As systems simultaneously become more powerful and cheaper, the range of practical application of video is broadened. Applications which could not afford video a few years ago may soon embrace it simply because it has become the most cost-effective approach. One class of application that may increase its use of video is education. However, the development of courseware is an important element in educational applications and the cost of doing that must also be brought in line with the lower hardware costs. That is a technical and marketing challenge.

An important component of video systems is the imager. The trends of solid-state technology also apply to CCD imagers and higher resolutions at lower costs will continue to occur. This is a driver that will result in improved image quality of video systems at all cost levels.

19.2.2 Trends in digital technology

Most features of the new digital video systems are based on principles that were developed in the computer industry years ago. They have simply become practical for digital video because of continued cost reduction of digital functionality. However, because digital video products are new, there is a lot of room for improvement and the process of refinement will go on for some time. An overwhelming breakthrough is unlikely in the near future, but there will be new applications introduced and nearly all analog functionality in video systems will be replaced by digital devices over the next five or ten years.

In the digital era, hardware engineers are even more likely to say "we can do *anything*," but the practicality and success of new products will depend more on their market suitability and presentation as much as or more than their technology (see Section 19.3.2).

19.2.3 Trends in optical technology

Communication, cameras, recorders, and displays depend on optical technology. Great improvement in materials and devices has been made, but optical technology is still expensive for many uses. The following paragraphs discuss each of the application areas listed above.

19.2.3.1 Optical communication As digital systems demand higher bandwidths, the alternatives to optical transmission technology begin fading away. This is resulting in many applications for optical communication despite the cost and it is increasing the momentum of optical development. Costs for fiber-optic cable and transducer devices will continue to reduce and their volume of use will increase. This will be one of the drivers of the information superhighway (see Section 19.4).

19.2.3.2 Camera optics Camera lenses and prisms are mature components that are simply undergoing slow refinement and cost reduction as the volume of use increases. CCD imagers are adopting their own optical features, such as the on-chip lenses that increase sensitivity of interline-transfer devices (see Section 8.3.3).

19.2.3.3 Optical recording Although the laser optical videodisc has been on the market for more than 20 years, it never became a mainstream product. The world waited until the digital audio Compact Disc (CD) to embrace an optically-recorded product. Since then, various spinoffs have been developed (see Section 9.18.1), mostly for computer applications. A standardizing effort is currently underway to create a CD spinoff for digital video program distribution. This would potentially compete with prerecorded videotapes for the home video play-only market. However, the video CD requires a wholly new infrastructure for program replication and consumer playback, and that could take a long time to appear (see Section 19.5.2).

In any case, the video CD brings with it a substantial advance in optical recording density, and that will be applicable to other uses such as a high-capacity CD-ROM, which does not require the massive consumer-market infrastructure to get started. It is easy to predict that optical recording density will increase by as much as 10-fold in the next few years, but it's harder to say where it will make the greatest market penetration.

19.2.4 Trends in display technology

In both computers and television the trend is to larger, higher-resolution displays. Although CRT technology is currently dominant, flat panel displays have highly desirable features for the market and much research and development is going into this area. The portable computer market is an effective proving ground for this work.

As HDTV comes on line in the next few years, the pressure on new display technologies will increase even further because the HDTV market will need displays that are at or beyond the practical limits of CRT size. This may be the opportunity for flat panels to move into the television market in a big way. At this point, many laboratories are addressing the problems of large, high-resolution flat panels. If they are successful, it could provide the breakthrough that the display market needs.

When display components are readily available in the HDTV 16:9 format, they also will be used for personal computers. Thus, the market for 16:9 flat panel displays will be even larger than the TV market.

19.2.5 Trends in audio technology

The audio market was the first consumer application of digital technology. The audio CD has nearly taken over the prerecorded audio market from records and cassettes. This success is a result of the superb quality and reliability of digital technology and the convenience and durability of the optical medium.

In audio production and postproduction, digital audio is also taking over as both digital tape and hard disk recorders become available. In the long run, hard disk recorders will probably prevail because of their fantastic editing features. Tape will become a backup and archiving medium, much as it has done in the computer business.

19.2.6 Trends in recording technology

Magnetic recording is the primary technology for video recording, but optical recording has become important for audio and may soon be an important video medium as well. As shown in Figure 9.9, magnetic recording density has steadily improved over the years, resulting in recorders that are smaller and less expensive to operate. Magnetic video recording has also made the transition from analog to digital and new video recorders in all fields should be digital from now on.

Further advances in both magnetic and optical recording are certainly expected—the density curve should continue to go up and optical recording costs should continue to come down.

19.2.7 Technology conclusions

The bottom line of technology trends is that all the technologies of video engineering have potential for improvement and none are up against any brick walls. In several places (recording and displays), there are even possibilities for a breakthrough. Thus, the deployment of video systems in the foreseeable future should not be limited by technology considerations. But there are other factors. . .

19.3 Video Systems

A video system is a set of equipment, software, and connections that provide a specified end-to-end service to its users. Examples of video systems are:

  television broadcasting

  cable television

  prerecorded videotapes

  CD-ROM multimedia

  satellite broadcasting

  picture-phone

Each of these systems includes all the ingredients shown in Figure 1.1 for creation, transmission, and reception. These elements are called the *infrastructure* of the system, which provides program content material, transmission paths, consumables (such as tapes), and the means whereby users acquire the necessary equipment. Systems also require *standardization* so that signals and equipment will be compatible between different sources and manufacturers. Video system considerations are discussed below.

19.3.1 Video systems engineering

There are two levels of video systems engineering. The first is the process of creating an entirely new system using available technology to provide a new service. This is discussed in the following sections. The second class of systems engineering is the specification of a video system using existing standards and specifications, for example, the design of a new broadcasting station. This type of systems engineering will not be discussed further.

19.3.2 Developing a new video service

The first step of creating a new video service is the specification of what that service will do and who the users will be. This is actually the most crucial part of the entire development project because it determines where the market is and the potential market value of the service. These two things have to be defined and understood at the outset and all subsequent steps of the development must be consistent with these objectives both in content and cost. Many projects are doomed to failure because this market focus doesn't exist or it gets lost; it's too easy to let the technological issues overwhelm the project to the detriment of the market objectives.

Once the service objective makes sense in market terms, technological choices can begin so that the entire development can be planned and cost estimated. All of the considerations of infrastructure must enter at this stage and be included in the projected cost. The cost of infrastructure can often be greater than the equipment development cost, so much so that no single entity can handle it. This leads to the "chicken and egg" syndrome, where equipment cannot be sold until the necessary infrastructure is in place, but no one wants to invest in infrastructure until equipment is selling. Many new systems face this dilemma.

A classic example of "chicken and egg" was NTSC color television, which was standardized in 1953 but it took more than 10 years after that before receivers began selling at a level that generated any profit. During that 10-year gestation period, investors in color TV were acting on faith because there were no profits. If they hadn't stuck it out, color TV as a worldwide service might never have happened.

Most new systems require a good strategy to overcome "chicken and egg." For example, the new home digital cassette system (DVC—see Section 10.9) is being introduced first in camcorders, which are usable by consumers even though there are no prerecorded tapes available. When the value of the system has been established by camcorder sales, DVC players can be introduced to play tapes made in camcorders and then prerecorded tape suppliers may begin offering the DVC format. Eventually, DVC may replace all analog home recorders.

19.3.3 Standards

The purpose of standards is to remove technology decisions from the marketplace. When a product is properly standardized, products, signals, and media from all manufacturers or sources will work interchangeably. The user does not have to worry about compatibility issues and he/she can base buying decisions on other considerations such as features, performance, and price. The value of the user's investment in equipment and media is also increased because there is a wider range of use for the system.

With standards, manufacturers also know exactly the requirements their product must meet to guarantee compatibility with other products in the field. Well-conceived standards define only what is necessary for compatibility and still allow manufacturers to develop their own competitive edges within the standard.

However, standards for new systems are not easy to achieve because the developers of new systems do not want to compromise their competitive position by sharing information with potential competitors in order to achieve standardization. But history has shown time and again that offering incompatible products to the market for the same service will, in the long run, delay market acceptance, alienate customers, and produce one or more losers in the manufacturing community. The battle in the initial VCR marketplace between Beta and VHS is a good example. When preparing to introduce a new service, it makes no sense to proceed without achieving a standard among potential competitors.

Fortunately, there are several industry organizations who will develop standards and the industry will also form an ad-hoc organization to create standards for an important new service. A recent example of the latter approach is the Grand Alliance HDTV system. The SMPTE has been very active in video recording standards since the original quadruplex days, and essentially all video recording formats are standardized by the SMPTE.

19.4 Communications

Video systems all involve some means to move video or other information from one location to another—communication. This may be by means of cable, broadcast, fiber optics, or by physically moving a recorded medium package. In the digital era, communication consists simply of moving bits and performance is equated to a sufficient bit rate and bit error rate for the intended service.

A digital communication path can be used for many types of information simply by changing the meaning of the bits being communicated. This opens the possibility of general-purpose digital communication services that will convey bits for any purpose and for anyone, just like the present telephone network offers voice communication to anyone who is connected. The idea of a high-speed two-way digital communication network connecting homes and offices around the world has been given the popular name *information superhighway* (ISH). It could become the universal medium for communication of voice, video, business data, and anything else worldwide. It is obvious that whoever could control such a network would have tremendous political and economic power and for this reason many companies and countries are pursuing ways to build such a system. The investment required is enormous and it is unlikely that a single entity could pull it off—it is going to take an international consortium of some type. At this time, it is not clear how that is going to happen, but the idea is so compelling that both industry and government are continuing to work on it.

One system already in place that partially fits the model of the ISH is the *Internet*, which was originally developed as a computer network for military and academic purposes but is now a public network. Some have proposed that the Internet might evolve into the ISH, but that is only part of the solution because the Internet is based on the existing telephone system for its communication paths, which severely limit data rates. Of course, telephone systems are beginning to offer high-speed digital services, but the physical network could not support widespread use of such service until upgrading to fiber optics is accomplished. That is being explored, but it will take many years to complete.

Another contender for the ISH is CATV service. Cable systems have analog bandwidths of up to 300 MHz, which would be capable of handling high data rate digital signals. The Grand Alliance HDTV proposal contains means for transmitting 38.6 Mb/s in each 6-MHz TV channel of a cable system, which would give a 50-channel cable system a total digital transmission capacity of 1.93 Gb/s and it is already connected to homes! But again, that is only part of the solution because cable systems are local entities that are not interconnected worldwide and the protocols are not developed for general digital communication via cable service.

Still other potential ISH providers are satellite communication companies, who have existing capability for high-speed digital communications worldwide. Again, however, the protocols for using satellite distribution for general digital communication are not developed, and the cost of connecting everyone to a satellite system is high, especially when two-way communication is considered.

The above discussion shows that no one has a complete handle on the ISH today—it is an exciting idea that does not yet have a means of fulfillment. However, there is much activity in industry and government and a solution will probably evolve in the next five to ten years.

19.5 New Video Systems

Although the march of technology will help the existing video systems to improve and become less expensive, it is more interesting to explore what new systems might become possible. Many possibilities have been discussed in the press and the literature, usually based on the technology aspects only. As mentioned above, the business and market considerations are much more difficult to develop and these will probably prevent some of the new ideas from ever becoming commercial successes. However, here are some of the ideas.

19.5.1 HDTV

A technical proposal for digital HDTV is available from the Grand Alliance in the United States. As described in Chapter 6, it offers exciting features and it can operate over existing broadcast and cable channels. Although it is offered as an eventual replacement for present analog television systems, it is not compatible in that the public must buy new receivers to enjoy the service. Likewise, program producers must buy new origination equipment to produce HDTV programs. This is worse than the "chicken and egg" situation that existed at the start of color TV in 1953, where at least there was enough compatibility that existing receivers could view color programs in monochrome. With digital HDTV, existing receivers can receive nothing on the HDTV channels, so HDTV program producers are addressing only viewers who have new receivers.

Existing analog program material can be converted to digital HDTV format for broadcasting but the picture and sound quality will not be much better than analog broadcasting. This will not create much incentive for consumers to buy new, more expensive, HDTV receivers. However, HDTV manufacturers will surely be very creative in building features into digital receivers that will be usable even without much actual HDTV program material. But the receiver base is likely to grow very slowly.

Program producers have been improving their picture quality recently by using digital component video systems in production and postproduction. This is advantageous to them because it allows programming to be converted to and released in various formats, such as NTSC, PAL, videotape, even film. That increases the market value of their programming. Similarly, programs produced in existing digital component formats can be converted to HDTV with improved picture quality. The problem with this is the aspect ratio, which cannot be fully converted. Material produced for 4:3 viewing must either be shown on HDTV with black areas at the edges or it must be enlarged to full width with loss of information at the top and bottom. Either way, some of the quality advantage of the component origination is lost.

A source of program material that is valuable for HDTV is motion picture film that has been produced for wide-screen theater viewing. With a digital HDTV telecine, this material converts beautifully to HDTV. To the extent that motion pictures can be shown on TV (due to rights considerations), films are a source of programming that can fully exploit the performance of HDTV. Some TV programming is actually produced in wide-screen format for possible showing in theaters; in this case, the 4:3 aspect ratio TV viewers have to take the disadvantage of aspect ratio conversion, but the material will show perfectly on HDTV.

HDTV is exciting, but it suffers from initially high costs and poor availability of full-quality program material. Market acceptance will occur, but it will be slower than most participants desire.

19.5.2 Video CD

Another new service that potentially can replace an existing one is the new video CD, which is currently facing standardization between two competing formats. The video CD is a prerecorded digital format that extends the video playing time and picture quality of the existing CD-ROM so that it could be a future replacement of prerecorded videotapes. It would be higher quality, lower cost, more durable, and more convenient than videotape, but it requires an entirely new infrastructure in that neither disc replication facilities nor players now exist in the marketplace.

The investment in disc replication facilities is major, but possible for some large companies, and existing program material can be reprocessed so that a lot of content can be made available when players are introduced. At the start, video CD players will probably be more expensive than videotape players, but early adopters and quality-conscious consumers will probably buy them, so the "chicken and egg" problem should not be too great. The situation can be compared to what existed at the introduction of the audio CD, which took over the recorded audio market in a matter of five years after introduction.

The video CD will likely spawn a new generation of CD-ROM players in the computer market, which will provide additional outlets for players and discs. Since the computer market usually adopts new technologies faster than the consumer market, this will further speed the acceptance of video CD. All of this depends on a single standard being adopted before product introduction—if that doesn't happen, market acceptance will be much slower.

19.5.3 The Information Appliance

With television becoming digital in the form of digital HDTV, a combined television and computer product is technically feasible. The same product could be used either for computing or TV viewing or, through on-screen windowing, both at the same time. This would be valuable to some customers, but for the bulk of the TV market, it is probably not desirable. That is because of the basic difference between the viewing environment for computers and television. Computers are operated at viewing ratios as low as 1:1 because the single user sits close to the screen. The typical television viewing situation has several people sitting across a room from the display, so viewing ratios are much higher—4:1 or more. This means that a TV display must be much larger so the viewers can see all the picture detail from across the room. (This statement assumes that HDTV and computers have similar resolution performance, which will be true in the HDTV era.) Because serious computing usually has only one user at a time and other materials such as books and papers are often required, computers are typically set up on a desktop rather than in a living room environment. A single display configuration cannot serve both uses.

There is another, broader, version of the information appliance that may have more potential, although it will be more difficult to develop and sell. That is the Home Information System (HIS), which is a digital network in the home that integrates video, audio, computer data, and home control into a single system.

Home networks have been marketed before with little success because they were proprietary systems to one manufacturer, they did only one of the above functions, and they were too difficult for the consumer to understand, install, and operate. The HIS faces the same problems, but its value to the user is greater because it integrates all home communications and with standardization it can be nonproprietary so multiple manufacturers can compete for the parts of the system. The understanding, installation, and use problems can be overcome with proper design of hardware and software.

Figure 19.1 shows a concept for the HIS, configured for a family of four.

19.5.3.1 The HIS server The heart of the system is a powerful PC that provides interface to the outside world, storage of local data, and network control—this is called a *server*. The server has its own large-screen flat-panel display—probably located in the kitchen. All displays are in HDTV 16:9 aspect format, although they may have different resolutions. The kitchen display might be fairly high resolution and it could serve as the family bulletin board (like the notes stuck on the front of the refrigerator), a local TV display, and the entry point for home control, system control, door and phone answering, etc. The server has a multitasking operating system so it keeps the rest of the system going while family members interact with its display, it could even support more than one person using it at a time by having multiple mice with different cursors operating on the same screen. If the server display is full HDTV resolution (1920×1080) and roughly 30 inches wide, separate windows could be simultaneously displayed for each of its functions.

19.5.3.2 ISH input/output The server accepts the home's connection to the ISH and processes data between it and other locations within the home. If telephone connections are separate from the ISH, they also are handled by the server. For telephone, the server performs answering service, voice mail, intercom, and routing of phone connections. If future phone connections are via the ISH, the server provides the same services from that source.

Video coming in over the ISH is either stored in the server or routed to another location for decompression and immediate display. Because the home network handles only compressed video, displays that show video must have decompression capability built in. The displays generally also have capability

Figure 19.1 Home information system of the future

to display data windows along with video, so the communication and control capabilities of the system can be accessed from any location.

At least one video digitizing input is available on the server. It has a video switcher so that the one digitizer can support cameras for door answering, room surveillance (baby's room, for example), and local camcorder use. The server stores the digital video or routes it over the home network to any location that requests display.

19.5.3.3 HIS displays Small video displays are used everywhere except the server and the living room. The small displays are lower resolution, such as 854 × 480, but the server and living room displays are 1920 × 1080 for showing full-resolution HDTV or multiple computer windows. All displays have a button that can be touched or clicked to gain access to the server menus for program selection, communication, or control. Displays also have touch-screen, wired mouse, infrared remote mouse, and/or keyboard as required. To support these operations, all displays must have a local CPU, I/O, and memory, but they rely on the server for mass storage.

If the home includes a home office computer, that is probably a stand-alone PC, but it should be connected to the home network to utilize the I/O and storage capabilities of the HIS server.

19.5.3.4 Developing HIS The HIS concept is something that will require concerted industry effort to accomplish because of the number of different products that are involved. That is unlikely to happen all at once, but it might begin with a standardizing task force that defines the interfaces needed by such a system. With that done, manufacturers can make their own decisions about providing those interfaces in their products, either built-in or as options. Software developers can work on approaches to create the server device using a standard computer as the platform and system marketers can begin promoting the system concept. This way, the system is achieved in an evolutionary way.

In today's terms, the HIS looks like an expensive concept and it is. However, with the steady cost reduction of system components that can be counted on in coming years, it won't look that expensive. Also, homeowners can begin with the server and add the peripherals over time to spread the cost over a number of years. In the end, it still depends on whether homeowners will buy the concept.

19.6 Conclusion

As shown in these examples, the digital video era offers many possibilities, including super-quality television (HDTV), high-quality home video production (DVC), worldwide video communication (ISH), high-quality prerecorded video (video CD), and computer and TV integration (HIS). The next five or ten years should be a very exciting and challenging time for video engineers as they help to bring these or other new developments to fruition.

Glossary

3:2 pull down A technique for presenting 24 frames/sec. motion picture film on a 30 frames/sec. interlaced television system. One film frame is shown for three television fields and the next film frame is shown for two TV fields.

A–B roll editing An editing method that uses two source recorders (A and B) running synchronously and combining their outputs through a video switcher to a third recorder in the Record mode.

A–D See *analog to digital conversion*.

active-matrix LCD A *liquid-crystal display* that has a transistor array behind the panel to substantially improve the speed and contrast of the display.

acuity, visual The ability of the human eye to distinguish fine detail. It is usually expressed in seconds of arc.

adaptive differential PCM (ADPCM) A form of *differential PCM* that adjusts its parameters dynamically according to the characteristics of the signal being encoded.

additive color A color reproduction system that mixes colored lights (usually red, green, and blue) to display an image. (See also *subtractive color*.)

address In computers, the use of a digital word to specify a location in a memory array. For example, 16-bit addressing can specify 65,536 locations.

ADPCM See *adaptive differential PCM*.

algorithm The specification of the steps that perform a process, such as drawing a circle or compressing an image.

aliasing An artifact produced by *sampling* a signal that contains frequency components higher than one-half the sampling frequency. In horizontally-sampled images, aliasing produces jagged lines.

alpha-wrap In a helical-scan tape recorder, a configuration where the tape wraps completely around the scanner in the manner of the Greek letter alpha (α).

amorphous silicon A noncrystalline form of silicon.

amplitude modulation (AM) Modulation that is accomplished by varying the amplitude of a carrier signal in accordance with the information to be transmitted.

analog The expression of a variable as a continuous function of time, amplitude, etc.

analog to digital conversion (A–D) The process of converting an analog signal to a digital bit stream. (See also *sampling* and *quantization*.)

aperture (1) In image *scanning*, the size of the sensitive spot that moves over the image. (2) In an antenna, an imaginary cylinder in space surrounding the radiating elements.

aperture correction In video cameras, an electronic process that corrects for some of the high-frequency losses due to *aperture response*.

aperture response In video cameras, the frequency response characteristic due to the effect of a finite-sized scanning *aperture*. It is usually specified in terms of the *modulation transfer function*.

application software In computers, the software that performs a specific end-user task, such as word processing or drawing.

artifact Any unnatural component. With respect to images, distortions of all types are called artifacts.

aspect ratio The ratio of the width to the height of an image. For example, analog television uses a 4:3 (1.33) aspect ratio, while HDTV has an aspect ratio of 16:9 (1.78).

assemble editing In video postproduction, single-recorder editing that is done by adding new material onto the end of a previous recording.

asynchronous transport mode (ATM) A data communication transport system based on *packetization*.

ATM See *asynchronous transport mode*.

azimuth recording In video recording, the technique of offsetting the azimuth of adjacent tracks to reduce cross talk between the tracks at short wavelengths. (The azimuth is the angle between the head gap and a perpendicular to the track edge.)

basic input/output system (BIOS) In computers, built-in software (usually in *ROM*) that performs the interface between the system's permanent hardware and the operating system software.

beam tilt In broadcasting antennas, the technique of directing an antenna's vertical pattern slightly below the horizontal to improve fringe-area signal strength.

beam width In broadcasting antennas, the angle between the half-power points of a directivity pattern.

BER See *bit error rate*.

binary A number system based on powers of 2. A binary digit is a *bit*.

BIOS See *basic input/output system*.

bit A digit of a *binary* system. It has only two possible values: 0 or 1.

bit error rate (BER) In digital transmission or recording systems, the average rate at which bit errors may occur. For example, a BER of 10^{-4} means that, on the average, there will be a bit error for every 10,000 bits received.

bitmapped graphics Graphics or images that are defined by specifying every pixel value in the image. For example, a 640 × 480 bit map will contain 307,200 pixel values. (See also *vector graphics*.)

black matrix In *cathode-ray tubes*, a technique to make the screen appear black when no pixels are illuminated. This improves the contrast by reducing the reflection of incident illumination.

blanking In an analog video signal, the periods between active lines and frames, intended to provide time for retrace of the display scanning beam. Blanking intervals are normally at black level or blacker-than-black and they also are used for transmitting synchronizing information.

blooming An artifact of a camera imager where an extreme highlight in the scene causes a spreading of a white area in the reproduced picture.

Boolean algebra A branch of mathematics named for the English mathematician, George Boole, that deals with the relationships between and manipulation of binary numbers.

burned-in time code In video tape recording, the technique of placing time code numbers in the picture for *off-line editing*.

bus In computers, a parallel interconnection of all the internal units of the system. It provides for transfer of data, address, and control information.

C-band The frequency band from 3.0 to 10.0 GHz.

cable television (CATV) A television system where programs are distributed to users over coaxial cable or fiber-optic cable. It provides for multichannel service in areas where broadcast service is limited or to augment broadcast service in other areas.

cache memory In computers, a block of fast memory used to temporarily hold data for use by the CPU. It provides a significant speedup when information from main memory or mass storage can be moved into cache memory before the CPU needs it because the CPU does not have to wait for access to slower memory.

camcorder An equipment that combines a video camera and recorder in one unit.

cassette A box that holds two reels of tape for use in an automatically-threading video cassette recorder (VCR).

cathode-ray tube (CRT) A video display device that operates by electron-beam bombardment of an electroluminescent phosphor in a vacuum. There are monochrome and color CRTs.

CATV See *cable television*.

CCD See *charge-coupled device*.

CD-ROM An acronym for Compact Disc–Read-Only Memory. (See also *Compact Disc*.)

central processing unit (CPU) In a computer, the main processor that performs data movement and manipulation. In a personal computer, the CPU is usually a microprocessor chip.

characteristic impedance In a transmission line, the ratio of the voltage to the current in the line. If the load impedance at the end of the line equals the characteristic impedance, all the energy coming down the line is absorbed. If the load impedance does not match the characteristic impedance, some energy is reflected at the load and travels back the line. (See also *voltage standing-wave ratio*.)

charge-coupled device A solid-state imaging device that stores an electrical charge representation of the optical image by means of *photoconductivity*. A readout mechanism converts the charge image to a video signal.

chroma keying A means for combining two video signals where a color in the foreground image specifies whether to show a second (background) image. For example, an announcer standing in front of a blue screen can have a second image replace all blue areas, giving the appearance that he is actually standing in the second image.

chromaticity diagram A standard diagram that shows the chromaticity coordinates of the visible spectrum and on which any color can be plotted. It was originally developed by the CIE (*Commission Internationale de l'Eclairage*) in 1931. (See also *trichromatic*.)

chrominance In color video systems, the signal component(s) that describe *color-difference* information.

CISC See *complex information-set computer*.

clamping In analog video systems, the process of setting some portion of the video signal to a fixed level to restore the DC component. Usually part of the blanking interval is used for clamping.

clock frequency (1) In a computer, the operating frequency of the *CPU*. (2) In digital transmission systems, the basic rate of data transmission.

CLUT See *color lookup table*.

coercivity A measure of threshold magnetizing force for a magnetic material.

color burst In an analog color video signal, a burst of the color subcarrier frequency placed in the blanking interval for subcarrier synchronization purposes.

color difference In a color video system, signals that represent the difference between specified colors and the luminance component. They have the property that they go to zero for monochrome images.

color lookup table (CLUT) In digital bit-mapped images, the bit map data value for each pixel is an index into a color lookup table that reads out the actual pixel value. For example, an 8 bits/pixel bit map can have only 256 different values in its bit map, but by using a CLUT containing 18-bit values, the 256 colors in the image can be selected from a palette of 262,144 colors. This improves color rendition if the CLUT colors are carefully selected for each image.

color-under In analog video recording, a system of color encoding that encodes the *chrominance* signal in a frequency band below the frequency range of the *luminance* FM signal. It is used in home recording systems.

colorimetry (1) The science of color. (2) In a video system, the specific colorimetric parameters of the system.

comb filter A *transversal filter* designed to produce a frequency response that has a series of nulls at equally-spaced frequencies. For example, a comb filter can be used to separate the interleaved chrominance and luminance sidebands in an NTSC video signal.

comet tails In video cameras, an overload artifact that shows tails behind moving highlights.

Compact Disc (CD) A 12-cm diameter digital optical disc used for audio or computer data storage and distribution. The computer format is called *CD-ROM*.

compiler In computer software, a program that reads a high-level language such as C, Pascal, or BASIC and generates microprocessor *instructions* in the form of an executable file.

complex instruction-set computer (CISC) A *central processing unit* that has a large *instruction* set of both low- and high-level operations. Typically, different CISC instructions take different numbers of CPU clock cycles to execute, proportional to their complexity. (See also *reduced instruction-set computer*.)

component signal Signals in a color video system where separate signals represent aspects of the picture—typically three. Examples are R,G,B; Y,I,Q; etc.

composite signal Signals in a color video system that multiplexes all the picture information into a single signal, such as NTSC, PAL, or SECAM.

compression In digital systems, the process of data manipulation to reduce the amount of data needed to represent something, such as a text file, a still image, motion video, or audio.

constellation diagram In a modulated digital transmission system, a diagram that shows the allowable modulated-signal states, usually shown in terms of modulated-signal amplitude and phase.

contrast In an image, the ratio of the brightest to the darkest areas. Sometimes called "contrast ratio."

contrast compression In a video camera, special processing to improve reproduction of a scene that has a contrast ratio higher than can be handled by the video system.

CONUS an acronym for "Continental United States."

CPU See *central processing unit*.

cross color In a composite color system, distortion caused by *luminance* components appearing in the *chrominance* channels.

cross luminance In a composite color system, distortion caused by *chrominance* signals appearing in the *luminance* channel.

crossword code In digital *error protection*, the application of the error protection *algorithm* to a two-dimensional data block. This is also called *product coding*.

CRT See *cathode-ray tube*.

cyclic redundancy check (CRC) In digital *error protection*, a form of *Hamming code* for error detection of data blocks.

D–A See *digital-to-analog conversion*.

dark current In an *imager*, the signal current that flows in the output when no light is present on the sensitive surface (lens capped).

DCT See *discrete cosine transform*.

decode The process of undoing an encoding algorithm to restore the original signal.

definition In an image, the property of sharpness. This is different than the *resolution* of an image.

dichroic surface An optical surface that has a special coating of a thickness that reflects light in one region of the optical spectrum and transmits the rest.

differential gain In a composite video system, a nonlinear effect that causes the luminance level to affect the subcarrier amplitude. It shows up as a variation of color *saturation* with image brightness.

differential phase In a composite video system, a nonlinear effect that causes the luminance level to affect the subcarrier phase. It shows up as a variation in color *hue* with image brightness.

differential pulse code modulation (DPCM) A form of *pulse code modulation* where the transmitted values represent the differences between adjacent sample values of the input signal.

digital A signal system where signals have values chosen from a specified set. For example, a *binary* digital system has only two valid signal values; similarly, an *octal* digital system has eight valid signal values.

digital-to-analog conversion (D–A) The process of converting digital values into analog values. It is a necessary step to display digital video signals on an analog display such as a CRT.

digital signal processor (DSP) A special microprocessor that has an *instruction* set optimized for signal processing tasks.

digitization Another word for *analog-to-digital conversion*.

dipole A basic radiating element of a broadcasting antenna.

discrete cosine transform (DCT) In image compression, a transform that converts blocks of pixel data into a form that is more suitable to run-length and statistical compression. It is a variation of the *Fourier transform*.

display refresh In digital displays, the process of continuously reading from display memory to deliver signals to the monitor.

dithering In a digital system, the introduction of small random variations that mask the effect of *quantization* errors. In representing digital images with a limited number of colors, dithering varies the colors of adjacent pixels to give the appearance of more colors.

dot pitch In a color display device, such as a *CRT* or *LCD* panel, the distance between adjacent dots of the same color. This parameter determines the *limiting resolution* of the display.

downlink In satellite communications, the transmission path from the satellite to the ground. See also *uplink*.

DPCM See *differential pulse code modulation*.

driver In computer software, a module that interfaces custom hardware to the system.

drop-out In magnetic video recording, a signal interruption caused by a surface defect or foreign particle that momentarily affects the head-to-tape contact.

drops In CATV systems, the cables that connect individual homes to the *feeder* cable.

DSP See *digital signal processor*.

edit decision list (EDL) In audio or video editing, a list of commands and *time code* locations that can control the final editing of a program.

editing In audio or video production, the process of assembling individual shots or clips into a complete program.

EDL See *edit decision list*.

effective radiated power (ERP) In a broadcast transmission system, the specification of radiated power that takes into account the transmitter power output, the transmission line losses, and the antenna gain.

EIA See *Electronic Industries Association*.

Electronic Industries Association (EIA) An organization of electronics manufacturers in the United States.

electronic news gathering (ENG) The use of portable video and audio equipment for collecting news stories for television.

encode The process of converting information to a different form for transmission or storage. (See also *decode*.)

ENG See *electronic news gathering*.

envelope delay distortion In an analog system, distortion caused by the system delay varying with frequency.

ERP See *effective radiated power*.

error protection In digital systems, the provisions made to detect and correct for errors.

exciter In a broadcast transmission system, the equipment that generates the modulated carrier.

fade margin In FM communication systems, the difference in dB between the receiver input power under nonfading conditions and the receiver's threshold (the point where the *signal-to-noise ratio* of the receiver begins to deteriorate rapidly with reduction of input power).

FDM See *frequency-division multiplex*.

feeders In a CATV system, the cables that run in front of homes, from which individual homes are connected by *drops*.

field In *interlaced scanning*, one vertical scan.

field sequential A composite color TV system that transmits different colors on successive scanning fields. In the 1950s, such a system was briefly adopted in the United States, but because it was incompatible with existing monochrome receivers, it was quickly replaced by the NTSC color system.

field strength In broadcasting, the measure of electric field strength at a receiving location, defined in millivolts/meter or μvolts/meter.

flash memory In computers, a form of nonvolatile solid-state memory that can be updated by applying a special procedure.

flying-spot scanner A form of image sensing that uses a scanning spot of light and a photocell pickup. In a light-transmission mode with a CRT as the light source, flying-spot scanners are sometimes used for motion picture film pickup. They can also be used in a reflective mode to scan physical objects, but this is not practical for television.

FM improvement factor In frequency-modulated (FM) transmission systems, the ratio (in dB) of the system *signal-to-noise ratio* to the signal-to-noise ratio of the unmodulated carrier.

Fourier transform A mathematical process that converts an electrical waveform (time domain) into its frequency components (frequency domain).

frame In video systems, the *scanning* of a complete picture. To convey motion, frames are repeated at a rate high enough that the eye sees a continuous image, typically 30 frames/second or higher.

frequency-division multiplex (FDM) A technique for combining multiple signals in the same transmission path by sending each signal at a different frequency. For example, a CATV system uses FDM.

front porch In an analog video signal, the part of the horizontal blanking interval ahead of the sync pulse. It exists to prevent video cross talk into the sync.

gamma correction In a video camera, the process of adjusting the amplitude *transfer characteristic* to obtain optimum reproduction of the brightness levels in a scene.

genlock In video systems, the process of synchronizing a video source to the signal from a remote source.

geosynchronous orbit A satellite orbit having the property that the orbital rotation is the same as the earth's rotation, causing the satellite to appear stationary relative to the earth. It is at an altitude of 22,300 miles.

ghost   In television broadcasting, interference due to *multipath* transmission that produces multiple images.

graceful degradation   In analog systems, the name given to the accumulation of distortions that occurs as system elements or processes are cascaded. The signal becomes steadily worse as the system is extended, but it may remain useful even though seriously impaired.

Grand Alliance   An organization formed in the United States in 1993 to develop a single proposal for an HDTV system. This work was completed in 1995.

graphical user interface (GUI)   In personal computers, a system for user-computer interaction that includes a video display and a pointing device such as a mouse. The user controls the system by pointing at objects (*icons*) on the screen and clicking a button.

graphics   In computers, the name given to computer-generated drawings and pictures. (See also *vector graphics* and *bit-mapped graphics*.)

graphics accelerator   In a PC video display adaptor, a special-purpose processor that speeds up common graphical operations such as displaying windows or drawing objects.

gray scale   Refers to the amplitude characteristic of the luminance channel of a video system.

GUI   See *graphical user interface*.

Hamming code   A digital *error protection* method that is able to detect single errors and point to the specific bit that is in error. It is easy to implement, but lacks the power of better algorithms, such as the *Reed-Solomon* code.

hard disk   In a computer, a mass storage device utilizing a magnetic coating on a rotating disk and moving heads to achieve random access.

HDTV   See *high-definition television*.

head end   In a CATV system, the equipment that creates the signals and modulates them to drive the cable.

header   In a digital data stream or data file, a specified block that defines the characteristics of the data that follows.

helical scan   In video recording, a rotating-head scanner that writes diagonal tracks across the tape to achieve the high head-to-tape speed needed for video recording.

hexadecimal   A digital number system based on 4-bit words, giving 16 values per word.

high-definition television Television systems of higher resolution, wider screens, and better sound compared to present NTSC or PAL systems. In the United States, the Grand Alliance all-digital system meets these goals.

horizontal detail In an image, the scene features that are affected by resolution in the horizontal direction, i.e., vertical lines.

hue The "color" of a pixel, such as red, blue, yellow, violet, etc.

Huffman coding In digital systems, a form of statistical coding for data compression that assigns word lengths in proportion to the frequency of occurrence of values. For example, frequently-occurring values get short words, seldom-used values get long words.

hypertext In computers, a system for presenting text that allows words or phrases to be marked so the user can select them and cause the display of additional or related information.

I/O See *input/output*.

icon In computers, a graphical symbol that represents objects or actions on the screen. As in *graphical user interface*, the user selects an icon to go to the represented object or perform the represented action.

image enhancement The act of modifying or processing an image to make it look better. Examples are *aperture correction* in video cameras or adjusting the brightness and contrast of an image in a computer.

image scanner A hardware device for digitizing a hard-copy document or picture into a computer.

imager A device that converts optical images to electrical signals, such as a *charge-coupled device*.

impedance In electrical circuits or devices, the ratio of voltage to current.

incidental phase modulation In a television transmitter, an undesirable variation of carrier phase with carrier amplitude. Depending on the type of modulation, this may cause distortion of the transmitted video signal.

index of refraction In a material used to transmit electromagnetic waves (including light), a parameter that defines the refractive properties of the material.

input/output (I/O) In computers, the part of the system that deals with interfacing to external devices for input or output, such as keyboards or printers.

insert editing In audio or video editing, a technique for replacing a segment of a previously recorded program.

instructions In microprocessors, the digital words that tell the microprocessor what actions to perform.

interframe In television systems, any action that makes use of the content of more than one *frame*. For example, a compression system that was based on how much change there was between adjacent frames would be interframe compression. (See also *intraframe*.)

interactivity In any system, provision for the user to control the action of the system and for he/she to view the results of his/her action on the system.

interlaced scanning In video systems, the technique of scanning part of the lines of the image in a first vertical scan and adding the in-between lines in a subsequent vertical scan or scans. In standard analog television, 2:1 interlace is used: half of the lines are scanned in each *field* to create a *frame* out of two vertical scans.

interleaving In digital systems, a technique that rearranges the data in a specified way, usually to cause bursts of transmission errors to be spread out in the data.

International Standardizing Organization (ISO) A worldwide cooperative organization responsible for standards in many fields.

interpreter In computer software, a program that reads commands from a file written in a programming language such as BASIC, and converts the language statements to *instructions* for the system's microprocessor, which are immediately executed.

intraframe In television systems, any action that uses only information from a single *frame*. (See also *interframe*.)

ISO See *International Standardizing Organization*.

isotropic radiator In antennas, a theoretical radiator that radiates uniformly in all directions. Such a radiator cannot be constructed.

Joint Photographic Expert Group (JPEG) A working group of a committee of the *ISO*. They developed the JPEG worldwide standard for compression of still images that is applicable to nearly all fields of electronic imaging.

JPEG See *Joint Photographic Expert Group*.

Kell factor In video systems employing horizontal and vertical scanning, a factor that expresses the ratio of the observed vertical resolution to the total number of scanning lines. For conventional analog television systems, the Kell factor is about 0.7.

lag In video cameras, an artifact where moving objects in the image tend to smear as they move.

LCD See *liquid-crystal display*.

limiting resolution In an image, the highest *resolution* that can be seen in the reproduced image. Limiting resolution is often measured by observing a pattern of converging lines (called a "wedge") and looking for the place on the wedge where the lines just fade out.

linearity In any system, the degree to which a property approximates a straight line. In video systems, scanning linearity and amplitude linearity are important.

lip sync In a system involving the simultaneous presentation of sound and motion video, the degree to which the audio and video are in sync. The name "lip sync" comes from the observation of a talking-head scene, which is one of the most critical cases for audio/video synchronization.

liquid-crystal display (LCD) An electronic display panel that is based on the properties of a liquid-crystal material sandwiched between two transparent plates and illuminated from behind. The transmittance of the liquid-crystal material changes with the applied electric field.

local bus In computers, a special *bus* that connects parts of the system that require the fastest data transfer, such as the *CPU*, *RAM*, and *video display adaptor*.

luminance In color video systems, a component signal that represents the brightness of the image. (See also *luminosity function*.)

luminosity function A function that expresses the apparent brightness of the spectral colors. It is used in video systems to calculate the *luminance* signal.

MAC See *multiplexed analog components*.

masking In color video systems, the process of mathematically mixing the color channels to produce controlled changes in the system *colorimetry*.

mass storage In computers, the nonvolatile, large-capacity storage elements of the system, such as *hard disks*, *CD-ROM*, or floppy disks.

mastering In optical recording, the process of creating a master recording to be used for replication, usually by pressing.

meander gate In a PAL video system, a circuit that controls the insertion of the color burst during the vertical blanking interval.

memory-mapping In computers, the technique of accessing hardware by assigning the hardware to memory addresses in the system memory space.

millions of instructions per second (MIPS) In microprocessors, the average speed at which the microprocessor executes instructions.

MIPS See *millions of instructions per second*.

mixing In audio or video systems, the process of combining controlled proportions of two or more source signals to create a new signal.

modulation In communication systems, the process of modifying the channel signal (often called the "carrier") to represent information to be transmitted. Typical modulation methods involve the frequency, amplitude, or phase of the carrier.

modulation index In frequency-modulation (FM) systems, the ratio of the frequency deviation to the highest modulating frequency. This parameter determines the extent of the spectrum generated by the FM process.

modulation transfer function (MTF) In video systems, the system's response to line patterns of varying spacing. The cross-sectional darkness of the lines varies sinusoidally. This is one measure of *aperture response*.

motion compensation In video compression systems, a method of compression that determines the degree of motion between adjacent frames and encodes only the changes.

Motion Picture Expert Group (MPEG) A working group of a committee of the *ISO*. They developed the MPEG worldwide standard for motion video compression and transport.

MPEG See *Motion Picture Expert Group*.

MTF See *modulation transfer function*.

multimedia The combining of different media (text, sound, graphics, video, animation, etc.) to create a presentation. This is done with computers or video systems.

multipath In radio transmission, the effect of signal reflection or refraction. This causes receivers to see several components of the original signal coming at different times due to the different path lengths of direct, reflected, or refracted signals. (See also *ghosts*.)

multiplexed analog components (MAC) In video systems, a system of component transmission where the components are transmitted sequentially in time, usually at TV line rate.

multitasking In computer operating systems, the ability to run two or more programs concurrently by rapid time-switching of the microprocessor between programs.

multithreading In computer operating systems, a form of *multitasking* that allows multiple tasks (called "threads") within the same program.

National Television Systems Committee (NTSC) The committee that decided on the compatible color television system for the United

States, adopted by the FCC in 1953. This system is called the NTSC system.

N$_e$ In a television system, a single-number representation of aperture response determined by defining a rectangle that has the same area as the area under the aperture response squared curve. The N$_e$ is the width of the defined rectangle.

nonvolatile memory In computer memory devices, a memory that retains its stored data when power is removed.

nonlinear editing In video editing, a technique for viewing edits in real time by rapidly selecting video sources on the fly. This is usually done with computers and hard disk storage—it cannot be done with magnetic tape.

NTSC See *National Television Systems Committee.*

null fill In broadcast antennas, a technique that partially fills in the signal nulls that would otherwise appear in the vertical pattern. This improves the close-in coverage.

Nyquist frequency In a sampled system, the sampling frequency needed to avoid aliasing. It is twice the highest signal frequency.

Nyquist limit In a sampled system, the highest signal frequency that can be sampled without aliasing with a given sampling frequency. It is one-half the sampling frequency.

octal A digital system that has eight valid values. In binary digital systems, octal representation uses three-bit words.

off-line editing In audio or video editing, an editing system that operates with copies of the master recordings (usually in a different format) in order to facilitate or lower the cost of edit decision making. As edits are chosen, an *edit decision list* is created with commands and *time codes* that can be subsequently used to perform the final edits using the master recordings.

offset binary In digital systems, a binary representation where signal zero is offset to the center of the word value range. This allows positive and negative signal values to be represented. For example, in an 8-bit offset-binary representation, signal zero would be at binary 128, maximum-positive signal would be at binary 255, maximum-negative signal would be at binary 0. (See also *two's complement.*)

omega-wrap In a helical-scan tape recorder, a scanner design where the tape wraps partially around the scanning drum, entering and leaving the drum and its guides along a radius. This is like the Greek letter omega (Ω).

operating system In computers, the program that provides the basic functions needed to run programs.

orbital slot In satellite communication using *geosynchronous* satellites an arc of the orbit allocated to a single satellite. Orbital slots are identified by their longitude.

packetization In digital transmission systems, the technique of breaking up a data stream into blocks to facilitate transmission, multiplexing, and error protection.

PAL See *phase-alternating line.*

passive-matrix LCD A *liquid-crystal display* panel that uses a pattern of row and column lines to provide access for refreshing the display. This is the lowest-cost type of LCD panel.

pay-per-view In video distribution systems, the technique of controlling access so that the customer can be charged on the basis of what programs he/she watches.

PCM See *pulse code modulation.*

permeability The property of a magnetic material that defines how easily it can be magnetized.

phase-alternating line (PAL) The analog color video composite system developed in Europe and used by certain countries around the world. It is similar to *NTSC* but it uses a subcarrier phase alternation technique that makes certain kinds of transmission errors appear to cancel.

phase-lock loop A circuit that synchronizes a local oscillator to an incoming signal. Phase-lock loops are used, for example, to synchronize an analog television receiver to the horizontal scanning and color subcarrier of the incoming signal.

photoconduction The process whereby a material exhibits light sensitivity by changing its electrical conductivity with light level. For example, photoconductive materials are used in vidicon and Plumbicon pickup tubes and in CCD imagers.

photoemission The process whereby a material exhibits light sensitivity by emitting electrons in the presence of light. This was used in the early pickup tubes such as the iconscope and the image orthicon.

pixel In an electronic image, a single point in the picture. Each pixel has unique values of *luminance* and color. In a digital image, pixels are individual points of the image, each represented by a certain number of bits.

polarization In electromagnetic waves, describes the orientation of the electric field vector. It may be horizontal, vertical, or elliptical. With elliptical polarization, the electric field vector rotates as the wave passes a fixed point.

poly-silicon Poly-crystalline silicon. Its electrical properties are between single-crystal silicon and *amorphous silicon*. Unlike single-crystal silicon, which must be grown, poly-silicon can be deposited on a substrate.

postproduction In the creation of audio and video programs, the process of assembling previously-recorded source material into the finished program. Postproduction includes tasks such as *image enhancement*, audio *sweetening*, *special effects*, titling, and *editing*.

primary color In a *trichromatic* color system, one of the three colors that are mixed to create all other colors in the system.

print-through In magnetic tape recording, the tendency for demagnetization of recordings to occur between layers of the tape due to magnetic cross talk when tape is rolled up on a reel.

product code See *crossword code*.

progressive scanning A *scanning* process where all the lines of the image are scanned by every vertical scan. (See also *interlaced scanning*.)

pulse code modulation (PCM) In digital audio and video systems, the representation of an analog signal by its direct digitized values.

quantization In *analog-to-digital conversion*, the process of fitting the sampled values into the digital bits/sample range. For example, if there are 8 bits/sample, the sample values must be quantized to no more than 256 levels.

quantum efficiency In an imaging device, the percent of incident light quanta that create useful electrical output.

radiating pattern In antennas, the variation of radiation intensity in different directions, both horizontally and vertically.

RAM See *random-access memory*.

random-access memory (RAM) In computers, the main system memory, usually consisting of (*volatile memory*) solid-state chips.

raster In video *scanning*, the pattern made by the scanning spot as it covers the entire picture. The usual raster is a pattern of equally-spaced horizontal lines.

read-only memory (ROM) In computers, *nonvolatile memory* that stores permanent programs. ROM usually consists of solid-state chips.

reduced instruction-set computer (RISC) A microprocessor architecture based on a simplified instruction set that allows all instructions to be executed in a single clock cycle or less. Currently, the fastest microprocessors use this approach.

Reed-Solomon code A digital *error protection* code based on blocks of data with added redundancy. It is capable of correcting burst errors up to a limit set at design time.

reflection coefficient In electrical transmission lines, a parameter that describes the degree of mismatch between the *characteristic imped-ance* of the line and its load impedance.

rendering In computer *graphics*, the process of creating a bit-mapped representation of a *vector graphics* image.

resolution (1) In an electronic image, a measure of the reproduction of fine detail. See also *limiting resolution*. (2) In a digital image, the total number of pixels in the horizontal and vertical directions. For example, 640 × 480.

RISC See *reduced instruction-set computer*.

ROM See *read-only memory*.

S/N See *signal-to-noise ratio*.

sampling In an electronic system, the process of taking equally-spaced readings of the instantaneous value of an analog signal. It is the first step of converting the analog signal to a digital representation. (See also *analog-to-digital conversion*.)

saturation (1) In color video systems, the measure of the intensity or purity of a color. (2) In a magnetic material, the maximum flux density that can exist in the material.

SAW filter An electrical filter using the surface acoustic wave (SAW) technology. SAW filters are capable of transfer functions that are not possible in lumped-element circuits.

scanning In electronic image reproduction, the process of reading or displaying all the points of the image by passing a moving spot across the image. (See also *raster*, *interlaced scanning*, and *progressive scanning*.)

SECAM See *Sequential Coleur et Memoire*.

Sequential Coleur et Memoire The analog *composite* color televi-sion system developed in France and used in several other countries. SECAM uses a luminance signal and two FM subcarriers that transmit the *color-difference* components.

server On a digital network, a computer that is devoted to providing network control, data storage, and/or input/output interfaces.

service area In broadcasting, the geographical area covered by a broadcast transmitter at a specified level of received signal quality.

set-top converter An electronic unit that connects a CATV system *drop* cable to a subscriber's TV receiver. It provides frequency conversion so the receiver can access all the cable channels and it also may include features for *pay-per-view* or other premium services.

shift register An electronic digital circuit that stores data that can be clocked out in sequence. If the shift register has a parallel input, it can function as a parallel-to-serial converter.

signal-to-noise ratio (S/N) In an electrical channel, the ratio of the signal amplitude to the random noise in the channel. For video signals, the ratio is between the peak-to-peak video signal and the root-mean-square (RMS) noise.

SMPTE Society of Motion Picture and Television Engineers.

special effects In video postproduction, the creation of dynamic transitions between scenes or other dynamic modification to video signals.

soft-decision decoder In a digital transmission or recording system, a technique for converting the channel signal into proper digital values by taking account of the values of adjacent *symbols*. This is known as "Viterbi" decoding.

software In computers, the modules of information that either already are or result in *instructions* to the system *CPU*. The software tells the computer what to do. Software modules are called "programs." (See also *interpreter*, and *compiler*.)

streaking In an electronic image, an artifact where objects in the image create black-level shifts that go entirely across the image.

sub-Nyquist sampling In a sampled system, *sampling* at a rate lower than the *Nyquist frequency*.

subcarrier In an electrical transmission system, a carrier embedded in the baseband signal to transmit additional information.

subsampling In a sampled system, the process of deleting certain samples for the purpose of data reduction. For example, every other sample may be deleted. Proper use of subsampling requires filtering to prevent aliasing from the subsampling.

subtractive color A color reproduction system that uses pigments or dyes added to a white background. The pigments or dyes subtract certain colors from the light reflected from the white background. (See also *additive color*.)

super VGA (SVGA) In personal computers, a generic name for scanning and color standards that are an extension of the *VGA* standard.

SVGA See *super VGA*.

sweetening In audio postproduction, any process applied to the signal to make it sound better.

symbol In digital transmission systems employing modulation, the time period for transmitting a single unique state of the carrier. Depending on the modulation method, each symbol can transmit one or more bits of data.

tape transport In magnetic recording, the mechanism that moves the tape and presents it to the heads.

telecine A video system specifically designed for transferring motion picture film to television signals.

time code In audio and video recording, a system that records a number with each video frame or section of audio. This is used to facilitate precise control during *postproduction*. (See also *edit decision list*.)

transfer characteristic In electrical circuits, the property that specifies the ratio of output to input signal amplitude. (See also *linearity*.)

transfer function In electrical circuits, the complete relationship between input and output signals, usually expressed mathematically as a function of frequency, amplitude, or other parameters.

transitions In audio or video editing, the aural or visual actions that occur while changing the program from one scene to another. Examples are wipes, dissolves, fades, etc. (See also *special effects*.)

transmission line A hard-wired electrical circuit that transmits a signal from one point to another. It may be a coaxial cable, a waveguide, twisted-pair wires, etc., depending on the system requirements for frequency response, distance, and power level.

transponder In satellite communications, the equipment on the satellite that receives signals from the *uplink* and retransmits them for the *downlink*.

transport layer In the Grand Alliance digital HDTV system, the part of the system that packetizes the data streams to prepare them for transmission.

transversal filter An electrical filter produced by sending a signal down a *transmission line* and creating the output by combining signals tapped from various points along the line.

trellis coding In digital transmission using multilevel modulation, a technique for error protection that improves error performance without increasing the required channel bandwidth. Among other things, it makes use of *soft-decision decoding*.

trichromatic A color system that uses three color components to represent all colors. Also called a "tristimulus" system.

troposphere The lower portion of the earth's atmosphere, from about 4 to 12 miles altitude at the equator. In this region, temperature generally decreases with altitude.

TRS-ID In serial digital transmission systems, a code for synchronization and identification. The acronym means "timing reference and identification signal."

trunks In CATV systems, the cables that carry the signals between the *head end* and the *feeder* break-out points.

truth table In digital systems, a table that specifies the outcome of a process for all possible combinations of values of the inputs to the process.

TV lines (TVL) In video systems, a unit of measurement of resolution based on a pattern of equally-spaced lines. The TVL number is obtained by counting all lines and spaces if the pattern is extended to reach a distance equal to the picture height.

TVL See *TV lines*.

two's complement In binary digital systems, a method of representing signed values that keeps signal zero at binary zero. (See also *offset binary*.)

uplink In satellite communications, the transmission path from the ground to the satellite. (See also *downlink*.)

vector graphics In computer *graphics*, the representation of objects by mathematical means. It makes possible the *rendering* of the object to be displayed at any size or format. (See also *bit-mapped graphics*.)

vertical detail In an image, the scene features that are affected by resolution in the vertical direction, i.e., horizontal lines.

vertical interval time code In video systems, a *time code* that is transmitted during the vertical blanking interval of the signal.

vestigial-sideband modulation A *modulation* system where one of the sidebands resulting from modulation is partially removed. The sideband is only partially removed in order to ease the filtering requirements in signal generation.

VGA See *video graphics array*.

video (1) The technology for the electronic capture, storage, transmission, and reproduction of images and motion pictures. (2) The signals generated by video technology.

video adaptor In personal computers, a hardware module including memory and control circuits to drive a video display monitor. It performs the tasks of video *display refresh* and *graphics acceleration*.

video graphics array (VGA) In personal computers, a *video adaptor* standard originally developed by IBM for the PC/AT. It has been greatly extended by others. (See also *Super VGA*.)

video overlay In personal computers, the technique of combining analog video from a VCR or laser disc with computer information on the same screen. This often is accomplished by a special *video adaptor*.

video RAM In personal computers, a special form of RAM chip that has a separate serial-output port for display refresh operations. This architecture speeds up *video adaptor* performance.

viewing ratio When viewing a reproduced image, the ratio between the distance between viewer and screen to the height of the screen.

VITC See *vertical interval time code*.

Viterbi decoder See *soft-decision decoder*.

volatile memory In computers, memory that loses its data when power is removed.

voltage standing-wave ratio In a *transmission line*, the ratio between the voltage maximum to the voltage minimum in the standing wave which may exist on the line due to reflections. It is a fundamental measurement of transmission line performance.

VRAM See *video RAM*.

VSB See *vestigial-sideband modulation*.

VSWR See *voltage standing-wave ratio*.

waveform monitor An oscilloscope specifically designed for viewing television video waveforms.

waveguide In broadcast transmitters, a *transmission line* that consists of a hollow rectangular or circular tube within which an electromagnetic wave transmits power. It is sometimes used to connect a UHF transmitter to the antenna.

weighted noise In recognition of the fact that high-frequency noise is less visible in a picture than low-frequency noise, a high-frequency rolloff is sometimes used during *S/N* measurement. This rolloff is called a "weighting filter" and it makes the noise measurement more indicative of how the picture might look to an observer. Weighted-noise measurements give somewhat larger numbers than unweighted measurements.

wireless cable A multichannel broadcast service where the multiple channels are transmitted with *frequency-division multiplex* in a manner similar to cable television.

Bibliography

General References

ANSI/EIA/TIA *Standard Electrical Performance for Television Transmission Systems*, Electronic Industries Association, Washington, D.C., 1990.

Benson, K. Blair and J. Whittaker, *Television Engineering Handbook*, McGraw-Hill, New York, N.Y., 1992.

Fink, Donald G. and Donald Christiansen, *Electronic Engineers' Handbook*, McGraw-Hill, New York, N.Y., 1988.

Inglis, Andrew F., *Electronic Communications Handbook*, McGraw-Hill, New York, N.Y., 1990.

Rorabaugh, C. Britton, *Communications Formulas and Algorithms*, McGraw-Hill, New York, N.Y., 1990.

Various authors, *Television Technology in the 1990s*, SMPTE, White Plains, N.Y., 1981.

Various authors, *Tomorrow's Television*, SMPTE, White Plains, N.Y., 1982.

Various authors, *Television Technology: A Look Toward the 21st Century*, SMPTE, White Plains, N.Y., 1987.

Various authors, *Television Technology in Transition*, SMPTE, White Plains, N.Y., 1988.

Various authors, *Television—Merging Multiple Technologies*, SMPTE, White Plains, N.Y., 1990.

Chapter 1, Video System Fundamentals

Fink, D. G., *Television Engineering Handbook*, McGraw-Hill, New York, N.Y., 1957.

Chapter 2, Color Video Fundamentals

FCC rules, Part 73.

CCIR, *Recommendations and Reports, CCIR 15th Plenary Assembly*, Vol. XI, Broadcasting Service (Television), ITU, Geneva, 1982.

EIA/TIA Standard, "Electrical Performance for Television Transmission Systems," EIA-TIA-250-C, Washington, D.C., 1990.

Hirsch, C. J., W. F. Bailey and B. D. Loughlin, "Principles of NTSC Compatible Color Television," *Electronics*, Vol. 52, February 1952.

Loughren, A. V., "Recommendations of the National Television Systems Committee for a Color Television Signal," *J. SMPTE*, Vol. 60, April–May 1953.

Various authors, *Television Image Quality*, SMPTE, White Plains, N.Y., 1984.

Wentworth, John, *Color Television Engineering*, McGraw-Hill, New York, N.Y., 1954.

Chapter 3, Introduction to Digital Technology

Luther, Arch C., *Digital Video in the PC Environment*, McGraw-Hill, New York, N.Y., 1991.

Pohlmann, Ken C., *Principles of Digital Audio*, McGraw-Hill, New York, N.Y., 1995.

Various authors, *Digital Video,* Vols. 1–2, SMPTE, White Plains, N.Y., Vol. 1 (1977), Vol. 2 (1979).

Watkinson, John, *The Art of Digital Audio*, Focal Press, London, 1994.

Chapter 4, Elements of Image Quality

Ardito, Maurizio, "Studies on the Influence of Display Size and Picture Brightness on the Preferred Viewing Distance for HDTV Programs," *SMPTE J.*, August 1994.

Herman, S., "The Design of Color Television Color Rendition," *J. SMPTE*, Vol. 84, 1975.

Various authors, *Better Video Images*, SMPTE, White Plains, N.Y., 1989.

Schade, Otto H., Jr., "Electro-Optical Characteristics of Television Systems," *RCA Review*, Vol. 9, March, June, September, December, 1948.

Schade, Otto H., Jr., *Image Quality, A Comparison of Photographic and Television Systems*, RCA Laboratories, Princeton, N.J., 1975.

"Engineering Aspects of Television Allocations," TASO Report to the FCC, March 1, 1977.

Chapter 5, Analog Video Systems

RJG Ellis, "The PALplus Project: Conception to Introduction," *IBC Convention 16–20 September 1994*.

Also see references for Chapter 2.

Chapter 6, Digital Video Systems—HDTV

Benson, K. Blair, and Donald G. Fink, *HDTV-Advanced Television for the 1990s*, McGraw-Hill, New York, N.Y., 1991.

Glenn, William E., and Karen G. Glenn, "High-Definition Transmission, Signal Processing, and Display," *J. SMPTE*, July 1990.

"Grand Alliance HDTV System Specification," Version 2.0, December 7, 1994.

Various authors, "HDTV and the New Digital Television," *IEEE Spectrum*, April 1995.

Chapter 7, Digital Video Systems—Computers

Ozer, Jan, *Video Compression for Multimedia*, AP Professional, Boston, MA, 1995.

Chapter 8, Video Cameras

"Camera Glossary," Sony Publication.

Hamalainen, J., T. Leacock and P. Westerlink, "Facts and Fiction: Some Aspects Regarding the Design of Digital Television Cameras Using CCD Image Sensors," Matsushita publication.

Thorpe, Laurence J., "HDVS CCD Camera—A Significant Advance in Real Time High Definition Imaging," Sony Publication.

Thorpe, Laurence J., "The Electronic Pursuit of Film Imaging," *IS&T's 47th Annual Conference*, 1994.

Chapter 9, Professional Video Recorders

Felix, M. and H. Walsh, "FM Systems of Exceptional Bandwidth," *Proc. IEEE*, Vol. 112, No. 9, September 1965.

Watkinson, John, *The Digital Videotape Recorder*, Focal Press, London, 1994.

Chapter 10, Home and Semiprofessional Video Recorders

Videomaker Magazine, published monthly by Videomaker, Inc., P.O. Box 4591, Chico, CA 95927.

Chapter 11, Video Postproduction Systems

Anderson, Gary H., *Video Editing and Postproduction: A Professional Guide*, Second Edition, Knowledge Industry Publications, 1993.

Ohanian, Thomas A., *Digital Nonlinear Editing*, Focal Press, London, 1993.

Chapter 14, Interactive Video Systems

Laurel, Brenda, *The Art of Human-Computer Interface Design*, Addison-Wesley, Reading, MA, 1990.

Luther, Arch C., *Authoring Interactive Multimedia*, AP Professional, Boston, MA, 1994.

Luther, Arch C., *Using Digital Video*, AP Professional, Boston, MA, 1995.

Chapter 15, Terrestrial Broadcasting Systems

HDTV Tech Brief Series; published by Comark
    "Digital HDTV and RF Amplifier Dynamic Range," May 1992
    "Digital HDTV and Amplifier Class of Operation," June 1992
    "WRC-TV, Washington D.C., AD-ADTV test," October 1992
    "HDTV Transmitter Performance Criteria—Linearity Versus Range," December 1992
    "Linearity Correction for ADTV Transmitters—What's Best?" September 1993
    "8-VSB—The ATV Select Standard, What Is It?" March 1994

"Is COFOM Dead?" May 1994
"RFP/RFI—A Serious Consideration for IOT Transmitters," June 1994

Chris McConnell, "MSTV Proposed Allocation Plan," *Broadcasting and Cable*, January 16, 1995.

Chris McConnell, "ATV Comes Out on Top of NTSC Field Tests," *Broadcasting and Cable*, September 26, 1994.

Chapter 16, CATV Systems

Bartlett, Eugene R., *Cable Technology and Operations*, McGraw-Hill, New York, N.Y., 1990.

Inglis, Andrew F. (Ed.), *Electronic Communications Handbook*, Chapter 17, McGraw-Hill, New York, N.Y., 1988.

Cicora, Walter S., "Inside the Set-Top Box," *IEEE Spectrum*, April 1995.

Large, David, "Creating a Network for Interactivity," *IEEE Spectrum*, April 1995.

Chapter 17, Satellite Video Communications

Ha Tri, T., *Digital Satellite Communications* (2nd edition), McGraw-Hill, New York, N.Y., 1990.

Inglis, Andrew F., *Satellite Technology, An Introduction*, Focal Press, Stoneham, MA, 1991.

Miya, K., *Satellite Communications Technology*, KDD Engineering and Consulting, Inc., Tokyo, 1981.

Baylin, Frank, *Miniature Satellite Dishes* (2nd edition), Baylin Publications, Boulder, CO, 1995.

McConnell, Chris, "The SNG Edge," [special report on satellites in 1994], *Broadcasting and Cable*, June 11, 1994.

FCC Docket 86-496, [Reduced orbital spacings].

Chapter 18, Fiber-Optic Transmission Systems

Kao, C. K., *Optical Fiber Systems: Technology, Design, and Applications*, McGraw-Hill, New York, N.Y., 1982.

Midwinter, J. E., *Optical Fibers for Transmission*, John Wiley & Sons, New York, N.Y., 1979.

Personick, S. D., *Fiber Optics Technology and Applications*, Plenum Press, New York, N.Y., 1985.

Chapter 19, The Future of Video Systems

Negroponte, Nicholas, *Being Digital*, Alfred A. Knopf, New York, N.Y., 1995.

Index

additive primaries, 35
algorithm, 76
aliasing, 91
   CCD, 104, 217
   frame, 103, 136
   scanning line, 102
analog-digital comparison, 27
analog-digital conversion, 26, 50, 74, 227
animation, 354
antennas, 391
   aperture, 391
   bandwidth, 393
   beam tilt, 395
   beam width, 391
   circularly polarized, 394
   directivity, 391 - 392
   earth station, 438
   gain, 391
   null fill, 395
   power ratings, 396
   radiation pattern, 391
   radiators, 393
   slot radiators, 394
   superturnstile, 393
   VSWR, 393
aperture correction, 222
aperture response, 95, 107, 110
aspect ratio, 13, 401
audio, 27
   bandwidth, 28
   cassette, 242
   digital, 28
   HDTV, 177
   home VCR, 283
   mixing, 306
   sweetening, 306
authoring, 355
automatic scan tracking, 257
azimuth recording, 282

bandwidth, 14, 105
binary arithmetic, 60
   addition, 60
   multiplication, 61
bit error rate, 66
bit mapped images, 353
black body radiator, 39
blanking interval, 18
Boolean logic, 61
byte, 49
cable television, 405
   channels, 406
   coaxial cable, 408
   digital transmission, 422
   downlinks, 447
   fiber optics, 423
   head end, 414
   history, 405
   line amplifiers, 410
   performance standards, 413
   set-top converters, 415
   signal-to-noise ratio, 411
camcorder, 291
camera
   aperture correction, 222
   beam splitters, 220
   camcorder, 291
   color correction, 223
   colorimetry, 40
   digital electronics, 226
   ENG, 219, 291
   gamma correction, 223
   image stabilization, 294
   imager, 109
   lens, 108, 220
   live cameras, 196
   operating controls, 229
   optical filter, 108, 221
   specifications, 229
   studio, 219

telecine, 197
triax cable, 225
viewfinder, 225, 293
cassettes, 242
CATV
  see cable television
CCD, 200
  aliasing, 104, 217
  aperture response, 109
  definition and resolution, 212
  frame interline architecture, 210
  frame transfer architecture, 207
  interline transfer, 209
  lag, 217
  MTF, 213
  on-chip lenses, 204
  performance, 212
  sensitivity, 215
  sensor construction, 201
  signal-to-noise ratio, 214
  smear and related defects, 218
  spectral response, 216
  transfer function, 217
  tri-color, 204
CD-ROM, 184
chroma keying, 305
chromaticity coordinates, 37
chrominance, 143
closed captioning, 322
coaxial cable, 408
  attenuation, 409
  characteristic impedance, 409
  loop resistance, 410
  return loss, 410
color difference components, 37
color distortions
  chroma/luminance gain, 153
  differential gain, 154
  differential phase, 154
  envelope delay distortion, 154
  subcarrrier cross talk, 155
color lookup table, 188
color temperature, 39
color-under, 278, 287
colorimetry, 31

camera, 40
comb filters, 155
communications theory, 75
Compact Disc, 270
component signal formats, 156
  MAC, 156
compression
  DBS, 428
  HDTV, 166
  JPEG, 191
  motion video, 192
  use in recorders, 272
computer
  analog video output, 331
  colorimetry, 46
  display monitors, 46, 334
  floppy disk, 268
  hard disk, 268
  scanning standards, 15
  video formats, 188, 193
  video overlay, 332
  video subsystem, 327
constellation diagram, 67
contrast ratio, 121
CONUS, 431
CPU, 182
cross color, 142
cross luminance, 142
cyclic redundancy check, 72
data
  interleaving, 73
  randomizing, 174
data headers, 25
dichroic surface, 220
differential gain, 142, 154
differential phase, 142, 154
diffraction, 387
digital
  audio, 28
  effects, 305
  modulation modes, 66
  postproduction, 308
  synchronization, 69
digital mirror device, 338
digital recording, 261

channel coding, 263
D-1, 262, 264
D-2, 262, 264
D-3 and D-5, 262, 265
D-6, 262, 268
error management, 263
formats, 262
home VCRs, 290
standardization, 262
Digital Satellite Service, 449
digital signal processing, 63
for display, 330
postproduction editing, 65
standards conversion, 64
time-base error correction, 63
direct broadcast service, 428
HDTV, 455
system architecture, 452
discrete cosine transform, 79, 167
display device primaries, 42
display memory size, 329
display refresh, 328
dithering, 189
DMD, 338
earth stations
CATV, 447
TVRO, 437
editing, 298
A-B roll, 301
assemble, 300
insert, 301
nonlinear, 309
off-line, 302
effective isotropic power, 428
effective radiated power, 370
electromagnetic radiation, 385
field strength, 389
polarization, 387
electronic editing, 298
encoding
offset binary, 57
pure binary, 57
two's complement, 57
equivalent line number, Ne, 98
error protection, 70

cyclic redundancy check, 72
detection, 71
effectiveness of, 73
encoding, 71
Hamming code, 72
product codes, 72
Reed-Solomon codes, 72
fade margin, 446
fiber optics, 423
fiber construction, 458
modulation modes, 463
optical transmitters, 465
photodetectors, 468
receivers, 467
transmission systems, 457
field sequential system, 141, 359
film, 132
frame aliasing, 136
MTF, 133
sensitivity, 137
signal-to noise ratio, 137
flat-panel displays, 335
electroluminescent, 339
field-emission, 338
liquid-crystal, 336
micromirror, 338
plasma, 338
television, 321
flicker, 11, 90, 130
flying spot scanners, 195
frame rate, 11
frequency modulation, 444
FM improvement factor, 445
signal-to-noise ratio, 444
frequency synthesis, 315
gamma, 123
correction, 124, 223
genlock, 224
geosynchronous orbit, 428
ghosts, 92
graceful degradation, 127
Grand Alliance, 159
graphics acceleration, 330
gray scale, 90, 119

half-wave dipole, 369
Hamming code, 72
hard disk, 268
HDTV, 159, 361, 401
    aperture response, 112
    assignments, 364
    audio, 177
    broadcasting, 398
    by satellite, 455
    camera processing, 227
    channels, 398
    compression, 166
    D-6 recorder, 268
    layered architecture, 163
    modulation modes, 68
    picture formats, 164
    production standard, 89
    receiver, 339
    sampling rates, 165
    system objectives, 161
    transmission layer, 172
    transmission standards, 400
    transmitters, 403
    transport layer, 168
    trellis coding, 68
helical-scan, 234
    ½-in. cassette recorders, 259
    alpha wrap, 241
    home VCR, 275
    omega wrap, 241
    tape transport, 241
home VCR, 275
    digital, 290
    frequency spectrum, 285
    record parameters, 280
    specifications, 276
hue, 32
Huffman coding, 80
human vision
    aperture response, 97
    brightness/contrast perception, 92
    characteristics, 92
    detail/sharpness perception, 93
hypertext, 352

iconoscope, 196
icons, 347
illuminants, 39
image
    capture, 189
    compression, 191
    definition, 89, 94, 116
    enhancement, 112
    gray scale, 119
    stabilization, 294
image orthicon, 196
imagers
    see also CCD
    history, 195
incidental phase modulation, 379
index of refraction, 386
information superhighway, 477
interactivity, 343
interlaced scanning, 11
interleaving data, 73
Internet, 478
isotropic radiator, 369
JPEG image compression, 81, 191
Kell factor, 105
kinescope recording, 298
kinescopes, 319
    projection, 321
lag, 91
lasers, 465
light emitting diodes, 465
limiting resolution, 104
liquid-crystal display, 336
    active-matrix, 337
    improvements, 338
    passive-matrix, 336
low noise amplifier, 442
luminance, 31, 143
MAC, 157
magnetic tape recording, 236
    azimuth recording, 265
    digital recording, 261
    editing, 299
    recording density, 248
    tape, 243
memory-mapping, 182

Microsoft Windows, 348
MIPS, 183
modulation transfer function, 96
motion compensation, 83, 167
motion video
    capture, 192
    compression, 192
Moviola, 302
MPEG compression, 82
    MPEG-1 and MPEG-2, 85
multimedia, 350
    animation, 354
    audio, 352
    authoring, 355
    graphics, 353
    hypertext, 352
    images, 353
    motion video, 351
    text, 352
multipath, 387
multitasking, 187
National Television Systems Committee
    see NTSC
Ne, 98
noise figure, 324
noise temperature, 441
nonlinear editing, 310
notch diplexer, 369
NTSC, 141, 361
    chrominance, 143
    encoder, 224
    spectral content, 149
    subcarrier, 145
number systems, 47
    binary, 48
    hexadecimal, 49
Nyquist limit, 51
offset binary, 57
operating system, 186, 354
optical fiber
    attenuators, 469
    connectors, 469
    construction, 458
    couplers, 469

graded-index, 458
modal dispersion, 460
multimode, 458 - 459
single-mode, 458 - 459
step index, 458
packetization, 65, 168
PAL
    chrominance, 144
    reference phase, 151
    spectral content, 152
    subcarrier, 150
pay TV systems, 417, 425
personal computer, 179
    architecture, 181
    BIOS, 183
    bus, 181
    CPU, 182
    input/output, 184
    mass storage, 183
    memory, 183
    multitasking, 187
    operating system, 186
    software, 185
photodetectors
    avalanche, 468
    PIN junction, 468
picture-in-picture, 323
pixels, 7, 9, 107
Planckian radiator, 39
Plumbicon, 199
polarization, 387
postproduction, 298
    editing, 65
    tasks, 304
primary colors, 34
    additive, 35
    camera, 40
    display device, 42
    NTSC/FCC, 43
    SMPTE 240 (Appendix), 44
    SMPTE Standard C, 43
    subtractive, 35
prism beam splitter, 220
production, 297
progressive scanning, 13

projection displays, 321
pulse code modulation, 28, 463
pure binary, 57
quantization, 10, 48, 54
    errors, 55
randomizing, 174
receivers, TV
    cable-ready, 314
    noise figure, 324
    picture-in-picture, 323
    projection, 321
    remote control, 322
    varactor tuners, 315
    video frequency response, 315
Reed-Solomon codes, 72
reflection coefficient, 380
rendering, 353
resolution, 6
    horizontal, 105
    limiting, 89, 104
    vertical, 89, 105
sampling, 7, 101
    aperture response, 101
    frequency domain, 8
    horizontal, 51
    subsampling, 77
    time domain, 7
satellite
    circuit, 427
    construction, 434
    downlink, 428, 440
    earth stations, 437
    fade margin, 446
    footprint, 428
    low noise amplifier, 442
    orbital slots, 429
    payload, 435
    solar eclipse, 431
    spectrum bands, 432
    system performance, 447
    threshold extender, 447
    transmission formats, 442
    transponder, 428, 436
    uplink, 428, 439
Saticon, 199

saturation, 32
SAW filter, 374
scanning, 4
    aperture response, 111
    computer, 15
    interlaced, 11
    progressive, 13
server, 481
set-top converters, 314, 415
shadow-mask kinescopes, 319
signal waveforms, 17
signal-to-noise ratio, 90, 444
    cable television, 411
    definition, 125
    in digital systems, 126
    standards, 127
    unweighted and weighted, 125
SMPTE Time Code, 307
SMPTE Type C recorders, 252
    automatic scan tracking, 257
    drop-out compensation, 258
    drum assembly, 252
    playback electronics, 255
    record electronics, 254
    record pattern, 253
    slow-scan and still-frame, 259
standardization
    digital recording, 262
standards conversion, 64
statistical coding
    Huffman, 80
subsampling, 77
subtractive primaries, 35
super VGA, 16, 328
symbols, 66, 173
synchronizing signals, 20
telecine, 230
television
    allocations, 362
    assignments, 362
    aural carrier, 368
    channels, 362
    history, 359
    interactive, 356
    receivers, 313

signal waveforms, 17
synchronizing signals, 20
transmitters, 371
visual carrier, 365
television broadcasting
   antenna height, 371
   effective radiated power, 370
   field intensity curves, 397
   performance standards, 376
   service area, 396
   transmission line, 379
   transmitter power ratings, 376
threshold extender, 447
time code, 307
   SMPTE, 307
   vertical interval, 307
time-base error correction, 63, 278
transform coding, 79
   discrete cosine, 79
   Fourier, 79
transmission
   digital video, 65
   fiber optics, 457
   parallel , 58
   serial, 58
   tropospheric, 386
transmission line
   characteristic impedance, 381
   coaxial line, 379
   diameters, 381
   power ratings and attenuation, 382
   reflection coefficient, 380
   voltage and current vectors, 380
transponder, 428
trellis coding, 68, 175
troposphere, 386
TRS-ID, 69
truth table, 61
U-Matic format, 281
user interfaces, 346
vector graphics, 353
vestigial sideband, 68, 176, 367
VGA, 16
video
   analog overlay, 332

DC component, 19
definition, 1
digital, 25, 47
digital transmission, 65
effects, 305
signal formats for color, 140
signal standards, 2
spectral content, 21
video compression, 75
   algorithms, 76
   DBS, 428
   motion compensation, 83
   MPEG, 82
video graphics array, 16
video recording, 233
   8-mm format, 281
   ¼-inch tape formats, 272
   analog signal modulation, 249
   audio tracks, 283
   azimuth recording, 282
   Beta format, 281
   cassettes, 242
   head materials, 240
   head performance, 245
   helical-scan, 234, 275
   Hi-8 format, 289
   history, 233
   home VCR, 275
   quadruplex, 233
   record density, 286
   S-VHS format, 289
   semiprofessional, 294
   SMPTE Type C, 252
   tape, 243
   time code, 307
   U-Matic format, 281
   VHS drum configurations, 280
   VHS format, 281
   video head design, 238
vidicon, 199
viewfinder, 293
viewing ratio, 6, 94
Viterbi decoder, 68
voltage standing wave ratio, 380

VSWR
  see voltage standing wave ratio
waveguides, 383

ABOUT THE AUTHORS

ANDREW F. INGLIS is a consultant to the communications industry and recently retired as president and CEO of RCA American Communications. During his 30-year career at RCA, he also served as vice-president and general manager of its Broadcast Equipment and Communications Systems Division. He is the author of *Behind the Tube* and *Satellite Technology: An Introduction*.

ARCH C. LUTHER consults and writes in the fields of television, computer multimedia, digital audio, and digital video. He is retired from the David Sarnoff Research Center, and is the author of *Digital Video in the PC Environment*, now in its Second Edition; *Designing Interactive Media*; *Authoring Interactive Multimedia*; and *Using Digital Video*.